Advances in plant phenotyping for more sustainable crop production

It is widely recognised that agriculture is a significant contributor to global warming and climate change. Agriculture needs to reduce its environmental impact and adapt to current climate change whilst still feeding a growing population, i.e. become more 'climate-smart'. Burleigh Dodds Science Publishing is playing its part in achieving this by bringing together key research on making the production of the world's most important crops and livestock products more sustainable.

Based on extensive research, our publications specifically target the challenge of climate-smart agriculture. In this way we are using 'smart publishing' to help achieve climate-smart agriculture.

Burleigh Dodds Science Publishing is an independent and innovative publisher delivering high quality customer-focused agricultural science content in both print and online formats for the academic and research communities. Our aim is to build a foundation of knowledge on which researchers can build to meet the challenge of climate-smart agriculture.

For more information about Burleigh Dodds Science Publishing simply call us on +44 (0) 1223 839365, email info@bdspublishing.com or alternatively please visit our website at www.bdspublishing.com.

Related titles:

Advances in crop modelling for a sustainable agriculture
Print (ISBN 978-1-78676-240-5) Online (ISBN 978-1-78676-242-9; 978-1-78676-243-6)

Advances in breeding techniques for cereal crops
Print (ISBN978-1-78676-244-3) Online (ISBN 978-1-78676-246-7; 978-1-78676-247-4)

Robotics and automation for improving agriculture
Print (ISBN 978-1-78676-272-6) Online (ISBN 978-1-78676-274-0; 978-1-78676-275-7)

Genome editing for precision crop breeding
Print (ISBN 978-1-78676-447-8) Online (ISBN 978-1-78676-449-2; 978-1-78676-450-8)

Chapters are available individually from our online bookshop:
https://shop.bdspublishing.com

climate•SMART•publishing
IN AGRICULTURAL SCIENCE

BURLEIGH DODDS SERIES IN AGRICULTURAL SCIENCE
NUMBER 117

Advances in plant phenotyping for more sustainable crop production

Edited by Professor Achim Walter, ETH Zurich, Switzerland

burleigh dodds
SCIENCE PUBLISHING

Published by Burleigh Dodds Science Publishing Limited
82 High Street, Sawston, Cambridge CB22 3HJ, UK
www.bdspublishing.com

Burleigh Dodds Science Publishing, 1518 Walnut Street, Suite 900, Philadelphia, PA 19102-3406, USA

First published 2022 by Burleigh Dodds Science Publishing Limited

Library of Congress Control Number: 2022934179

British Library Cataloguing in Publication Data
A catalogue record for this book is available from the British Library

ISBN 978-1-78676-856-8 (Print)
ISBN 978-1-78676-859-9 (PDF)
ISBN 978-1-78676-858-2 (ePub)
ISSN 2059-6936 (print)
ISSN 2059-6944 (online)

DOI 10.19103/AS.2022.0102

Typeset by Deanta Global Publishing Services, Dublin, Ireland

Contents

Part 2 Sensor types

Series list

Title	Series number
Achieving sustainable cultivation of maize - Vol 1 From improved varieties to local applications Edited by: Dr Dave Watson, CGIAR Maize Research Program Manager, CIMMYT, Mexico	001
Achieving sustainable cultivation of maize - Vol 2 Cultivation techniques, pest and disease control Edited by: Dr Dave Watson, CGIAR Maize Research Program Manager, CIMMYT, Mexico	002
Achieving sustainable cultivation of rice - Vol 1 Breeding for higher yield and quality Edited by: Prof. Takuji Sasaki, Tokyo University of Agriculture, Japan	003
Achieving sustainable cultivation of rice - Vol 2 Cultivation, pest and disease management Edited by: Prof. Takuji Sasaki, Tokyo University of Agriculture, Japan	004
Achieving sustainable cultivation of wheat - Vol 1 Breeding, quality traits, pests and diseases Edited by: Prof. Peter Langridge, The University of Adelaide, Australia	005
Achieving sustainable cultivation of wheat - Vol 2 Cultivation techniques Edited by: Prof. Peter Langridge, The University of Adelaide, Australia	006
Achieving sustainable cultivation of tomatoes Edited by: Dr Autar Mattoo, USDA-ARS, USA and Prof. Avtar Handa, Purdue University, USA	007
Achieving sustainable production of milk - Vol 1 Milk composition, genetics and breeding Edited by: Dr Nico van Belzen, International Dairy Federation (IDF), Belgium	008
Achieving sustainable production of milk - Vol 2 Safety, quality and sustainability Edited by: Dr Nico van Belzen, International Dairy Federation (IDF), Belgium	009
Achieving sustainable production of milk - Vol 3 Dairy herd management and welfare Edited by: Prof. John Webster, University of Bristol, UK	010
Ensuring safety and quality in the production of beef - Vol 1 Safety Edited by: Prof. Gary Acuff, Texas A&M University, USA and Prof. James Dickson, Iowa State University, USA	011
Ensuring safety and quality in the production of beef - Vol 2 Quality Edited by: Prof. Michael Dikeman, Kansas State University, USA	012
Achieving sustainable production of poultry meat - Vol 1 Safety, quality and sustainability Edited by: Prof. Steven C. Ricke, University of Arkansas, USA	013
Achieving sustainable production of poultry meat - Vol 2 Breeding and nutrition Edited by: Prof. Todd Applegate, University of Georgia, USA	014

Introduction

Plant phenotyping is rapidly developing technology that involves the quantitative analysis of structural and functional plant traits. It is widely recognised that phenotyping needs to match similar advances in genetics if it is to not create a bottleneck in plant breeding.

This volume reviews the wealth of research on advances in plant phenotyping for more sustainable crop production, focusing on new technologies such as optical and thermographic sensors, as well as alternative carrier systems such as field robots and unmanned aerial vehicles (UAVs). The book is split into five parts: Part 1 examines the development of phenotyping as a research field, chapters in Part 2 describe the different sensor types that can be used and Part 3 chapters look at the use of carrier and delivery systems in crop phenotyping. Chapters in Part 4 review data analysis in crop phenotyping systems. The three final chapters of the book in Part 5 provide case studies on using phenotyping techniques in different settings.

Part 1 The development of phenotyping as a research field

The book opens with a chapter that focuses on origins and drivers of crop phenotyping. Chapter 1 begins by outlining how plant phenotyping has developed over recent decades, driven by factors such as advances in optical sensors, image analysis and automation as well as multidisciplinary cooperation in establishing facilities for high throughput plant phenotyping. The chapter also describes successful uses of plant phenotyping and stresses the importance of collaboration in further development, particularly to address emerging potential bottlenecks such as management of data to enable interoperability. It highlights the development of networks across centres and why plant phenotyping will always require multi-disciplinary and multi-site resources and expertise to improve the knowledge about plant-environment interactions.

Chapter 2 examines the evolution of trait selection in breeding. It first outlines a historical basis for plant selection and examines the areas in which sensor technology has evolved from our eyes to the application of proximal and remote sensing. The chapter also discusses the use of none-invasive methods that can aid with selection of crop characteristics critical for improving yield potential, such as photosynthesis and partitioning-related traits, as well as the detection of traits that help protect yield, related to disease and lodging resistance. It also examines how trait selection could look in the future.

Part 2 Sensor types

The first chapter of Part 2 reviews advances in optical analysis for crop phenotyping. Chapter 3 begins by focusing on optical sensors and the recently published literature in this area. The chapter also provides a comparison of advantages and disadvantages between the different sensor options. Moreover, several challenges related to the application of optical sensors are discussed. These include the distribution of stresses, compounds and colour across the canopy, but also the complexity of the canopy's 3D structure, diurnal changes in plants and the impacts of rapidly changing environmental conditions. The chapter also provides potential solutions to these challenges through the inclusion of case studies and it discusses relevant technology trends.

The subject of Chapter 4 is advances in the use of thermography in crop phenotyping. The chapter first describes the foundational theory of thermography and then goes on to highlight the principles of thermography measurements. The chapter discusses the different technologies and procedures currently in use and what their limitations are. Two case studies demonstrate the use of thermography phenotyping in different settings. The first focuses on thermography phenotyping in plant breeding and the second case study focuses on the application of thermography phenotyping in Australian environments. Sections on the main challenges and future trends of thermography in crop phenotyping conclude this chapter, highlighting the relevance of airborne imaging because this can facilitate simultaneous analyses of a high number of plots.

Chapter 5 examines advances in the use of X-ray computed tomography in crop phenotyping. The chapter addresses the physical basis and the most relevant hardware components of a computed tomography system used for crop phenotyping. The chapter reviews different system setups and ways of acquiring a computed tomography dataset. It also discusses the post processing of the generated data using specialized algorithms for segmentation. For above- and belowground phenotyping applications two examples are shown to present the utilization of X-ray computed tomography for crop phenotyping: Belowground phenotyping of potato tubers and aboveground phenotyping of wheat ears.

Part 3 Carrier/delivery systems

Part 3 begins with a chapter that focuses on the use of field robots for plant phenotyping. Chapter 6 first provides an overview of current research on different morphologies of plant-phenotyping robots and state-of-the-art sensors and technologies for automatic plant phenotyping. The chapter summarises the advantages of and problems associated with this research field and presents prospects for future development. Platforms are developed in many variations, with some being capable of carrying high loads and allowing simultaneous measurements with multiple sensors. A section on sensors and technologies for phenotyping field robots is also provided, which is then followed by an analysis of using robotic arms for fruit phenotyping and harvesting, before providing a summary of how using field robots for phenotyping can improve the overall collection of phenotypic information from crops.

Chapter 7 examines the advances in high-throughput crop phenotyping using unmanned aerial vehicles (UAVs). The chapter first discusses approaches to UAV remote sensing and data analysis for high-throughput field phenotyping and ecophysiological research. The chapter reviews the use of UAV remote sensing to measure key plant traits through a plant's lifetime: canopy cover; leaf area index; plant emergence, density and tillering; plant height, growth and lodging; biomass; flowering, fruiting and yield; senescence; and canopy temperature. The chapter concludes with reflections on the need for plant scientists and remote sensing specialists to interact closely with each other for unlocking the full potential of their data in the interest of a more sustainable agriculture.

Part 4 Data analysis

Part 4 opens with a chapter that focuses on the advances in computer vision (CV) and crop phenotyping through data challenges. Chapter 8 begins by reviewing the key dimensions that should be considered in CV applications in plant phenotyping, focusing specifically on dimensions such as trait or scenario, plant type and organ and scale. The chapter also examines the technical issues such as hardware constraints and various data analysis needs. An analysis of the Computer Vision Problems in Plant Phenotyping (CVPPP) Workshop is also included, focusing on how the CVPPP workshop not only helped to attract experts already interested in the area, but also attracted new researchers. The chapter then goes on to discuss the various data challenges and how these can be used to accelerate progress in CV techniques, such as leaf counting and segmentation. This is followed by a review of recent agriculture-related CV challenges.

The subjects of Chapter 9 are digital phenotyping and how genotype-to-phenotype (G2P) models are implemented to predict complex traits in cereal crops. The chapter first reviews the use of digital phenotyping as a tool to support breeding programmes, then goes on to examine various G2P models and how the data collected from phenomics and envirotyping can be used in predictive breeding. The current status of statistical models to incorporate secondary traits in univariate and multivariate models, as well as how to better handle longitudinal traits - such as light interception, biomass accumulation or canopy height - is also reviewed.

Chapter 10 examines the role of crop growth models in crop improvement: integrating phenomics, envirotyping and genomic prediction. The chapter shows how crop models can be used to explore knowledge of phenotype, environment and genotype to better understand and predict their complex interrelationships. It describes how crop growth models have been used in envirotyping and in setting targets for phenotyping programmes. It also describes current limitations in modelling, how these are being addressed as well as emerging trends.

Part 5 Case studies

The final part of the book begins with a chapter that addresses using phenotyping techniques to analyse crop functionality and photosynthesis. Chapter 11 first reviews the relation of photosynthesis to crop growth and stress response. It then goes on to examine phenotyping photosynthesis in varying environmental conditions, focusing specifically on climate effects and photosynthesis. A section on using gas exchange to analyse photosynthesis is also included, which is then followed by a review of using porometry and thermal imaging of stomatal conductance and hyperspectral techniques. The chapter also provides a review of using chlorophyll fluorescence and analyses how photosynthesis can be affected by heat stress, drought stress and elevated carbon dioxide levels in the atmosphere. Four case studies analysing these effects are also provided.

Chapter 12 examines the use of phenotyping techniques to predict and model grain yield: translating phenotyping into genetic gain. The chapter describes the importance of boosting genetic gain in grain yield by focusing on phenotypic traits such as plant height, flowering time, aboveground biomass and ear density. A section on stomatal conductance is also provided, which is then followed by a review of the functional 'stay green' trait. A case study that analysed the difference between the data obtained from advanced sensors, digital cameras and UAV technology assessing grain yield in wheat in Spain is also provided.

The final chapter of the book focuses on the automated assessment of plant diseases and traits by sensors, as well as how digital technologies can support smart farming and plant breeding. Chapter 13 begins by providing an overview of digital plant disease detection, focusing on aspects such as non-invasive technologies, motivation, optical sensors and the combination with data analysis and carrier platforms at different scales. The chapter moves on to address the complexity of host-pathogen interactions as well as the complexity in a crop stand. A case study that examines the application of deep learning to foliar plant diseases is also included, which is then followed by a summary of why developing digital technologies is important in improving plant disease detection.

Acknowledgement

Cover image copyright

- Cover image: A hexacopter acquiring thermal and color images of wheat plots on a field site in Switzerland in July 2020. © Norbert Kirchgessner, Helge Aasen and Achim Walter.

Part 1

The development of phenotyping as a research field

Chapter 1

Origins and drivers of crop phenotyping

Roland Pieruschka and Ulrich Schurr, Institute for Bio- and Geosciences (IBG), IBG-2: Plant Sciences, Forschungszentrum Jülich, Germany

1 Introduction

Understanding the phenotype of plants is essential in the context of food or biomass production from crops, for efficient use of resources such as water or nutrients or in understanding plant ecological performance. All these depend on the interaction between plant genetic makeup and the prevailing environment. Understanding multidimensional plant–environment interactions has a long history in the eco-physiological sciences. The subject gained new momentum when genomics technologies became available about three decades ago. An increasing number of plant genome projects were initiated to analyse the genetic makeup of plants. Within the last few decades, about 600 genome assemblies from different plant species have been made available in public repositories (Kersey, 2019). Crop species dominated initially but a wider range of plants, including non-domesticated species, have now been analysed. In parallel with these developments, there have been advances in technologies to modify plant genetics. Recent progress in genetic engineering -specifically CRISPR/CAS9 - provides '...enormous power in this genetic tool, which affects us all. It has not only revolutionised basic science but also resulted in innovative crops and will lead to ground-breaking new medical treatments', to quote Claes Gustafsson, chair of the Nobel Committee for Chemistry. (https ://www.nobelprize.org/prizes/chemistry/2020/press-release/).

http://dx.doi.org/10.19103/AS.2022.0102.01

However, genetic makeup provides only one element that determines the phenotype of a plant. Genes are the toolbox with which a plant can work when the plant is exposed to challenges - especially abiotic and biotic environment - in real life. Plant phenotyping is the process of quantitatively characterising the structural and functional properties of a phenotype within a dynamic environment. It is the next step beyond genotyping towards an integrated understanding of plant–environment interactions (Fig. 1). Based on a wealth of research, in the late 1990s plant phenotyping started to integrate genomics and eco-physiological perspectives into a more holistic approach. New methods for high-throughput phenotyping were based on technologies like computers, optical sensors and automation. It is worth remembering that computers only became generally accessible in the 1990s - a fact that we often forget given their ubiquity today. High-throughput genotyping revolutionised our understanding of the genetic makeup of plants. This development then increased the need for quantitative assessment of the phenotype as a basis for understanding plant–environment interactions and for breeding applications.

The challenges posed by the complexity of plant structures and functions and their environmental plasticity continue to drive research. This research requires dynamic assessment of individual plant organs, developmental stages, entire canopies or plant processes. These components are interconnected. A change in one has an impact on the others and, in turn, the whole plant and its interactions with the environment. Continuous measurements with non-invasive technology have provided the key to connecting phenotypic dynamics, molecular properties and environmental dynamics to create models to visualise and predict these complex interrelationships and transfer them into applications such as breeding.

Phenotypic data are pivotal to understanding quantitative traits which are under polygenic control. Quantitative description of a phenotype is extremely challenging. It ranges from subcellular, cellular, tissue and organ levels up to the level of the whole organism or even the arrangement of individual plants in a stand. These different levels must be analysed in the context of dynamic

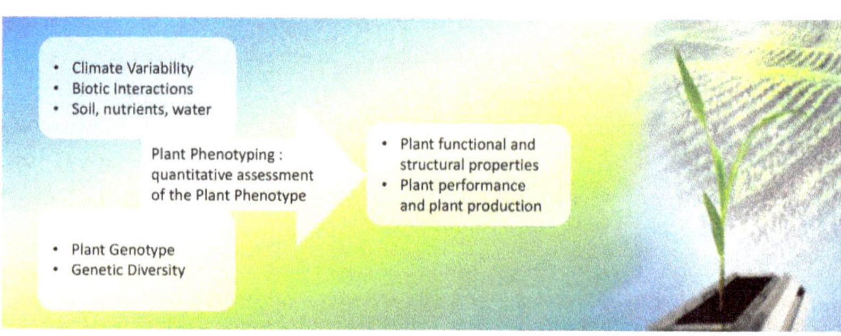

Figure 1 Multidimensional system addressed by plant phenotyping.

environmental conditions in space and time (Pieruschka and Poorter, 2012). Houle et al. (2010) described this relation using genotype-phenotype maps to determine the phenotypic state that organisms occupy within the overall space of possible phenotypes, taking into account phenotype response to variation in environmental conditions within a complex multidimensional system (Fig. 1). This level of analysis requires robust and accurate measurement technologies (technology development), integration of technologies in phenotyping infrastructures and applications (phenotyping infrastructures), use of phenotyping systems in interaction with plant material which has been properly characterised in its genetic setup (genotype analysis) and with analysis of its often dynamic (bio-)chemical composition (chemical phenotype). Only by combining all these dimensions in adequate temporal and spatial domains, can a mechanistic understanding of interactions of genetics, structural, physiological and environmental processes be made. Since this is the basis for crop improvement, a concerted effort is essential to further advance the quantitative understanding of crops in a dynamic environment. This is particularly important with multiple grand challenges related to feed and food for a growing population in times of climate change (FAO, 2017).

Selection based on phenotypic traits has been the basis for crop improvement for thousands of years and has been fundamental for breeding from the beginning of crop domestication until genetic tools became available. Wheat is an outstanding example where phenotypic selection based on traits such as large seeds, plants with low toxicity and reduced seed dispersal was essential for adaptation of plants to crop cultivation. The success of using early, non-quantitative phenotyping was vital to the shift from hunter-gatherer to agricultural societies, and it stimulated the development of cities and modern civilisations. The increased understanding of genetics, which became available during the Green Revolution, allowed genetic improvement towards much higher-yielding varieties through the introduction of dwarfing traits with huge increases in wheat yield (Vergauwen and De Smet, 2017). The development of various DNA/gene marker technologies shifted the emphasis of breeders from selection based on phenotypes to marker-assisted methods based on access to cheap sequencing methods (Collard and Mackill, 2008). Phenotypes were linked to DNA-markers which could then be used to assess crossing material without expressing the phenotype at each step of the selection process. This has been critical to the success of modern breeding. One study has shown that, on average and across all major arable crops cultivated in Europe, quantitative phenotyping contributes approximately 74% to overall productivity growth (Noleppa, 2016).

Plant phenotyping has been recognised as one of the major limitations to further advances in breeding and pre-breeding (Watt et al., 2020). Progress in phenotyping requires improved technology and access to state-of-the-art methods and facilities, as well as dissemination of novel technologies to phenotype an increasing diversity of traits under controlled and field conditions above and belowground. These must be combined with integrated data management techniques to allow reuse of data.

This chapter outlines how plant phenotyping has developed over recent decades, driven by factors such as advances in optical sensors, image analysis and automation as well as multidisciplinary cooperation in establishing facilities for high throughput plant phenotyping. The chapter describes successful uses of plant phenotyping and stresses the importance of collaboration in further development, particularly to address emerging potential bottlenecks such as management of data to enable interoperability.

2 Technological progress in plant phenotyping

The terms phenotype and genotype were first mentioned over a hundred years ago and quantitative plant phenotyping has been a central element of eco-physiological research ever since (Johannsen, 1903, 1911). Over the past two decades, however, there has been a dramatic increase in the accuracy and speed of phenotyping with new technologies in high-throughput based on huge volumes of genetic information from next-generation sequencing. Since then, plant phenotyping has made impressive progress, developing novel sensors and imaging techniques able to quantitatively measure a wide range of traits at many spatial scales and in many different temporal dimensions. Quantitative phenotyping to understand the interaction between plants and environmental conditions is now possible, using low and high-throughput phenotyping for quantitative screening of a high number of genotypes under well-defined conditions at cellular, organ, plant and canopy levels under a variety of environmental conditions.

This progress in phenotyping can be seen in the massive increase in the number of publications addressing plant phenotyping from around 2010 (Costa et al., 2019). This period coincides with the development of tools for quantitative assessment of plant traits that allowed the matching of phenotypic analysis with genetic information from DNA sequencing.

Non-invasive imaging sensors and computer vision technologies have significantly improved measurement of plant traits. These were based on the rapid increase in affordable and robust imaging systems in the 1990s following the development of digital imaging sensors (Blais, 2004). Non-invasive technology initially focused on recording time series at the level of single organs in single (or a few) plant(s). Schmundt et al. (1998) were able to demonstrate the temporal and localised character of growth in leaves of *Ricinus communis* and tobacco with growth restricted to the base of the leaf and to a few hours at the end of the night and the start of the day. The ability to obtain

Published by Burleigh Dodds Science Publishing Limited, 2022.

high-resolution time-lapse videos, paired with development of algorithms in image analysis, represented a substantial advance in understanding growth processes. The progress was possible by linking image analysis expertise with plant physiology. Similar examples include characterisation of growth of primary roots (Walter et al., 2002) and functional imaging of photosynthesis (Genty and Meyer, 1995).

Newly developed imaging systems and the increasing number of image analysis programs led to the next step in the evolution of phenotyping. These phenotyping approaches were integrated into automated systems allowing increasingly high throughput. Automated systems first addressed phenotyping of small plants such as *Arabidopsis* to measure growth dynamics (Granier et al., 2006) or *Arabidopsis* and tobacco seedlings to measure growth and photosynthetic properties (Jansen et al., 2009). These marked an important step towards measuring genetic variability. The importance of imaging approaches in plant phenotyping can be seen in the growth in publications using the terms plant phenotyping with different types of imaging over the last decade (Fig. 2). This analysis shows that functional imaging (spectral, hyperspectral, thermal) also grew over the same decade. Merging these different imaging modes (sensor fusion) may become an important tool in the future, enabling simultaneous monitoring of the dynamics of structural and functional properties by combining measurement of 3D plant structure with spectral measurements (Paulus, 2019). Increased use of imaging can be expected with the development of affordable, mobile technologies providing high-quality images (Mueller-Linow et al., 2019; Reynolds et al., 2019).

Many of the ongoing developments in plant phenotyping are driven by developments in non-invasive sensors providing information from the entire electromagnetic spectrum. A wealth of information can be obtained from data from reflected, transmitted or emitted radiation. The magnetic resonance

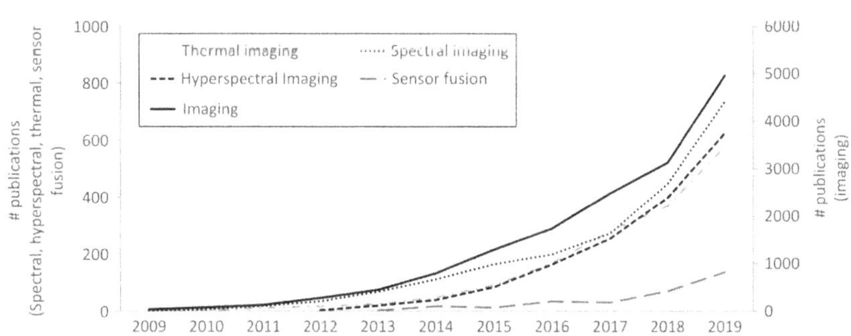

Figure 2 Number of publications using imaging technology. Web of Science (Clarivate Analytics) was used for a simple evaluation of publications within the last 10 years with search terms: phenotyping x imaging, spectral imaging, hyperspectral imaging, thermal imaging and sensor fusion.

Published by Burleigh Dodds Science Publishing Limited, 2022.

properties of nuclei provide information for instance on *in situ* root properties (Fiorani et al., 2012). Two-dimensional or three-dimensional representations of plant features show structural properties that can be combined with functional imaging techniques in providing unique information on structural-functional or agronomic traits (Paulus, 2019). Advances in automation, robotics and computing capacity have led to advances in: i) combining experimental design with growth facilities to simulate relevant environments at a sufficient capacity to support large-scale phenotyping; ii) the integration of non-invasive technologies into standardised experimental protocols (key for best practice in phenotyping); and finally, iii) data analysis, management and storage capacities (Fiorani and Schurr, 2013).

Plant phenotyping will always require a wide range of technologies, expertise and knowledge about plant systems and the environments. In this context, the diversity of technologies and infrastructures mirrors the diversity of traits and crops. This is in strong contrast to the 'genomics world', where the unifying principle of nucleic acids provides a common focus for analysis. While analysing nucleic acids of crops in greenhouses and field conditions uses the same techniques, phenotyping in field or controlled environment conditions requires very different approaches. Therefore, plant phenotyping will always be a field of activity, which requires strong consideration of how to bring together the necessary infrastructure and expertise for the multitude of relevant traits. It is a very challenging task to understand the diversity of phenotypes in response to different environmental cues. This requires addressing a diversity of traits that change dynamically during the lifetime of a plant in a dynamic environment. To decipher this interaction, we need to interpret various structural and functional traits, spectral ranges, the development of various organs, etc. The number of outcomes (phenotypes) from the potential combinations of genotype, environment and phenotype is enormous. It is essential to identify the relevant traits, environmental scenarios and technologies that advance our understanding, improve our models and provide data that are reusable.

The approach needs to be complemented by the optimisation of the costs of plant phenotyping tools and facilities. Low-cost phenotyping will mostly be limited to a few common traits that can be measured with affordable, often manual tools. In contrast, high quality and high throughput analyses of functional and structural traits require higher initial investment costs for instance in sensors, vehicles and automation technologies, but may deliver more robust, higher quality and relevant results with lower operational costs over the long term (Reynolds et al., 2019). Complex phenotyping also requires investment in expertise and software. Image analysis instruments, e.g. need to be tailored to their experimental setting to address specific phenotyping problems which differ according to crop, environment and genotype. One software solution often cannot be transferred from one crop or treatment to another. This means that acquiring images is often easier (and less costly) than quantitative analysis of images, with the latter becoming a potential bottleneck (Choudhury et al., 2019; Minervini et al., 2015). The use of artificial intelligence in image analysis problems may help to address this challenge. Machine learning approaches have emerged as a tool to analyse and find patterns in large amounts of data in plant phenotyping (Singh et al., 2018). Machine learning techniques may play a prominent role in the future of image-based phenotyping in, e.g. detection of biotic and abiotic stress (Ubbens and Stavness, 2017). These approaches, however, require choices about target features to be extracted. They depend on the availability of high-quality reference data sets to provide the learning algorithm with the optimal description of the data and problem in question (Pound et al., 2017). Variability in methods may also create standardisation problems.

Despite substantial progress, plant phenotyping still faces challenges which are inherent in the complexity and diversity of plants and traits and the nearly countless environmental scenarios that need to be quantified. There is an increasing range of technologies available with the potential to address these challenges but it still require significant development in software solutions and data analysis.

While analysing gene sequences require essentially one basic approach across all plants or living organisms, with advances in one area potentially benefitting all others, the gap between phenotyping and genotyping may remain large. Pooling of infrastructures and expertise is essential for plant phenotyping to increase efficiency. This requires utilising synergies by pooling investment in new facilities, sharing specialised technology, software solutions and knowledge on physiology. This also includes the development of technological and scientific expertise in conjunction with best phenotyping practices to benefit knowledge and technology transfer (Fig. 3).

The development of common standards to enable interoperability will reduce costs and help close the gap between phenotype and genotype. Interoperability of data and model development will become integral elements in designing phenotyping experiments to address knowledge gaps. These considerations show the importance of the phenotyping community organising broad networks to integrate knowledge and facilities. The challenges of quantifying the diversity of relevant plant phenotypes cannot be addressed in a single research institution or country. Open collaborative approaches between plant phenotyping centres are key to address gaps in plant research and translation of this knowledge into applications.

3 Community integration in plant phenotyping

The first phenotyping centres started to emerge in the early 2000s. From the beginning, these centres needed to build on strong multidisciplinary expertise.

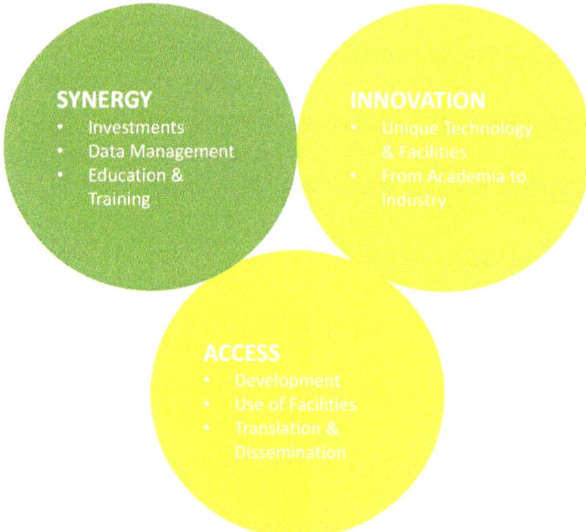

Figure 3 Community integration to foster synergies, innovation and access (modified from EMPHASIS: (https://emphasis.plant-phenotyping.eu/).

These included integration of (novel) sensors, establishing facilities simulating relevant environmental conditions (often in contrast to constant conditions used in in many experimental facilities) and specialised platforms for deep and high-throughput phenotyping under controlled and field conditions. The resulting research had to link genetic, biochemical and phenotypic data. Challenges remain in the need to address different plants and a diversity of environmental conditions. It is particularly challenging to balance a high level of flexibility with harmonisation of experimental procedures and data management to allow reusability of data. This is another important role of centres with high phenotyping expertise. They often form important nuclei for knowledge and technology transfer. This task cannot be addressed by a single research group or organisations and requires community integration across different levels (Fig. 3). This integration has been achieved in several countries from the early 2000s with the establishment of national infrastructures to integrate the expertise of different institutions. These national infrastructures pool both technology and expertise to address the needs of users from both academia and industry. An important aspect is the support by funders and policymakers to optimise research investment, reduce duplication and enable sustainable operation of the phenotyping infrastructure within a country.

One of the first national infrastructures established in 2009 was the Australian Plant Phenomics Facility (APPF, https://www.plantphenomics.org.au/).

This combined The Plant Accelerator in Adelaide as the first automated high-throughput phenotyping system, the High Resolution Plant Phenomics Centre (HRPPC) at Commonwealth Scientific and Industrial Research Organisation (CSIRO, Australia) which combined expertise in plant science and engineering to develop and build phenotyping technologies and, finally, the Australian National University (ANU) with expertise in phenomics, bioinformatics, hardware and software development and data visualisation. The German Plant Phenotyping Network (DPPN, https://dppn.plant-phenotyping-network.de/index.php?index=6) started in 2012 with three complementary sites to combine infrastructure and methods, using a range of technology platforms to address different plant phenotyping approaches. PHENOME (France, https://www.phenome-emphasis.fr/phenome_eng/) was established in 2013 as a federated network of phenotyping platforms at five different sites in France. Since then, many countries have initiated building of a national plant phenotyping infrastructure to integrate national research. Table 1 lists national and regional plant phenotyping initiatives. The challenge is to integrate the activities across borders.

In Europe, in parallel with the establishment of national centres, discussions were initiated about developing cooperation between centres. The process of integration was facilitated by the EC-funded projects EPPN (2012-2015) and EPPN2020 (2017-2021), which developed best phenotyping practice and enabled excellence-driven access to phenotyping facilities under controlled environmental conditions (EC, 2016). These projects provided 200 opportunities for access to enable experiments, resulting in over 80 publications (https://eppn2020.plant-phenotyping.eu/EPPN_Publiations). Most users were from academic institutions addressing a wide range of scientific questions. Facilities providing access within EPPN2020 also have a strong interaction with industry. On average, about 30% of the experiments within the plant phenotyping facilities include industry partners – in most cases based on bilateral agreements (EPPN2020, personal communication). Since projects such as EPPN or EPPN2020 usually have a limited duration, and national infrastructure efforts are usually limited to a certain country or region, it is important to develop ways to allow long-term and sustainable development of a regional infrastructure. The European Strategy Forum for Research Infrastructures (ESFRI) is a Europe-wide instrument to improve scientific integration across Europe (and beyond), to strengthen its international outreach and enable long-term sustainable operation across borders (ESFRI, 2017). The integration of plant phenotyping within Europe has been developed within the framework of European Infrastructure for Multi-scale Plant Phenotyping and Simulation (EMPHASIS), added to the ESFRI Roadmap in 2016 and expected to become operational during 2022 or 2023 (ESFRI, 2016). The main objectives of EMPHASIS (see Fig. 4) are to:

- develop an infrastructure to coordinate and enable access to facilities and services across borders, make investments in new infrastructure more efficient;
- link data acquisition to a pan-European information system and modelling by developing best phenotyping practice; and
- develop, evaluate and disseminate knowledge and technology to support the commercial sector as well as provide new technological solutions both for phenotyping and related disciplines such as precision agriculture.

Table 1 Plant phenotyping networks/Initiatives

National networks/Initiatives:	
APPF: Australian Plant Phenotyping Facility	http://www.plantphenomics.org.au/
APPN: Austrian Plant Phenotyping Network	http://www.appn.at/
BePPN: Belgian Plant Phenotyping Network	-
CPPN: China Plant Phenotyping Network	-
CzPPN: Czech Plant Phenotyping Network	http://www.czppn.cz/
DPPN: German Plant Phenotyping Network	http://www.dppn.de/
NaPPI: Finland National Plant Phenotyping Infrastructure	http://blogs.helsinki.fi/nappi-blog/about/
Phen-Italy: Italian Plant Phenotyping Network	http://www.phen-italy.it/
PHENOME-EMPHASIS	https://www.phenome-emphasis.fr/
Phenom-UK	https://www.phenomuk.net/
Regional networks / Initiatives:	
EMPHASIS	https://emphasis.plant-phenotyping.eu/
EPPN: European Plant Phenotyping Network	https://www.plant-phenotyping-network.eu/
EPPN2020: European Plant Phenotyping Network 2020	https://eppn2020.plant-phenotyping.eu/
LatPPN: Latin American Plant Phenotyping Network	(Camargo and Lobos, 2016)
NaPPNord American Plant Phenotyping Network	https://nappn.plant-phenotyping.org/ (Carroll et al., 2019)
Nordplant	https://www.nordplant.org/
Global networks / Initiatives:	
IPPN: International Plant Phenotyping network	https://www.plant-phenotyping.org/

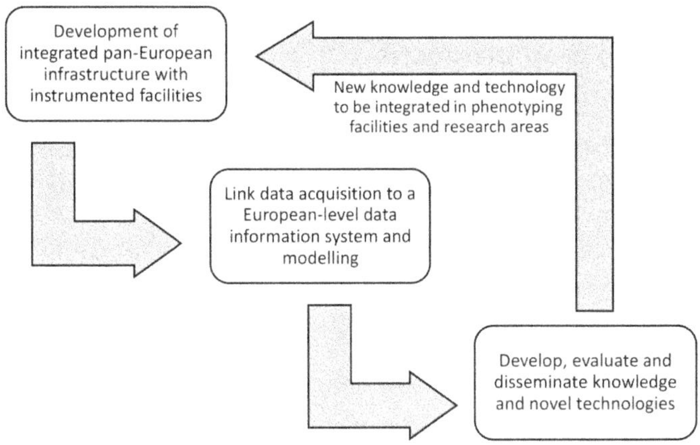

Figure 4 Integration of distributed plant phenotyping research infrastructure (Example from EMPHASIS, modified from: https://emphasis.plant-phenotyping.eu/).

EMPHASIS has a strong European focus. However, cooperation beyond Europe is important and an integral part of the ESFRI internationalisation strategy (see RI-VIS project: https://ri-vis.eu).

There is also a strong need for development of phenotyping networks outside Europe. Latin American Plant Phenomics Network (LatPPN), e.g. focuses on Latin America and the Caribbean countries covering a wider range of climatic conditions and potentially any crop (Camargo and Lobos, 2016). Another example is the North American Plant Phenotyping Network (NaPPN) to promote collaboration between groups in North America to support agricultural and other sciences (Carroll et al., 2019). National and international networks are listed in Table 1. With the increasing number of national and continental networks, there is an opportunity to integrate at a global scale. This role is taken by the International Plant Phenotyping network (IPPN). The IPPN has the goal of linking different regional networks and providing a communication and exchange platform. The IPPN includes academic members from phenotyping centres across the globe as well as key companies involved in phenotyping. The network addresses different topics with dedicated working groups organising workshops, round table discussions, training, etc. A key initiative is the organisation of the International Plant Phenotyping Symposia, starting in 2009 and bringing the global plant phenotyping community together to exchange information and ideas. The symposia are often complemented by publication of the latest research in dedicated special issues in different journals (Pieruschka and Lawson, 2015; Pommier et al., 2020; Reynolds and Schurr, 2019). Working groups develop integrated views on particular areas of research and support the exchange between experts and users of plants phenotyping.

4 Plant phenotyping as a tool for enhanced and sustainable crop production

Sustainable intensification of crop production is a major challenge to ensure the amount and quality of plant biomass needed for human nutrition and other uses. There is an urgent need to design new varieties which are high yielding and adapted to contrasting environmental conditions, including those associated with climate change, which makes efficient use of resources and are compatible with new agricultural management systems. We highlight three research areas that have substantial potential in advancing the development and application of phenotyping solutions and play a key role in advancing plant breeding:

- utilisation of genetic resources;
- phenotyping of roots; and
- phenotyping under field conditions.

In each of those areas, we summarise the benefits plant phenotyping is delivering and discuss potential future developments.

4.1 Phenotyping to support utilisation of genetic resources

Genetic resources for food and agriculture are the raw materials on which the world relies to improve the productivity and quality of crops. The conservation and sustainable use of genetic resources for food and agriculture is therefore at the centre of food security and nutrition. Modern, high-yielding agricultural crop varieties have resulted in reduced genetic diversity, including the potential of crop adaptation to future environmental conditions compared to natural populations (Jump et al., 2009). Utilisation of genetic resources in gene banks with future value is essential in preserving the evolutionary potential of crops. There are about 7.4 million accessions locked in storage rooms in more than 1750 gene banks worldwide (FAO, 2010). These resources are substantially underutilised because of a lack of access to information. In particular, lack of phenotyping data about accessions is an important limiting factor in the use of genetic resources for crop improvement and basic research.

An essential step towards the utilisation of plant genetic resources is the informed selection of the most promising genetic resource. Passport records of accessions usually include provenance, genotypic data and phenotypic observations if available. Quantitative phenotyping during successive rounds of seed multiplication is rarely performed because of limited resources and capabilities. However, it may contribute substantially to complementing passport data with a relevant description of trait, which would improve the development of core populations and consequently the translation of novel traits (genes, alleles) into pre-breeding and breeding programmes (Mascher et al., 2019; Rebetzke et al., 2019).

New, more affordable techniques for single seed phenotyping will facilitate description of relevant traits of accessions, which would make the translation of novel traits into pre-breeding and breeding programmes more effective. Single seed phenotyping is often based on image analysis of 2D shape parameters such as seed length, width and surface area, which can be recorded using high throughput methods and used for quantitative trait locus (QTL) analysis as demonstrated with rice (Tanabata et al., 2012). Assessing seed properties using 3D techniques such as stereoscopic imaging is more challenging, especially for small seeds, but provides useful additional information (Roussel et al., 2016). Seed imaging connected to a robotised platform including weighing of seeds provides a useful tool to measure a number of traits (Jahnke et al., 2016). This may then be linked to an X-ray CT scanner to investigate the internal structure of seeds (Gargiulo et al., 2019). Irregularly shaped seeds with rough surfaces are difficult to assess accurately with image-based tools, though seed volume can be measured acoustically (Sydoruk et al., 2020).

To complement seed phenotyping, high throughput phenotyping of plants during regeneration cycles may provide useful additional information on traits to add to accession passport data. Low-cost solutions based on smartphone applications are widely available with increasing resolution and processing capabilities. The Plant Screen Mobile, e.g. provides a portfolio of segmentation options that allows simple geometric calibration to extract traits such as projected leaf area and shape parameters (Mueller-Linow et al., 2019). Those data can be used as a proxy for leaf area, biomass or growth when recorded over time and, when combined with environmental data, they provide a flexible and easy tool in providing quantitative phenotypic data (Jun and Park, 2017).

There is therefore a range of increasingly affordable technologies that can be used in gene bank pipelines to characterise single seeds and plant traits during regeneration cycles. Those approaches have the potential to substantially enhance the utilisation of genetic resources and make translation of novel traits into pre-breeding or breeding programmes more effective. Collaboration between gene bank curators and technology developers is important to identify the relevant steps in preservation and utilisation of genetic resources.

4.2 Exploring the hidden part of the plant: root phenotyping

Quantitative assessment of root structural and functional properties relates directly to properties such as water and nutrient use efficiency which are key to plant productivity. While changes in shoot properties have been extensively studied, analysis of roots has been neglected. However, the development of new crop cultivars with improved root properties is an important strategic goal for global agriculture (Kuijken et al., 2015).

High throughput analysis can be achieved with seedlings grown in agar-filled Petri dishes where root system architecture and growth can be quantified with a good temporal resolution, including traits such as branching angles and frequency of lateral roots (Nagel et al., 2009; Nagel et al., 2020). Growth in cylinders filled with transparent gel allows measurements of 3D root properties (Clark et al., 2011). While growth in agar-filled Petri dishes is limited to a few days, pouches using germination paper (with roots growing along the paper) allow experiments over weeks (Gioia et al., 2017; Le Marie et al., 2016). Root growth in artificial systems provides an opportunity to simulate aspects of the root environment such as nutrient availability. However, it can be challenging when using the root traits from one growth system to predict those in another growth system (Liu et al., 2017). Well-defined experimental settings to study root properties in soil are essential to allow comparison.

In controlled conditions plants can be grown in soil-filled pots (Shi et al., 2018) rhizotrons (Nagel et al., 2012) or rhizotubes (Jeudy et al., 2016) with transparent walls that allow measurement of root traits. The root system fractions visible from the transparent parts of the pots or rhizotrons are species dependent. However, the variation of roots growing along the transparent part of the pot or rhizotron correlates well with a number of traits such as root biomass or total root length (Nagel et al., 2012). These experiments require careful preparation but provide the opportunity to measure root and shoot properties simultaneously. Analysing the growth of maize under drought conditions, e.g., revealed a significant reduction in total root length as well as rooting depth and width, together with significantly reduced leaf growth parameters. Such an approach has the potential to identify key candidate traits for future breeding programmes (Avramova et al., 2016).

New approaches for root phenotyping have recently been developed to study 3D root structure and function *in vivo* in soil-filled pots using X-ray computed tomography (CT) or magnetic resonance imaging (MRI) (Metzner et al., 2015; Mooney et al., 2012). Widely used in medicine and the life sciences, both methods involve 3D reconstruction of scanned objects (Kherlopian et al., 2008). These methods produce high-quality 3D images of root systems and an opportunity for deep phenotyping of root responses to a range of relevant environmental conditions. MRI can also be combined with positron emission technology (PET) where carbon transport into roots can be measured by tracing the short-lived radioactive carbon isotope (11)C (Jahnke et al., 2009).

Root phenotyping often starts using controlled conditions with the goal to transfer knowledge to the field (Kuijken et al., 2015). There are several approaches to phenotyping of roots under field conditions. Digging out the roots (shovelomics) or coring are crude but effective methods and can be used to reveal biologically important variation and genome regions affecting root architecture (Burridge et al., 2017) or identify variation in deep root traits to speed up selection of genotypes for breeding programs (Wasson et al., 2014). Deep rooting genotypes can also be identified using indirect methods such as a tracer or herbicide placed at a certain depth in the soil: root activity can be assessed by measuring the responses of shoots above ground (Chen et al., 2019). An interesting approach is root observation through long transparent tubes that are placed diagonally in the soil at differing depths, allowing measurement of the time when roots are visible at different depths. The approach can be combined with water and nutrient stress analysis, providing direct measurements of both root growth and stress symptoms in the canopy (Svane et al., 2019).

There is therefore a range of methods to assess root properties in a detailed and systematic way and then translate a better understanding of root properties into crop improvement. These approaches are essential for the development and validation of models on how root phenotypes interact dynamically with the environment, acquire resources and thus influence plant growth and yield (Postma et al., 2017). The capacity for high throughput phenotyping of roots is currently rather limited. It is essential to use available data and models to identify the main gaps in knowledge that need to be addressed experimentally. Phenotyping facilities and technology that simultaneously measure roots and shoots must be developed for a better understanding of global plant productivity and translation of this understanding into applications (Tracy et al., 2020; Watt et al., 2020).

4.3 Plants under dynamic environment: field phenotyping

Plant performance is strongly affected by environmental conditions. Plants grown under controlled conditions can be manipulated in a well-defined way that cannot be replicated in the field. The translation of results from controlled to field conditions is challenging because of the highly dynamic environment in the field (Poorter et al., 2016). There are some facilities where some aspects of the environment can be controlled such as the CO_2 concentration in Free Air CO_2 Facilities (FACE, (Muller et al., 2018)) or control of water and nutrient availability in rain-out shelter facilities (Beauchêne et al., 2019). However, in most cases field phenotyping focuses on plant canopies growing in a highly dynamic environment usually over extended areas where both the environment and the traits of interest must be assessed in a quantitative way. Here we focus on summarising the current development of field and airborne platforms as well as sensor networks in experimental field sites and discuss potential requirements for future developments.

Requirements for quantitative assessment of plant traits in the field can be easily described but are challenging to achieve in practice. There is a need for a dedicated ground and airborne carrier system for sensors complemented by environmental monitoring (Araus et al., 2018; Cendrero-Mateo et al., 2017; Jin et al., 2020). Ground-based phenotyping systems may be categorised as phenopole-based, phenomobile and stationary systems. Phenopoles are simple fixed or mobile 'sticks' with an attached sensor that must be fixed manually. Phenomobiles may involve automatic, human or tractor-based movement. Both systems may include an integrated GPS system to deliver high-quality data. Stationary systems include cable-suspended sensors (Kirchgessner et al., 2017) or dedicated gantry systems (Virlet et al., 2017) which provide high-resolution spatial and temporal data but are often restricted to a specific field-side because the carrier system is usually not transferable. Unmanned aerial vehicles (UAVs) have recently become increasingly important in field phenotyping. This has become possible because of increasing flight times, increased payload capacity and decreasing costs. An additional factor is the development of miniaturised sensors covering a wide range of the electromagnetic spectrum, as well as using thermal imaging and hyperspectral imaging to assess a diversity of functional and structural traits in the field at high spatial and temporal resolution. This has resulted in the increasing use of this technology (Yang et al., 2017).

Assessing genetic variability of crops under field conditions usually requires multiple experiments across different sites and countries, which need to take into account year-to-year and site-to-site climatic variability. This requires comparable environmental characterisation of individual experiments in addition to harmonised trait assessment over multiple years to assess the performance of genotypes in crops such as maize (Millet et al., 2016) and allow prediction of yield under a wide range of environmental scenarios (Millet et al., 2019). Further development of multi-site experimental capacity is essential and requires harmonisation of experimental procedures so that data are comparable and reusable. Further advances will come from multi-scale monitoring of plants and the environment using autonomous robotics combined with big data analysis through artificial intelligence. With the G5 technology on the horizon, it will be possible to transfer data to servers for complex analysis in real time, providing opportunities for use in precision farming.

There is currently substantial development in technology for plant phenotyping including ground and airborne systems with a wide range of sensors to analyse a diversity of traits. Further development in robotised systems will boost plant phenotyping and related disciplines such as precision farming. Assessment of genetic variability of crops requires multi-site experiments and, despite the availability of phenotyping technology and environmental sensors, there is a need to harmonise experimental procedures, data acquisition and management, particularly considering that experiments often need to be performed on multiple sites over multiple years.

5 Future trends

Plant phenotyping is becoming an integral tool in the plant sciences and related disciplines, including in breeding and precision farming. Phenotyping can exploit advances in vision systems, machine learning and robotics to improve efficiencies and accuracy. There is no one-size-fits-all solution. Achieving sustainable plant productions requires three key elements:

- phenotyping and environmental observations of diverse crops and traits of interests in dedicated facilities;
- hardware and software solutions to address the diversity of required solutions; and
- multi-scale systems.

These are summarised in Fig. 5.

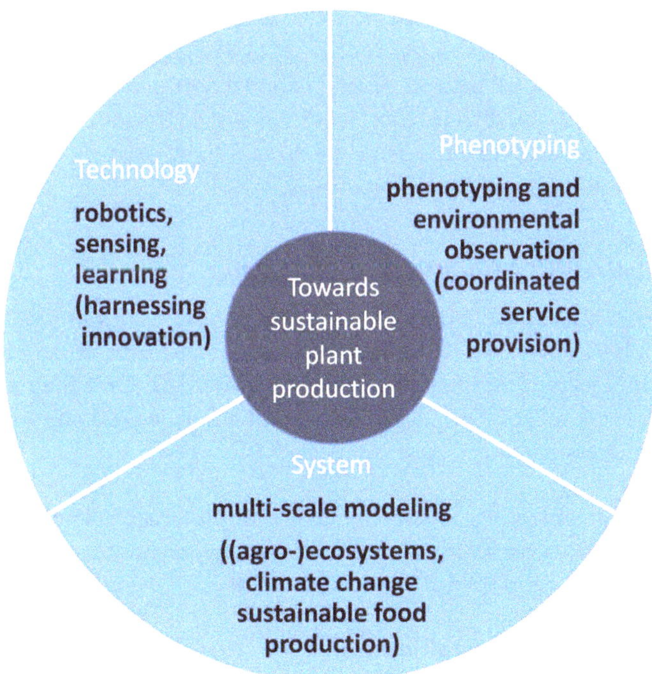

Figure 5 Phenotyping, systems and technology as three integral elements required to advance sustainable plant production.

Plant phenotyping is an important tool in understanding plant-environment interactions. Given the diversity of facilities, tools and methods, a coordinated approach is needed to effectively utilise these resources for sustainable plant production. This includes sharing facilities to minimise duplication of costs and resources, with investment focussed on developing new facilities which will have the most impact. The integration of robotics and data science will help to achieve higher throughput and accuracy (Watt et al., 2020). Coordination is essential to enable harmonisation from experimental design to data acquisition and reusability of data in models (Bolger et al., 2019; Tardieu et al., 2017). Analysis of genotype-phenotype relationships in diverse environments is not feasible through experiments alone (Houle et al., 2010). There is therefore a need for interoperable phenotyping data for modelling to identify gaps that dedicated experiments can then fill.

6 Where to look for further information

6.1 Conferences

The IPPN organises biannual International Plant Phenotyping Symposia in different regions across the globe, usually hosted by a dedicated plant phenotyping centre in a specific region. The symposia are often linked to special issues (Pieruschka and Lawson, 2015; Pieruschka and Poorter, 2012; Pommier et al., 2020; Reynolds and Schurr, 2019). There are also several regional events organised by regional networks such as the North American Plant Phenotyping Network (NAPPN, https://nappn.plant-phenotyping.org/) or research infrastructures such as EMPHASIS (https://emphasis.plant-phenotyping.eu/).

6.2 Journals addressing plant phenotyping

According to a latest bibliometric study plant phenotyping was mentioned in 174 journals (Costa et al., 2019). The top five journals were: Frontiers in Plant Science, Journal of Experimental Botany, Plant Methods, Plant Theoretical and Applied Genetics, PLoS ONE. Specific plant phenotyping journals addressing plant phenotyping are:

- Plant Phenomics (https://spj.sciencemag.org/journals/plantphenomics/).
- The Plant Phenome Journal (https://www.wiley.com/en-us/The+Plant+Phenome+Journal-p-9780JRNL79529).

6.3 Plant phenotyping initiatives

There are several plant phenotyping networks and initiatives on a country, regional and global level as summarised in Table 1.

7 Acknowledgements

The authors acknowledge support from the Helmholtz Association for the Forschungszentrum Jülich. The writing of this publication was made possible by funding from the German Ministry for Education and Research (BMBF) to the German Plant Phenotyping Network (DPPN; BMBF 031A053A), by funding from the European Union to the European Plant Phenotyping Network (EPPN2020; Grant Agreement 731013), from EMPHASIS-PREP (Grant Agreement 739514), from EOSC-Life (Grant Agreement: 824087) and from RI-VIS: (Grant Agreement: 824063). Particularly, we thank all of our colleagues in the various phenotyping networks for intensive discussions in recent years, which enriched the content of this publication significantly.

8 References

Araus, J. L., Kefauver, S. C., Zaman-Allah, M., Olsen, M. S. and Cairns, J. E. (2018). Translating high-throughput phenotyping into genetic gain. *Trends in Plant Science* 23(5), 451–466. doi:10.1016/j.tplants.2018.02.001.

Avramova, V., Nagel, K. A., AbdElgawad, H., Bustos, D., DuPlessis, M., Fiorani, F. and Beemster, G. T. S. (2016). Screening for drought tolerance of maize hybrids by multi-scale analysis of root and shoot traits at the seedling stage. *Journal of Experimental Botany* 67(8), 2453–2466. doi:10.1093/jxb/erw055.

Beauchêne, K., Leroy, F., Fournier, A., Huet, C., Bonnefoy, M., Lorgeou, J., de Solan, B., Piquemal, B., Thomas, S. and Cohan, J. P. (2019). Management and characterization of abiotic stress via PhénoField®, a high-throughput field phenotyping platform. *Frontiers in Plant Science* 10(904), 904. doi:10.3389/fpls.2019.00904.

Blais, F. (2004). Review of 20 years of range sensor development. *Journal of Electronic Imaging* 13(1). doi:10.1117/1.1631921.

Bolger, A. M., Poorter, H., Dumschott, K., Bolger, M. E., Arend, D., Osorio, S., Gundlach, H., Mayer, K. F. X., Lange, M., Scholz, U. and Usadel, B. (2019). Computational aspects underlying genome to phenome analysis in plants. *The Plant Journal: For Cell and Molecular Biology* 97(1), 182–198. doi:10.1111/tpj.14179.

Burridge, J. D., Schneider, H. M., Huynh, B. L., Roberts, P. A., Bucksch, A. and Lynch, J. P. (2017). Genome-wide association mapping and agronomic impact of cowpea root architecture. *TAG. Theoretical and Applied Genetics. Theoretische und Angewandte Genetik* 130(2), 419–431. doi:10.1007/s00122-016-2823-y.

Camargo, A. V. and Lobos, G. A. (2016). Latin America: a development pole for phenomics. *Frontiers in Plant Science* 7, 1729. doi:10.3389/fpls.2016.01729.

Carroll, A. A., Clarke, J., Fahlgren, N., Gehan, M. A., Lawrence-Dill, C. J. and Lorence, A. (2019). NAPPN: who we are, where we are going, and why you should join us! *The Plant Phenome Journal* 2(1), 1–4. doi:10.2135/tppj2018.08.0006, 180006.

Cendrero-Mateo, M. P., Muller, O., Albrecht, H., Burkart, A., Gatzke, S., Janßen, B., Keller, B., Körber, N., Kraska, T., Matsubara, S., Li, J., Müller-Linow, M., Pieruschka, R., Pinto, F., Rischbeck, P., Schickling, A., Steier, A., Watt, M., Schurr, U. and Rascher, U. (2017). Field phenotyping: concepts and examples to quantify dynamic plant traits across scales in the field. *Terrestrial Ecosystem Research Infrastructures*, pp. 53-80. CRC Press Taylor and Francis Group.

Chen, S., van der Graaff, E., Ytting, N. K. and Thorup-Kristensen, K. (2019). Evaluation of deep root phenotyping techniques in tube rhizotrons. *Acta Agriculturae Scandinavica, Section B - Soil and Plant Science* 69(1), 62-74. doi:10.1080/09064710.2018.1500635.

Choudhury, S. D., Samal, A. and Awada, T. (2019). Leveraging image analysis for high-throughput plant phenotyping. *Frontiers in Plant Science* 10, 508. doi:10.3389/fpls.2019.00508.

Clark, R. T., MacCurdy, R. B., Jung, J. K., Shaff, J. E., McCouch, S. R., Aneshansley, D. J. and Kochian, L. V. (2011). Three-dimensional root phenotyping with a novel imaging and software platform. *Plant Physiology* 156(2), 455-465. doi:10.1104/pp.110.169102.

Collard, B. C. Y. and Mackill, D. J. (2008). Marker-assisted selection: an approach for precision plant breeding in the twenty-first century. *Philosophical Transactions of the Royal Society of London. Series B, Biological Sciences* 363(1491), 557-572. doi:10.1098/rstb.2007.2170.

Costa, C., Schurr, U., Loreto, F., Menesatti, P. and Carpentier, S. (2019). Plant phenotyping research trends, a science mapping approach. *Frontiers in Plant Science* 9, 1933. doi:10.3389/fpls.2018.01933.

EC (2016). European charter for access to research infrastructures: principles and guidelines for access and related services. Available at: https://ec.europa.eu/research/infrastructures/pdf/2016_charterforaccessto-ris.pdf.

ESFRI (2016). Strategy report on research infrastructures. Available at: https://www.esfri.eu/sites/default/files/20160308_ROADMAP_single_page_LIGHT.pdf.

ESFRI (2017). Long-term sustainability of research infrastructures. Available at: https://www.esfri.eu/sites/default/files/ESFRI_SCRIPTA_SINGLE_PAGE_19102017_0.pdf.

FAO (2010). The second report on the state of the world's plant genetic resources for food and agriculture. Available at: http://www.fao.org/3/i1500e/i1500e00.htm.

FAO (2017). The future of food and agriculture - trends and challenges. Retrieved from Rome. Available at: http://www.fao.org/3/a-i6583e.pdf.

Fiorani, F., Rascher, U., Jahnke, S. and Schurr, U. (2012). Imaging plants dynamics in heterogenic environments. *Current Opinion in Biotechnology* 23(2), 227-235. doi:10.1016/j.copbio.2011.12.010.

Fiorani, F. and Schurr, U. (2013). Future scenarios for plant phenotyping. *Annual Review of Plant Biology* 64(1), 267-291. doi:10.1146/annurev-arplant-050312-120137.

Gargiulo, L., Grimberg, Å, Repo-Carrasco-Valencia, R., Carlsson, A. S. and Mele, G. (2019). Morpho-densitometric traits for quinoa (Chenopodium quinoa Willd.) seed phenotyping by two X-ray micro-CT scanning approaches. *Journal of Cereal Science* 90. doi:10.1016/j.jcs.2019.102829.

Genty, B. and Meyer, S. (1995). Quantitative mapping of leaf photosynthesis using chlorophyll fluorescence imaging. *Functional Plant Biology* 22(2), 277-284. doi:10.1071/PP9950277.

Gioia, T., Galinski, A., Lenz, H., Muller, C., Lentz, J., Heinz, K., Briese, C., Putz, A., Fiorani, F., Watt, M., Schurr, U. and Nagel, K. A. (2017). GrowScreen-PaGe, a non-invasive, high-throughput phenotyping system based on germination paper to quantify crop phenotypic diversity and plasticity of root traits under varying nutrient supply. *Functional Plant Biology: FPB* 44(1), 76-93. doi:10.1071/FP16128.

Granier, C., Aguirrezabal, L., Chenu, K., Cookson, S. J., Dauzat, M., Hamard, P., Thioux, J. J., Rolland, G., Bouchier-Combaud, S., Lebaudy, A., Muller, B., Simonneau, T. and Tardieu, F. (2006). PHENOPSIS, an automated platform for reproducible phenotyping of plant responses to soil water deficit in Arabidopsis thaliana permitted the identification of an accession with low sensitivity to soil water deficit. *New Phytologist* 169(3), 623-635. doi:10.1111/j.1469-8137.2005.01609.x.

Houle, D., Govindaraju, D. R. and Omholt, S. (2010). Phenomics: the next challenge. *Nature Reviews. Genetics* 11(12), 855-866. doi:10.1038/nrg2897.

Jahnke, S., Menzel, M. I., van Dusschoten, D., Roeb, G. W., Bühler, J., Minwuyelet, S., Blümler, P., Temperton, V. M., Hombach, T., Streun, M., Beer, S., Khodaverdi, M., Ziemons, K., Coenen, H. H. and Schurr, U. (2009). Combined MRI-PET dissects dynamic changes in plant structures and functions. *Plant Journal: For Cell and Molecular Biology* 59(4), 634-644. doi:10.1111/j.1365-313X.2009.03888.x.

Jahnke, S., Roussel, J., Hombach, T., Kochs, J., Fischbach, A., Huber, G. and Scharr, H. (2016). phenoSeeder–a robot system for automated handling and phenotyping of individual seeds. *Plant Physiology* 172(3), 1358-1370. doi:10.1104/pp.16.01122.

Jansen, M., Gilmer, F., Biskup, B., Nagel, K. A., Rascher, U., Fischbach, A., Briem, S., Dreissen, G., Tittmann, S., Braun, S., De Jaeger, I., Metzlaff, M., Schurr, U., Scharr, H. and Walter, A. (2009). Simultaneous phenotyping of leaf growth and chlorophyll fluorescence via GROWSCREEN fluoro allows detection of stress tolerance in Arabidopsis thaliana and other rosette plants. *Functional Plant Biology: FPB* 36(11), 902-914. doi:10.1071/FP09095.

Jeudy, C., Adrian, M., Baussard, C., Bernard, C., Bernaud, E., Bourion, V., Busset, H., Cabrera-Bosquet, L., Cointault, F., Han, S., Lamboeuf, M., Moreau, D., Pivato, B., Prudent, M., Trouvelot, S., Truong, H. N., Vernoud, V., Voisin, A. S., Wipf, D. and Salon, C. (2016). RhizoTubes as a new tool for high throughput imaging of plant root development and architecture: test, comparison with pot grown plants and validation. *Plant Methods* 12, 31. doi:10.1186/s13007-016-0131-9.

Jin, X., Zarco-Tejada, P. J., Schmidhalter, U., Reynolds, M. P., Hawkesford, M. J., Varshney, R. K., Yang, T., Nie, C., Li, Z., Ming, B., Xiao, Y., Xie, Y. and Li, S. (2020). High-throughput estimation of crop traits: a review of ground and aerial phenotyping platforms. *IEEE Geoscience and Remote Sensing Magazine* 9(1), 200-231. doi:10.1109/MGRS.2020.2998816.

Johannsen, W. (1903). *Ueber Erblichkeit in Populationen und in reinen Linien*. Jena: Gustav Fischer.

Johannsen, W. (1911). The genotype conception of heredity. *The American Naturalist* 45(531), 129-159.

Jump, A. S., Marchant, R. and Penuelas, J. (2009). Environmental change and the option value of genetic diversity. *Trends in Plant Science* 14(1), 51-58. doi:10.1016/j.tplants.2008.10.002.

Jun, K. O. and Park, T. S. (2017). Applications of smartphone cameras in agriculture, environment, and food: a review. *Journal of Biosystems Engineering* 42(4), 330-338. doi:10.5307/jbe.2017.42.4.330.

Kersey, P. J. (2019). Plant genome sequences: past, present, future. *Current Opinion in Plant Biology* 48, 1-8. doi:10.1016/j.pbi.2018.11.001.

Kherlopian, A. R., Song, T., Duan, Q., Neimark, M. A., Po, M. J., Gohagan, J. K. and Laine, A. F. (2008). A review of imaging techniques for systems biology. *BMC Systems Biology* 2, 74–74. doi:10.1186/1752-0509-2-74.

Kirchgessner, N., Liebisch, F., Yu, K., Pfeifer, J., Friedli, M., Hund, A. and Walter, A. (2017). The ETH field phenotyping platform FIP: a cable-suspended multi-sensor system. *Functional Plant Biology : FPB* 44(1), 154–168. doi:10.1071/FP16165.

Kuijken, R. C. P., van Eeuwijk, F. A., Marcelis, L. F. M. and Bouwmeester, H. J. (2015). Root phenotyping: from component trait in the lab to breeding. *Journal of Experimental Botany* 66(18), 5389–5401. doi:10.1093/jxb/erv239.

Le Marie, C., Kirchgessner, N., Flutsch, P., Pfeifer, J., Walter, A. and Hund, A. (2016). RADIX: rhizoslide platform allowing high throughput digital image analysis of root system expansion. *Plant Methods* 12(1), 40. doi:10.1186/s13007-016-0140-8.

Liu, Z. G., Gao, K., Shan, S., Gu, R. C., Wang, Z. K., Craft, E. J., Mi, G., Yuan, L. and Chen, F. J. (2017). Comparative analysis of root traits and the associated QTLs for maize seedlings grown in paper roll, hydroponics and vermiculite culture system. *Frontiers in Plant Science* 8, 436. doi:10.3389/fpls.2017.00436.

Mascher, M., Schreiber, M., Scholz, U., Graner, A., Reif, J. C. and Stein, N. (2019). GeneBank genomics bridges the gap between the conservation of crop diversity and plant breeding. *Nature Genetics* 51(7), 1076–1081. doi:10.1038/s41588-019-0443-6.

Metzner, R., Eggert, A., van Dusschoten, D., Pflugfelder, D., Gerth, S., Schurr, U., Uhlmann, N. and Jahnke, S. (2015). Direct comparison of MRI and X-ray CT technologies for 3D imaging of root systems in soil: potential and challenges for root trait quantification. *Plant Methods* 11(1), 17. doi:10.1186/s13007-015-0060-z.

Millet, E. J., Kruijer, W., Coupel-Ledru, A., Prado, S. A., Cabrera-Bosquet, L., Lacube, S., Charcosset, A., Welcker, C., van Eeuwijk, F. and Tardieu, F. (2019). Genomic prediction of maize yield across European environmental conditions. *Nature Genetics* 51(6), 952–956+. doi:10.1038/s41588-019-0414-y.

Millet, E. J., Welcker, C., Kruijer, W., Negro, S., Coupel-Ledru, A., Nicolas, S. D., Laborde, J., Bauland, C., Praud, S., Ranc, N., Presterl, T., Tuberosa, R., Bedo, Z., Draye, X., Usadel, B., Charcosset, A., Van Eeuwijk, F. and Tardieu, F. (2016). Genome-wide analysis of yield in Europe: allelic effects vary with drought and heat scenarios. *Plant Physiology* 172(2), 749–764. doi:10.1104/pp.16.00621.

Minervini, M., Scharr, H. and Tsaftaris, S. A. (2015). Image analysis: the new bottleneck in plant phenotyping [Applications Corner]. *IEEE Signal Processing Magazine* 32(4), 126–131. doi:10.1109/MSP.2015.2405111.

Mooney, S. J., Pridmore, T. P., Helliwell, J. and Bennett, M. J. (2012). Developing X-ray computed tomography to non-invasively image 3-D root systems architecture in soil. *Plant and Soil* 352(1–2), 1–22. doi:10.1007/s11104-011-1039-9.

Mueller-Linow, M., Wilhelm, J., Briese, C., Wojciechowski, T., Schurr, U. and Fiorani, F. (2019). Plant screen mobile: an open-source mobile device app for plant trait analysis. *Plant Methods* 15, 2. doi:10.1186/s13007-019-0386-z.

Muller, O., Keller, B., Zimmermann, L., Jedmowski, C., Kleist, E., Pingle, V., Rascher, U. (2018). Field phenotyping and an example of proximal sensing of photosynthesis Under elevated CO2. IGARSS, 2018-2018 IEEE International Geoscience and Remote Sensing Symposium, 8252–8254.

Nagel, K. A., Kastenholz, B., Jahnke, S., Van Dusschoten, D., Aach, T., Muehlich, M., Truhn, D., Scharr, H., Terjung, S., Walter, A. and Schurr, U. (2009). Temperature responses of roots: impact on growth, root system architecture and implications for phenotyping. *Functional Plant Biology : FPB* 36(11), 947-959. doi:10.1071/FP09184.

Nagel, K. A., Lenz, H., Kastenholz, B., Gilmer, F., Averesch, A., Putz, A., Heinz, K., Fischbach, A., Scharr, H., Fiorani, F., Walter, A. and Schurr, U. (2020). The platform GrowScreen-Agar enables identification of phenotypic diversity in root and shoot growth traits of agar grown plants. *Plant Methods* 16(1), 89. doi:10.1186/s13007-020-00631-3.

Nagel, K. A., Putz, A., Gilmer, F., Heinz, K., Fischbach, A., Pfeifer, J., Faget, M., Blossfeld, S., Ernst, M., Dimaki, C., Kastenholz, B., Kleinert, A. K., Galinski, A., Scharr, H., Fiorani, F. and Schurr, U. (2012). GROWSCREEN-Rhizo is a novel phenotyping robot enabling simultaneous measurements of root and shoot growth for plants grown in soil-filled rhizotrons. *Functional Plant Biology: FPB* 39(11), 891-904. doi:10.1071/FP12023.

Noleppa, S. (2016). The economic, social and environmental value of plant breeding in the European Union. An *ex post* evaluation and *ex ante* assessment. Available at: http://www.plantetp.org/system/files/publications/files/hffa_research_paper_plant_breeding_eu.pdf.

Paulus, S. (2019). Measuring crops in 3D: using geometry for plant phenotyping. *Plant Methods* 15(1), 103. doi:10.1186/s13007-019-0490-0.

Pieruschka, R. and Lawson, T. (2015). Preface. Phenotyping in plants: preface. *Journal of Experimental Botany* 66(18), 5385-5387. doi:10.1093/jxb/erv395.

Pieruschka, R. and Poorter, H. (2012). Phenotyping plants: genes, phenes and machines. *Functional Plant Biology: FPB* 39(11), 813-820. doi:10.1071/FPv39n11_IN.

Pommier, C., Garnett, T., Lawrence-Dill, C. J., Pridmore, T., Watt, M., Pieruschka, R. and Ghamkhar, K. (2020). Editorial: phenotyping; from plant, to data, to impact and highlights of the international plant phenotyping symposium - IPPS 2018. *Frontiers in Plant Science* 11, 618342. doi:10.3389/fpls.2020.618342.

Poorter, H., Fiorani, F., Pieruschka, R., Wojciechowski, T., van der Putten, W. H., Kleyer, M., Schurr, U. and Postma, J. (2016). Pampered inside, pestered outside? Differences and similarities between plants growing in controlled conditions and in the field. *New Phytologist* 212(4), 838-855. doi:10.1111/nph.14243.

Postma, J. A., Kuppe, C., Owen, M. R., Mellor, N., Griffiths, M., Bennett, M. J., Lynch, J. P. and Watt, M. (2017). OPENSIMROOT: widening the scope and application of root architectural models. *New Phytologist* 215(3), 1274-1286. doi:10.1111/nph.14641.

Pound, M. P., Atkinson, J. A., Townsend, A. J., Wilson, M. H., Griffiths, M., Jackson, A. S., Bulat, A., Tzimiropoulos, G., Wells, D. M., Murchie, E. H., Pridmore, T. P. and French, A. P. (2017). Deep machine learning provides state-of-the-art performance in image-based plant phenotyping. *GigaScience* 6(10), 1-10. doi:10.1093/gigascience/gix083.

Rebetzke, G. J., Jimenez-Berni, J., Fischer, R. A., Deery, D. M. and Smith, D. J. (2019). Review: high-throughput phenotyping to enhance the use of crop genetic resources. *Plant Science: An International Journal of Experimental Plant Biology* 282, 40-48. doi:10.1016/j.plantsci.2018.06.017.

Reynolds, D., Baret, F., Welckere, C., Bostrom, A., Ball, J., Cellini, F., Lorence, A., Chawade, A., Khafif, M., Noshita, K., Mueller-Linow, M., Zhou, J. and Tardieu, F. (2019). What is cost-efficient phenotyping? Optimizing costs for different scenarios. *Plant Science: An International Journal of Experimental Plant Biology* 282, 14-22. doi:10.1016/j.plantsci.2018.06.015.

Reynolds, M. and Schurr, U. (2019). The 4th international plant phenotyping symposium. *Plant Science: An International Journal of Experimental Plant Biology* 4th International Plant Phenotyping Symposium 282, 1. doi:10.1016/j.plantsci.2019.01.025.

Roussel, J., Geiger, F., Fischbach, A., Jahnke, S. and Scharr, H. (2016). 3D surface reconstruction of plant seeds by volume carving: performance and accuracies. *Frontiers in Plant Science* 7, 745. doi:10.3389/fpls.2016.00745.

Schmundt, D., Stitt, M., Jähne, B. and Schurr, U. (1998). Quantitative analysis of the local rates of growth of dicot leaves at a high temporal and spatial resolution, using image sequence analysis. *The Plant Journal* 16(4), 505-514. doi:10.1046/j.1365-313x.1998.00314.x.

Shi, R. L., Junker, A., Seiler, C. and Altmann, T. (2018). Phenotyping roots in darkness: disturbance-free root imaging with near infrared illumination. *Functional Plant Biology: FPB* 45(4), 400–411. doi:10.1071/FP17262.

Singh, A. K., Ganapathysubramanian, B., Sarkar, S. and Singh, A. (2018). Deep learning for plant stress phenotyping: trends and future perspectives. *Trends in Plant Science* 23(10), 883–898. doi:10.1016/j.tplants.2018.07.004.

Svane, S. F., Jensen, C. S. and Thorup-Kristensen, K. (2019). Construction of a large-scale semi-field facility to study genotypic differences in deep root growth and resources acquisition. *Plant Methods* 15(1), 26. doi:10.1186/s13007-019-0409-9.

Sydoruk, V. A., Kochs, J., van Dusschoten, D., Huber, G. and Jahnke, S. (2020). Precise volumetric measurements of any shaped objects with a novel acoustic volumeter. *Sensors* 20(3). doi:10.3390/s20030760.

Tanabata, T., Shibaya, T., Hori, K., Ebana, K. and Yano, M. (2012). SmartGrain: high-throughput phenotyping software for measuring seed shape through image analysis. *Plant Physiology* 160(4), 1871–1880. doi:10.1104/pp.112.205120.

Tardieu, F., Cabrera-Bosquet, L., Pridmore, T. and Bennett, M. (2017). Plant phenomics, From sensors to knowledge. *Current Biology: CB* 27(15), R770–R783. doi:10.1016/j.cub.2017.05.055.

Tracy, S. R., Nagel, K. A., Postma, J. A., Fassbender, H., Wasson, A. and Watt, M. (2020). Crop improvement from phenotyping roots: highlights reveal expanding opportunities. *Trends in Plant Science* 25(1), 105–118. doi:10.1016/j.tplants.2019.10.015.

Ubbens, J. R. and Stavness, I. (2017). Deep plant phenomics: a deep learning platform for complex plant phenotyping tasks. *Frontiers in Plant Science* 8(1190), 1190. doi:10.3389/fpls.2017.01190.

Vergauwen, D. and De Smet, I. (2017). From early farmers to Norman Borlaug–the making of modern wheat. *Current Biology: CB* 27(17), R858-R862. doi:10.1016/j.cub.2017.06.061.

Virlet, N., Sabermanesh, K., Sadeghi-Tehran, P. and Hawkesford, M. J. (2017). Field Scanalyzer: an automated robotic field phenotyping platform for detailed crop monitoring. *Functional Plant Biology: FPB* 44(1), 143–153. doi:10.1071/FP16163.

Walter, A., Spies, H., Terjung, S., Küsters, R., Kirchgeßner, N. and Schurr, U. (2002). Spatio-temporal dynamics of expansion growth in roots: automatic quantification of diurnal course and temperature response by digital image sequence processing. *Journal of Experimental Botany* 53(369), 689-698. doi:10.1093/jexbot/53.369.689.

Wasson, A. P., Rebetzke, G. J., Kirkegaard, J. A., Christopher, J., Richards, R. A. and Watt, M. (2014). Soil coring at multiple field environments can directly quantify variation in deep root traits to select wheat genotypes for breeding. *Journal of Experimental Botany* 65(21), 6231-6249. doi:10.1093/jxb/eru250.

Watt, M., Fiorani, F., Usadel, B., Rascher, U., Muller, O. and Schurr, U. (2020). Phenotyping: new windows into the plant for breeders. *Annual Review of Plant Biology* 71(1), 689–712. doi:10.1146/annurev-arplant-042916-041124.

Yang, G. J., Liu, J. G., Zhao, C. J., Li, Z. H., Huang, Y. B., Yu, H. Y., Xu, B., Yang, X., Zhu, D., Zhang, X., Zhang, R., Feng, H., Zhao, X., Li, Z., Li, H. and Yang, H. (2017). Unmanned aerial vehicle remote sensing for field-based crop phenotyping: current status and perspectives. *Frontiers in Plant Science* 8, 1111. doi:10.3389/fpls.2017.01111.

Chapter 2

The evolution of trait selection in breeding: from seeing to remote sensing

Matthew Reynolds, Francisco Pinto, Liana Acevedo, Francisco J. Pinera-Chavez, and Carolina Rivera-Amado, International Maize and Wheat Improvement Center (CIMMYT), Mexico

1 Introduction

Crops fall into six categories: food, feed, fiber, oil, ornamental, and industrial crops (e.g. timber and pharmaceuticals). This chapter will focus on the traits considered to improve food crops, especially cereals since they predominate in the human diet (Fig. 1). However, of around 400 000 known plant species, an estimated 300 000 have potentially edible components in the form of seeds, fruits, roots, tubers, leaves, stems, or flowers. It is therefore surprising that just a little over 100 species are cultivated by humans, while nine crop species supply over 75% of human dietary calories, with 60% coming from just four crops: rice, wheat, soy, and maize (Warren, 2015; Milla and Osborne, 2021). This suggests that domestication of wild plants - that presumably started in the Neolithic age if not before - must have involved considerable trial and error in terms of plant selection to arrive at the 100+ species we cultivate today and that as far as we know have been cultivated since prehistory. It is notable that no new plant domestication or crop speciation event has been achieved in the modern era - indicating the challenge of selection for genetic improvement - with the exception of Triticale (Ammar et al., 2004). Even this twentieth-century crop

http://dx.doi.org/10.19103/AS.2022.0102.02

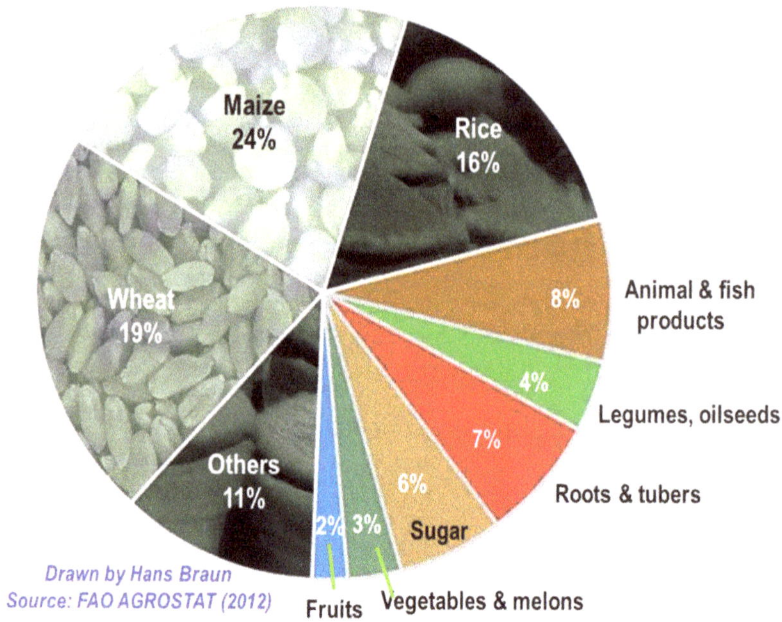

Figure 1 The proportions of the crops produced globally as a percent of their total dry matter (approximately 3 billion tons annually). Drawn by Hans Braun.

invention was a hybridization of other domesticated species, namely wheat and rye, and has proven to be quite difficult to market in spite of its many stress adaptive traits, wide adaptation, and high radiation use efficiency (Mergoum and Gómez-Macpherson, 2004). Clearly humans are quite conservative about the characteristics of our food crops.

Nonetheless, the principles of trait selection are similar for all categories of food and other crops. This is because they typically face similar types of challenges when it comes to resisting or tolerating biotic stresses (e.g. diseases, pests, and weeds), and since the majority of crops are grown in the field, they also must face environmental challenges associated with variable temperatures, water supply, light, and soil conditions.

Crop management - which was presumably also a fledgling discipline in neolithic times - can in theory optimize the crop environment through fertilization, control of weeds, pests and diseases, and choice of planting systems including sowing dates, crop rotations, and irrigation, where feasible. However, the fact that significant yield gaps exist still in most annual cropping systems (Lobell et al., 2009) underscores the importance of selection for heritable traits - that is, plant breeding. This is because once acquired by a farmer, a new cultivar can usually be relied on to express the desired expression of traits - including yield expectations and quality, as well as their robustness under stresses – across the target population of environments which must encompass seasonal weather fluxes. That given, understanding and improving trait selection is one of the most effective ways to improve crop productivity and changing cultivars is one of the easiest interventions a farmer can make. Furthermore, trait measurement is a prerequisite to the genetic dissection of yield and other traits (Reynolds et al., 2020) and the knowledge that can be used to facilitate the selection and accumulation of complementary alleles (Godwin et al., 2019).

1.1 Plant selection in prehistory

Domestication of plants is estimated to have started 9000–11 000 years ago and is believed to have occurred incrementally over periods of thousands of years through visual selection. Most domesticated plants used today were domesticated within their centers of origin, mainly in Asia and the Americas (Harlan, 1972; Milla and Osborne, 2021). Based on the characteristics that have been passed down through time and in comparison to wild ancestors, we can deduce that the early plant breeder/farmers would have selected for three main characteristics: first, the preferential growth of edible tissues to achieve bigger organs, whether seeds, fruits, roots, or tubers; essentially they were selecting for partitioning to maximize harvest index. Second – though not necessarily in that order – they were selected for palatability and nutritional value. It is perhaps counterintuitive that the edible target of many crop progenitors was toxic, such as the seeds of legumes, the tubers of some wild potatoes, and the roots of cassava: cultivated cassava still contains glycosides (linamarin) that are degraded to cyanide if not processed adequately, some beans such as those of Lathyrus are toxic at high doses, and potato skins left in the light produce solanine, a toxic glycoalkaloid. Fungi consumption dates back at least 13 000 years and their cultivation at least 2500 years despite the fact that some types that are toxic resemble edible species. It is a matter of speculation on how Neolithic man bred for nontoxic crops (one can only hope it was not largely a matter of trial and error). Third, adaptation to a range of stresses would have been key selection criteria when domesticating the species, especially biotic stresses which breeders continue to struggle with to the present day (Singh et al., 2015). In summary, while modern-day selection criteria have been refined, they are qualitatively the same kinds of traits that our ancestors selected for, suggesting that we are still domesticating our crop species (Denison et al., 2003).

1.2 Plant selection in the industrial age

Mendel's experiments with plant hybridization led to the first scientific documentation of hereditary principles, although this must have been understood intuitively by our ancestors for them to have domesticated so many plants and animal species. Nonetheless, the new discipline of genetics stimulated research in plant science with the objective of boosting crop productivity through selection of desirable traits, and the discipline of plant breeding was reborn as a scientific and commercial pursuit.

Gartons Agricultural Plant Breeders established in the 1890s in the UK was one of the first to commercialize new higher yielding crop cultivars, while William Farrer in Australia is credited with breeding the first rust fungus-resistant strain of wheat called Federation. In the same time frame Nazareno Strampelli in Italy created a number of higher yielding wheat lines between the two World Wars, some of which were exported to the Americas and even China (Salvi et al., 2013). In 1908, George Harrison Shull started to select for hybrid vigor in maize (Britannica, 2021). A decade later, the newly invented discipline of statistics enabled traits to be dissected genetically enabling for the first time a quantitative distinction between heritable variation and that caused by environment (Fisher and Mackenzie, 1923).

These efforts laid the groundwork for the Green Revolution which essentially led to a paradigm shift in plant selection and agricultural productivity, kick started by the successful introduction of semi-dwarf genes in wheat and other cereals in the 1960s. Before their adoption, cereal yields were limited by structural failure (lodging) that occurred when inputs such as fertilizer and irrigation caused plants to grow taller than most people. The source of the new allele in wheat (*Rht1*) was Norin 10 originally developed by Gonjiro Inazuka in Japan in 1935. It took more than 10 years to achieve its effective introgression, but its pleiotropic effects improved not only lodging resistance but also harvest index (HI) and nitrogen use efficiency (NUE), spearheading the Green Revolution that more than tripled yield potential of major cereal crops (Reynolds and Borlaug, 2006).

Shorter crops – that is, plant stands that can be easily observed from above – are also more amenable to measurement and selection than taller ones. So in a sense, the Green Revolution also facilitated a revolution in phenotyping. This chapter will discuss the different classes of phenotyping and how they can be applied in crop breeding.

1.3 The central role of phenotypic selection in plant breeding

Plant hybridization and selection has achieved not only steady yield gains (Fischer et al., 2014) but also resistance to a wide range of pests and diseases (Agrios, 2005) and high standards of end-use quality in many cases (Guzmán et al., 2017). Plant breeding of field crops still depends to a large extent on selection of a few agronomically important traits and this review will present these approaches as well as the potential to expand the inventory of traits and phenotyping tools that can be and in some cases are already being used in breeding programs. The advent of the 'gene revolution' led to much speculation that field breeding would shortly become redundant, something which has not happened to date. Certainly progress is being made, some markers can be applied routinely in breeding for genes of major effect and as part of genome-wide selection schemes (Bernardo, 2013). Gene editing can also be used to achieve targeted mutation where genes of major effects have been cloned (Metje-Sprink et al., 2019). The impact of such technologies will increase with better genome sequence and diversity information (Hu et al., 2018). However, the measurement and use of phenotypic traits are not only a useful stop-gap measure while more comprehensive genomic tools are developed but also complementary to marker-assisted selection (Rutkoski et al., 2016) and essential prerequisite to understanding the genetic bases of traits that molecular technologies are necessarily based on.

Apart from yield *per se*, different selection traits have shown value in breeding for both biotic factors (Singh et al., 2020) and abiotic ones (Reynolds, 2010; Sadras and Richards, 2014). Furthermore, public and private breeding programs have invested in high-throughput phenotyping (HTP) with the expectation of increasing efficiency and selection accuracy (Roitsch et al., 2019). However, precision phenotyping of traits – that cannot be estimated at high throughput – is also necessary in order to be able to make comprehensive assessment of candidate parents with the view to designing strategic crosses that stack complementary traits (Reynolds and Langridge, 2016). This is a natural progression in breeding based on improved phenomics capability, and also because as environments become more challenging, physiological and genetic models must encompass new traits and trait interactions.

While most breeders are reticent to apply physiological breeding approaches for complex traits into extant pipelines, new knowledge about physiological and genetic bases of yield and other commercially important traits coupled with advances in phenomics make this more feasible now. Nonetheless, wholesale adoption of advanced phenotyping methods must be affordable, add significant value, and be user friendly.

The degree to which new selection approaches are considered 'Breeder-friendly' will depend in part on scale, which in turn is a function of at what stage of the breeding pipeline plant or plot selection takes place (Fig. 2). We previously described three main classes of field-based phenotyping traits (Reynolds et al., 2020): 'Handy' phenotyping traits including those either routinely measured, or at least readily accessible for phenotyping at a breeding scale with relatively little capital investment; (2) 'High throughput phenotyping (HTP)' traits which require some training or specialization to measure and require the construction or purchase of specialized equipment; such data may also provide derived traits via models; and (3) 'Precision' phenotyping traits which are generally

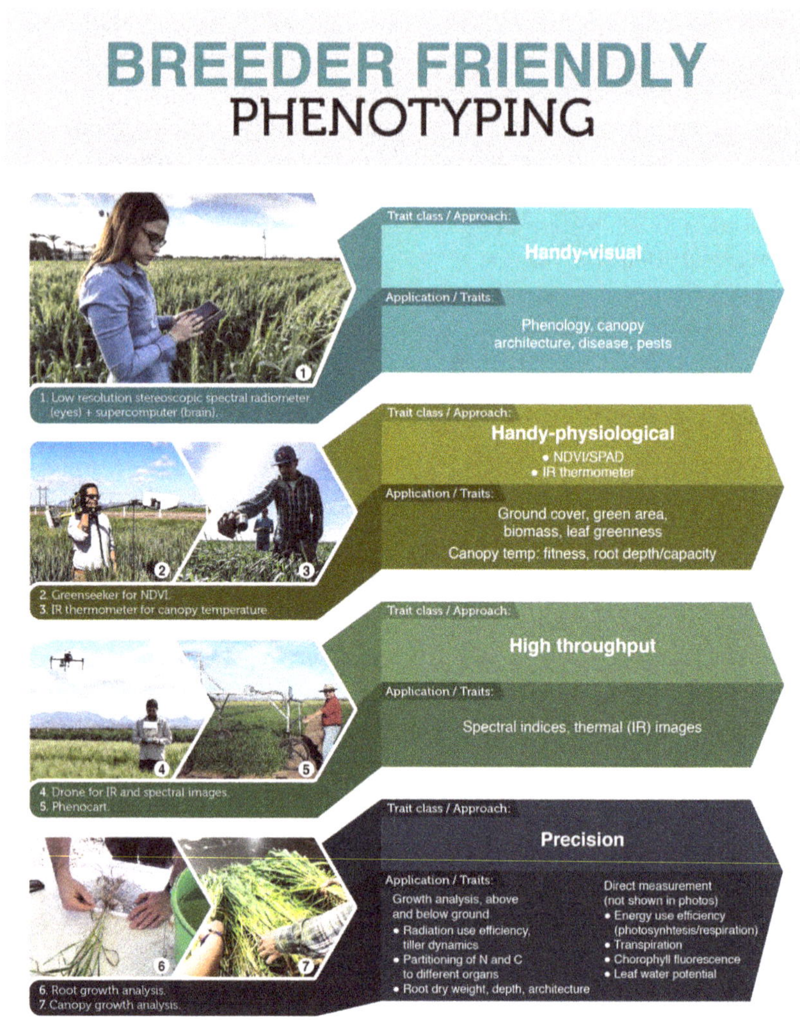

Figure 2 Examples of different classes and applications of breeder friendly phenotyping. Source: Adapted from Reynolds et al. (2020).

lower throughput and/or expensive but offer the potential for high pay-off in terms of opening current bottlenecks to yield improvement. Readers are also referred to other recent reviews, especially on high-throughput approaches (Araus et al., 2001, 2018; Araus and Cairns, 2014; Fahlgren et al., 2015; Araus and Kefauver, 2018; Furbank et al., 2019).

However, in this review we consider phenotyping interventions for different stages in the crop breeding pipeline. For example, selection of progeny - which may be in the millions of plants for a large breeding program - requires a much higher throughput than characterization of candidate parents. A trait taking several minutes or more to measure is not user-friendly for progeny selection. However, when comparing a few hundred candidate parents with the view to stacking complementary traits or for building prediction models based on panels of representative genotypes, more comprehensive and time consuming trait measurement is acceptable, and is more likely to fit in the context of translational research and pre-breeding (Richards, 2006; Reynolds et al., 2017, 2019; Molero et al., 2019) if not in a mainstream breeding context (Rutkoski et al., 2016).

2 Selection of progeny and large-scale genetic resources

2.1 Traits amenable to visual scores and proximal sensing

Practical plant breeders typically consider a minimum data set for agronomic evaluation. Such traits are useful for preliminary screening of large populations of uncharacterized lines such as segregating progeny or new genetic resources and are also generally required at the level of national yield trials to characterize potential new varieties. The minimum data set usually includes height, lodging tendency, time to heading/flowering and physiological maturity, disease and pest resistance, and seed characteristics such as size and color. The former are for the most part relatively simply inherited (Mendelian) traits and therefore show robust expression in terms of genotypic ranking across a wide range of environments. However, two other traits that show complex inheritance - and therefore may interact considerably with environments - are also essential, namely end-use seed quality characteristics and of course yield. In the last decade or so, a few handy physiological screens have started to be adopted that can help indicate the adaptive potential of a genotype. Some examples include assessing above-ground vigor using the Normalized Difference Vegetation Index (NDVI) (Rouse et al., 1974), assessing leaf or canopy N status using SPAD meter or NDVI, and estimating root capacity or relative gas exchange rate using canopy temperature under stress (Fischer et al., 1989; Pantuwan et al., 2002; Lopes and Reynolds, 2010; Pinto and Reynolds, 2015). These handy phenotyping tools have relatively low cost.

2.2 Traits amenable to remote sensing

Optical remote sensing (RS) of vegetation is a discipline that uses noncontact sensors to estimate structural and functional properties of plants, mainly through the measurement of their spectral reflectance. Although RS started to be used for studying the vegetation in the 1960s (Houborg et al., 2015), it was not until the recent massification of digital technology in the last two decades that the prospects of using RS expanded to plant phenotyping in the breeding context (Costa et al., 2019). Indeed, limitations of conventional phenotyping for screening large populations, especially for physiological traits, only started to be overcome with the advent of technologies capable to provide sufficient data resolution and that could increase our capacities for data collection (in space and time) and processing. The development of and increased accessibility to different measuring platforms, such as unmanned aerial vehicles (UAVs), aircrafts, or autonomous ground vehicles, have also contributed to facilitate the adoption of RS for plant phenotyping while reducing errors associated to data collection (Atzberger, 2013).

Any trait or process that affects the optical properties of the leaf or canopy can potentially be measured with a sensor. The biophysical and biochemical characteristics of vegetation determine the fraction of the incident radiation that is reflected, absorbed, or transmitted at a specific wavelength. At the leaf level, the main factors controlling the spectral reflectance are leaf pigment content, intracellular spaces in the mesophyll, and quantity of water in leaf tissue (Jensen, 2000). Pigments such as chlorophyll, carotenoids, or anthocyanins are usually the main factors shaping the reflected radiation in the visible part of the spectrum (VIS; 400–700 nm; Jensen, 2000; Blackburn, 2007), whereas water content, cell wall constituents, and intercellular spaces do it in the near infrared (NIR; 700–1100 nm; Knipling, 1970; Curran, 1989). At the canopy level, the interpretation of the spectral reflectance becomes more challenging. The canopy structure determines multiple scattering and reabsorptions of light within the vegetation (Kobayashi and Iwabuchi, 2008; Govind et al., 2013).

Spectral measurements and derived spectral reflectance indices (SRI) have been used to track changes and differences between genotypes in pigment-sensitive traits such as biomass (Yue et al., 2018; Robles-Zazueta et al., 2021), nitrogen content (Fava et al., 2009), stay green (Lopes and Reynolds, 2012), leaf area index (Yue et al., 2018), light interception (Pellegrini et al., 2020), photosynthetic capacity (Zhang et al., 2016; El-Hendawy et al., 2019), stover biochemical composition (Hames et al., 2003), and disease detection (Ashourloo et al., 2014). The NDVI is arguably best-known and most widely used. SRIs related to water content (i.e. in NIR and short-wave infrared part of the spectrum) have been used as indicators of stress adaptation (Gutierrez et al., 2010a; Wang et al., 2015) and also as indirect estimation of green biomass (Babar et al., 2006). Changes in time of some of these indices can also serve as proxy for tracking phenology (Zheng et al., 2016), changes in structural properties of the canopy (Jin and Eklundh, 2014), and evolution or intensity of plant diseases (Zhang et al., 2019).

Although the use of SRI is widespread and relatively easy to implement, these approaches are quite limited, for they use only a small part of the spectral information and in many occasions they lack sensitivity (Baret and Buis, 2008; Berger et al., 2018). With the development of machine learning algorithms, the analysis and construction of models based on the full spectral signature is expanding. Machine learning approaches are being used for the quantification of more complex traits such as radiation use efficiency (Fu et al., 2014; Zhang et al., 2017), photosynthesis (Serbin et al., 2012; Silva-Perez et al., 2018), or yield (Pantazi et al., 2016; Chlingaryan et al., 2018). Spectral measurements can also be combined with leaf or canopy radiative transfer models (RTMs) to retrieve relevant biophysical and biochemical parameters of the vegetation (Thorp et al., 2015; Yu et al., 2017). While machine learning algorithms are basically empirical models that require site-specific calibrations, RTMs rely on understanding of the radiative transfer functions within the plant tissue or canopy while considering plant metabolism. Since such models would account for different factors influencing the spectral reflectance, it is expected that they are more robust across different measuring conditions (Berger et al., 2018). However, the construction and validation of these models is complex and there are still challenges for their adoption in the field (Berger et al., 2018).

Spectroscopy can be used to measure not only reflected or transmitted radiation but also emitted radiation resulting from metabolic processes. The emission of infrared (IR) radiation by the plants is directly related to their temperature and consequently serves as a good proxy for the assessment of physiological changes in the vegetation resulting from its interaction with the environment (Reynolds et al., 2020). A good example is the use of canopy temperature as an adaptive trait for drought and heat environments (Lopes and Reynolds, 2010), where genotypes with cooler canopies are a consequence of their capability to keep stomata open for longer periods of time, ensuring more photo-assimilates and consequently resulting in higher yields. Likewise, canopy temperature has been shown to estimate water use efficiency in cotton (Conaty et al., 2015) and rooting capacity in wheat under heat and drought stress (Lopes and Reynolds, 2010; Pinto and Reynolds, 2015). Plants also emit a small fraction of radiation as chlorophyll fluorescence. Approaches designed to quantify the amount of sun-induced fluorescence (i.e. under the prevailing conditions; Cendrero-Mateo et al., 2018), as well as those using active fluorometry (e.g. LIFT; Keller et al., 2019), have great potential as proxies to identify genotypic differences in photosynthetic capacity and radiation use efficiency.

With the advent of higher-resolution imaging sensors and the rapid increases of computing capacities, computer vision has become one of the cornerstones for high-throughput phenotyping approaches nowadays. Image processing algorithms not only make use of the spectral information but also they are capable of analyzing shapes, patterns, and textures within the images. Deep learning (DL) is opening new opportunities for quantifying morphological and structural traits at unprecedented pace. The use of DL enables the accurate image-based identification and localization of different vegetation elements, which combined with other photogrammetric techniques provide powerful tools capable of quantifying traits such as stand density (Zhang et al., 2020), density of harvestable organs in cereals (Ghosal et al., 2019; Sadeghi-Tehran et al., 2019), internode length or diameter (Yamamoto et al., 2016; Xiang et al., 2020), phenology (Rasti et al., 2020), root architecture (Yasrab et al., 2019), and detection/identification of disease infections (Camargo and Smith, 2009; Neumann et al., 2014).

Although the list of traits that can potentially be assessed using RS tools is increasing due to new technologies available, the successful use of sensors for HTP to detect differences between genotypes is still challenged by confounding factors that need to be accounted for, especially under field conditions (Reynolds and Pinto, 2019). Correct interpretation of HTP data requires an understanding of how environmental factors affect the expression of a particular trait at the time of measurement (Reynolds et al., 2020). For instance, ambient temperature, cloud cover, wind gusts, or irrigation schedules can affect, almost instantaneously, the measurement of canopy temperature and SRI related to water content (Gutierrez et al., 2010b). Similarly, the growth stage also interacts with most traits, and therefore morphological or anatomical changes may have an impact in the signal captured by sensors (Gutierrez et al., 2015). Well-defined measurement protocols and adequate experimental designs can partly account for these confounding factors while improving heritability.

Good characterization of the environment can assist in the interpretation of sensor-based phenotypic data, whereas signal processing techniques help to improve the accuracy and repeatability of the measurements, increasing the capacity to discriminate phenotypic differences in time and between genotypes.

3 Characterization of parents and gene discovery panels: increasing throughput with sensors

Detailed characterization of candidate parents for strategic crossing, and within panels of selected lines for gene discovery research, typically involve destructive or invasive measurements. Nonetheless, less invasive techniques involving sensor technologies are becoming more feasible for such precision phenotyping.

3.1 Sensing photosynthesis

Despite improvements in light interception and partitioning of assimilates to harvested organs, little improvement has been made to light conversion efficiency in the last half century through modern plant breeding techniques. Therefore, improving photosynthesis presents an opportunity to significantly increase crop productivity, create more climate-resilient crops, and improve resource-use efficiencies. Utilizing transgenic means, it is estimated that improving photosynthesis could result in an increase in the yields of up to 60% (Long et al., 2015). However, through more conventional breeding approaches gains of 15–30% are still estimated for improving net photosynthesis (Long et al., 2015). Photosynthesis has not been a primary focus of breeding programs due to the low-throughput nature of infrared gas exchange measurements used to measure CO_2 assimilation. A single measurement can take upward of 3 minutes and utilizes equipment that is expensive and requires specialized knowledge to use. Consequently, great effort has recently been placed into developing methods to measure photosynthesis and photosynthesis-related traits in a high-throughput manner.

3.2 Improving plant response to fluctuating light and photoprotection

When a photon reaches a leaf, it can either be used to drive photochemistry, be reemitted as chlorophyll fluorescence, or be dissipated as excess heat through non-photochemical quenching (NPQ). Energy that is not quenched by the leaf, either through photochemistry or excitation transfer, can go on to form reactive oxygen species (ROS) that can damage the photosynthetic machinery of the leaf, particularly Photosystem II. The last process – the dissipation of heat through NPQ – acts in response to excess light to protect the leaf from light damage. An often-overlooked source of abiotic stress in plants is excess light. Prolonged excess light can result in the inhibition of photosynthesis or photoinhibition. Photoinhibition at early stages is reversible, however, if the leaf is exposed to excessive light for a prolonged period it can result in nonreversible photodamage of the photosynthetic machinery.

Excess light energy is dissipated by a leaf through the xanthophyll cycle. The xanthophyll cycle is comprised of three carotenoids – violaxanthin, antheraxanthin, and zeaxanthin – that are seen in greater quantities depending upon light intensity. When a plant is exposed to high light, violaxanthin is converted first into antheraxanthin and then zeaxanthin (Taiz and Zeiger, 2015). Upon a return to low light, zeaxanthin is converted back into violaxanthin in a process known as NPQ relaxation. Zeaxanthin is more prevalent under high-light conditions and reduces in abundance under low-light conditions.

In addition to its direct role in photoprotection, much interest has been dedicated recently to the improvement of NPQ and its relaxation as a method to increase overall plant productivity (Murchie and Niyogi, 2011; Long et al., 2015; Kromdijk et al., 2016). A faster NPQ response would allow leaves to respond more quickly to changes in irradiance and associated stress. Recently, increasing the rate of NPQ relaxation in tobacco resulted in an increase of

17% in biomass production (Kromdijk et al., 2016). Additionally, recent studies have found significant variation for NPQ and NPQ relaxation by characterizing the natural variation present in cultivated wheat and wild wheat relatives (McAusland et al., 2020). Photoprotection, or NPQ, can be measured through chlorophyll fluorescence measurements. Leaves must be dark adapted first to obtain F_m, or the maximum possible yield of fluorescence (Maxwell and Johnson, 2000; Murchie and Lawson, 2013). NPQ can then be calculated at a single point or through time as $(F_m - F_m')/F_m'$, where F_m' is the maximum possible yield of fluorescence in the light (Maxwell and Johnson, 2000). Measuring NPQ upon a transition from high-light to low-light, as when mimicking the passing of a sunfleck, can allow for the quantification of NPQ relaxation.

While fluorescence measurements that are coupled with gas exchange offer a more complete picture as to what is occurring in the leaf, gas exchange measurements can be time consuming and present a significant bottleneck to understanding photosynthesis-related traits on a larger scale. Chlorophyll fluorescence measurements can be deployed in the field through handheld devices, such as the MultispeQ (PhotosynQ, Lansing, Michigan USA) or LI-600 porometer/fluorometer (LI-COR Environmental, Lincoln, Nebraska USA). Measurements performed with these kinds of light-weight equipment provide insight into plant photoprotection and stress in real-time without altering the environment in which the plants are being grown. However, measurements can also be performed utilizing larger chlorophyll fluorescence imagers (Technologica Ltd., Colchester, United Kingdom; PSI Instruments, Drásov, Czech Republic), which allow for the screening of multiple samples or whole plants simultaneously. With adequate planning, these kinds of equipment can allow for the screening of hundreds of genotypes in one day, making it amenable for screening larger panels used in breeding programs. Beyond screening for photoprotection-related traits, chlorophyll fluorescence measurements can also be used to screen for stress from abiotic and biotic sources in the field, which may be of use to breeders.

4 Traits related to spike fertility and partitioning to yield

Grain yield in crops depends on the amount of assimilate supply, known as 'source' and reflected in biomass production and in the capacity of the grains to store it, known as 'sink' and reflected in grain set, final weight and biomass partitioning to grain (Sadras and Connor, 1991). In wheat, for example, recent evidence has shown trade-offs between biomass production and harvest index (HI, proportion of aboveground dry matter allocated in grains) (Rivera-Amado et al., 2019; Sierra-Gonzalez et al., 2021) and non-linear associations between HI and grain yield (Aisawi et al., 2015), indicating that maintaining or ideally increasing assimilate partitioning to grain is crucial in breeding for yield potential (Foulkes et al., 2011). The application of proximal and RS to phenotype biomass, greenness, or water content is moving fast in wheat (Tattaris et al., 2016; Lu et al., 2019; ten Harkel et al., 2019) and other crops such as maize and rice (Cen et al., 2019; Han et al., 2019; Niu et al., 2019) with promising results for canopy photosynthetic-related traits too (Robles-Zazueta et al., 2021) as described in the above sections. In contrast, phenotyping biomass partitioning and fertility-related traits rely mostly on manual or visual assessments that are generally susceptible to variability, error and are time-labor consuming and not easily assessed using HTP (Dreccer et al., 2019). The development of HTP protocols for 'precision phenotyping' traits related to biomass partitioning and spike fertility would benefit their application in segregant selection, mining genetic resources, or understating genotype by environment interactions (Dreccer et al., 2019; Reynolds et al., 2020).

HI is by nature a trait almost as complex as yield and generally low throughput. However, the relevance of this trait in breeding programs has stimulated research on estimations using crop modeling and traits related to source and sink physiological characteristics as well as associations between HI and the proportion of total water used during grain filling (Richards and Townley-Smith, 1987). A simple model based on the relationship between HI and the fraction of water transpired after anthesis accounted for 75% ($P<0.001$) of the variance of HI in wheat and 81% ($P<0.001$) in sunflower plants distinguishing between source and sink limitations (Sadras and Connor, 1991). Kemanian et al. (2007) followed an approach to estimate HI in wheat, barley, and sorghum based on the assumption that HI responds linearly or curvilinearly to fractional post-anthesis growth. Results showed the model was able to predict HI better under source or source-sink co-limited grain filling but in sink limited conditions the extent of the limitation needed to be added to the model. Recently, Campoy et al. (2020) proposed a RS approach to estimate HI in wheat based on variants of the two references above by estimating absorbed photosynthetically active radiation and crop transpiration from time series satellite images and meteorological data. Comparisons with empirical data showed the methodology was able to reproduce HI values across different management conditions but genotype effects were not tested.

Although dry matter (DM) partitioning to yield is normally assessed as HI, it is a continuous process involving the distribution of DM between plant organs (Marcelis, 1996), where early DM partitioning events can be decisive in the determination of fertility rates and yield. Through the last century, improvements in HI in grain crops have been a consequence of increased grain number per unit area (grains m^{-2}) and stable grain weight (Hay, 1995). In wheat, grains per unit area are a product of (1) the duration of the early reproductive phenophase, determining spike subunits, floret and tiller number and the late reproductive phenophase, determining the number of fertile florets simultaneously with the active growth of stems; (2) biomass production during that time; (3) the amount of biomass partitioned to the spikes at flowering; and (4) fruiting efficiency (FE, the ratio between final grain number and spike DM at anthesis) (Fischer, 2011). Further increases in grains per unit area in wheat may be achieved by increasing partitioning to spikes during the late reproductive phase (Foulkes et al., 2011). Recent research has proposed reductions in DM partitioning from middle stem internodes (i.e. peduncle -1 and -2) to favor increased DM partitioning to spikes at flowering, grains per unit area and HI in elite spring wheat (Rivera-Amado et al., 2019; Sierra-Gonzalez et al., 2021).

These studies have shown that internode length is a good proxy for internode DM, providing opportunities for HTP developments to screen germplasm for lines that allocate less assimilate to middle internodes. On the other hand, DM partitioning to lower internodes is a key trait in lodging resistance as discussed below. Xiang et al. (2020) reported a HTP field-based system to estimate stem diameter in Sorghum plants based on high-fidelity 3D models reconstructed from stereoscopic images. Although the potential to measure internode morphology at canopy level from image analysis can be limited by the occlusion of other parts of the plant such as spikes and leaves in wheat and other crops with a high tiller density, DL techniques able to estimate morphological parameters of stem subunits from field samples ex-situ could aid current protocols involving oven-drying and weighing. Image-analysis-based protocols have been successful in detecting nodes and measuring internode length in controlled environment experiments in tomato (Yamamoto et al., 2016) and in chopped stems in corn and switchgrass (Pothula et al., 2014). High-throughput characterization of spike dry-weight area or volume could aid estimations of DM partitioning to spikes and FE. However, due to the complex three-dimensional spike structure of wheat and the presence of awns, the determination of its area or volume by non-destructive methodologies represents a major challenge especially aiming for HTP (Dreccer et al., 2019). Estimation of area and volume in wheat ears based on images has been successful only in controlled environments or by destructive approaches with no field applications (Dreccer et al., 2019; Korohou et al., 2020).

As mentioned above, FE is the ratio between final grain number and spike DM at anthesis and the most common methodology to measure spike DM at anthesis in field plots is through direct samplings of aboveground biomass and spike partitioning (Slafer et al., 2015). Alternatively, Abbate et al. (2013) proposed a simpler method based on associations between spike DM at anthesis and spike chaff DM at harvest and between plot level (both components of FE expressed by unit area) and individual spike level. Based on strong linear associations in both comparisons ($R^2 > 0.7$), they concluded it was possible to produce sensible estimations of FE from a small sample of individual spikes at harvest. This methodology has been used recently showing promising results in the use of FE as a selection criterion for increasing yield genetic progress and stability (Alonso et al., 2018). However, selecting for FE - even from a small sample at harvest - might still be unrealistic or difficult in large breeding programs with thousands of selection plots per season (Slafer et al., 2015). Alonso et al. (2018) showed that spikes from a cultivar with low FE expressed lax and long with voluminous chaff structures and high spikelet numbers and a cultivar with high FE showed a shorter and compact spike with thinner chaff structures. In this setting, investigating non-destructive high-throughput approaches to estimate spike DM through volume or size could be valuable in screening for this trait, as discussed above. Considering the other numerical component of FE (grain number, a product of spike number per unit area and grains per spike) (Slafer et al., 1996), neural network approaches to count plants (Liu et al., 2017), or spikes per unit area (Velumani et al., 2017; Fernandez-Gallego et al., 2018; Sadeghi-Tehran et al., 2019) from RGB images in wheat plots have shown promising results. On the contrary, grain counting (per spike or unit area) in the field represents a bigger challenge and is best estimated or derived from samples taken after harvest (Dreccer et al., 2019).

Water soluble carbohydrate (WSC) accumulated in wheat stems can contribute to yield under favorable conditions from 10% to 20% (Bidinger et al., 1977; Schnyder, 1993; Gebbing et al., 1999; Shearman et al., 2005) or act as a buffer under water limitations or heat stress contributing to yield as much as 50% (Blum, 1998; van Herwaarden et al., 1998). Under drought, in particular, a greater accumulation of stem WSC seems to be associated to with better maintenance of HI (Foulkes et al., 2002). Stem WSC shows wide genetic variation across environments and large transgressive segregation but also weak correlations between concentration and amounts by unit area and a very complex genetic control (Ruuska et al., 2006; Rebetzke et al., 2008), although positive departures in understanding its genetic control have been identified more recently (McIntyre et al., 2012; Li et al., 2015; Dong et al., 2016). Stem WSC are traditionally evaluated from destructive sampling by a wet chemistry method (Yemm and Willis, 1954) or by NIR reflectance spectroscopy calibrated against the chemistry method (Wang et al., 2011). Non-destructive high-throughput estimations of stem WSC would allow not only for the trait to be selected in segregant populations but also as an input to crop model simulation to estimate the buffer capacity and improve yield predictions under unexpected seasonal changes (Dreccer et al., 2014). Robust nondestructive predictions of stem WSC concentrations and amounts in wheat canopies in the field were possible using hyperspectral data where more relevant regions were 350-988 nm and 1010-1290 nm (Dreccer et al., 2014). In perennial ryegrass, nondestructive field spectroscopy combined with machine learning allowed infield predictive models to measure nutritive value of individual plants model that included WSC (Smith et al., 2020).

5 Traits to improve lodging resistance in cereals

Lodging, the permanent displacement of crop stems from their vertical position, is a problem affecting the yield of a wide range of crops - wheat, barley, rice, oats, maize, oilseed rape, and so on - that involves the interaction of plants with abiotic (e.g. wind, rain, and soil) and biotic (e.g. diseases) factors and the crop management (e.g. fertilizers, irrigation schemes, and plant growth regulators) (Berry et al., 2004; Joseph et al., 2020). Reductions of grain yield have been estimated from 7% to 80% for wheat (Weibel and Pendleton, 1964; Fischer and Stapper, 1987; Easson et al., 1993; Berry et al., 2004; Tripathi et al., 2005; Acreche and Slafer, 2011; Berry and Spink, 2012; Peake et al., 2014), 28-65% for barley (Berry et al., 2004), 40% for rice (UKRI, 2021), 37% for oats (Berry et al., 2004), 14-48% for maize (Berry et al., 2021), and 16-52% for oilseed rape (Kendall et al., 2017). Additionally, lodging reduces grain quality and increases time to harvest and drying costs for wheat (Pinthus, 1974; Berry, 1998; Berry et al., 2004), maize (Berry et al., 2021), rice (UKRI, 2021), and oilseed rape (Kendall et al., 2017). The economic cost of lodging has been estimated at £105-188 million in the UK and US$4.7 billion at a global scale for wheat (Berry, 1998; Baker et al., 2014; Piñera-Chavez et al., 2016), US$7.5-30 billion per year at a global scale for maize (Berry et al., 2021), and £47-120 million per year in the UK for oilseed rape (Kendall et al., 2017).

Plant breeders commonly rely on the occurrence of natural lodging to visually score wheat and maize cultivars for lodging resistance (Berry et al., 2003a; Hu et al., 2012; Sekhon et al., 2020). In wheat, this limits the continuous identification and removal of lodging susceptible cultivars throughout the breeding cycle (Berry et al., 2003a). Consequently, this has driven the search for other alternatives to approach the lodging phenomenon of crops which includes the developing of lodging models and identification of traits, chemical composition of structural biomass, digital pushomics (Fig. 3), and RS.

5.1 Lodging models and traits

The developing of lodging models with engineering and mechanistic approaches applied at the crop level has been one promising avenue to understand the lodging phenomenon and to identify key traits that can aid crop improvement to increase lodging resistance. Lodging models for wheat, maize, barley, oat, and oilseed rape have been developed to determine lodging risk based on failure wind speeds for stem base (determinant of stem lodging) and root anchorage (determinant of root lodging) (parameters calculated from soil, rainfall, plant, and wind properties) (Baker et al., 1998, 2014; Berry et al., 2003b, 2006, 2021). From these models, a set of biophysical characteristics related to stem/stalk and anchorage strength have been identified. For instance,

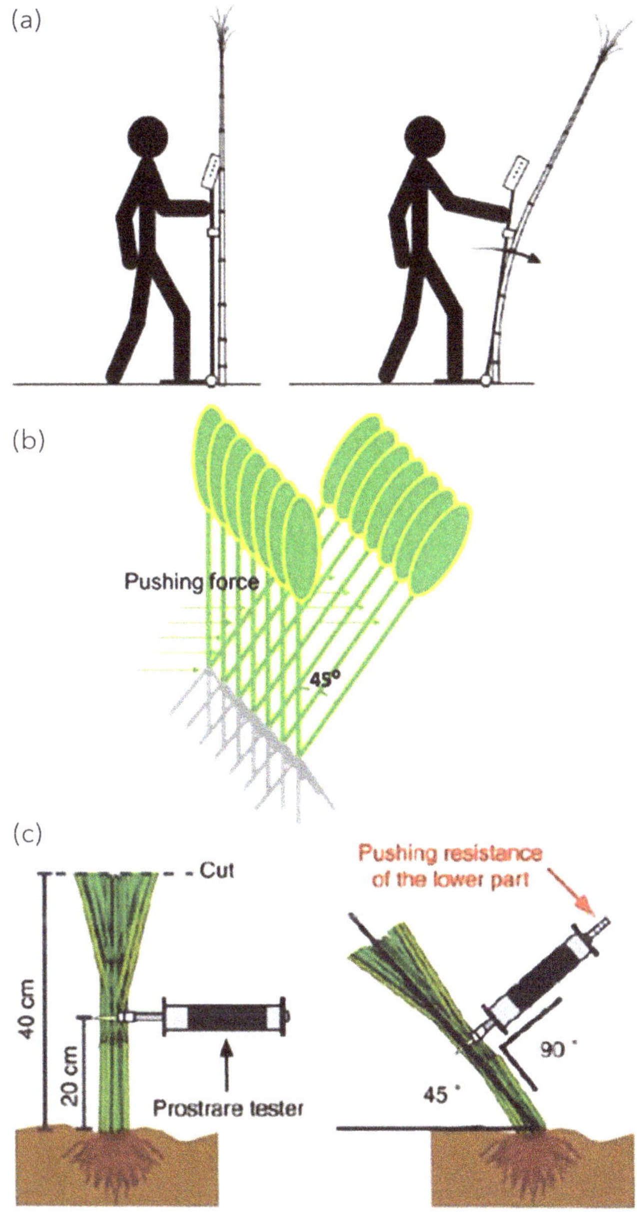

Figure 3 Schematic for instruments based on pushomics for maize (a), wheat (b), and rice (c). Source: Adapted from Kashiwagi and Ishimaru (2004); Reynolds et al. (2020); Sekhon et al. (2020).

in wheat key stem/root, biophysical characteristics that can help to enhance stem and anchorage strength (Berry et al., 2000, 2003b) can be used to screen germplasm for lodging resistance (Berry et al., 2007; Piñera-Chavez et al., 2016; Dreccer et al., 2020), enabling breeders to classify genotypes by type of lodging (root/stem lodging) and, more importantly, in the absence of natural lodging.

5.2 Chemical composition of structural biomass

Structural biomass of crop straw/stalk/stem/culm is mainly composed of primary and secondary cell walls of plants. Cell walls are predominantly composed of lignin, cellulose, and hemicellulose and one of the primary functions is the mechanical and structural support of the plant cell (Taiz, 1984; CCRC, 2007; Bosch et al., 2011; Komatsu and Yanagawa, 2013). Thus, cell wall composition plays an important role in stem/stalk strength in wheat and maize, respectively (Knapp et al., 1987; Bosch et al., 2011; Sekhon et al., 2020). In both, maize and wheat, a review of several studies suggested that there is no consensus on the key component of the cell wall conferring strength to the stalks/stems (Knapp et al., 1987; Ching et al., 2006; Ma, 2009; Bosch et al., 2011; Wiersma et al., 2011; Wang et al., 2012; Kong et al., 2013; Peng et al., 2014). For instance, in wheat, lignin has been long proposed to be associated with wheat stem strength (Ma, 2009; Wiersma et al., 2011; Peng et al., 2014). Possibly, because there is evidence that the regulation of the expression of a gene involved in the lignin synthesis contributes to the stem strength and lodging resistance (Ma, 2009). Nevertheless, a histochemical evaluation of different wheat stems differing in lodging resistance performance identified the cellulose component as the main factor conferring these differences in lodging resistance (Wang et al., 2012). Moreover, Knapp et al. (1987) concluded that the arrangement and interrelationships of the lignin and structural carbohydrates (cellulose and hemicellulose) in wheat cell walls may be more important than their simple concentrations. In maize, changes in cellulose and/or lignin concentrations can influence stalk strength (Ching et al., 2006; Bosch et al., 2011). Using cell wall compound concentrations (requiring complex laboratory measurements) (Liu et al., 2010) might not be the most practical way to rank cultivars for lodging resistance, thus, more rapid methods should be implemented. For instance, a study targeting renewable energy biomass developed a fast method for analyzing the chemical composition of corn stover and switchgrass with potential for chemical prediction of biomass in both crops (Liu et al., 2010). This method was based on the Fourier transform near-infrared (FT-NIR) spectroscopy coupled with multivariate analysis and aimed to analyze diverse types of biomass feedstocks.

5.3 Digital pushomics

Approaches to faster phenotype lodging resistance directly in the field for wheat, maize, and rice have been reported and showed promising results (Berry et al., 2003a; Kashiwagi and Ishimaru, 2004; Niu et al., 2012; Reynolds et al., 2020; Sekhon et al., 2020). These approaches are based on an instrument where the operational principle is to measure stem/stalk/culm strength when an external pushing force is applied to a certain degree angle from the vertical position (Fig. 3). Berry et al. (2003a) tested a lodging instrument based on this operational principle and was able to predict lodging with moderate precision in winter wheat. Similarly, for spring wheat a lodging instrument - pushometer - is in its third year of testing and has shown promising results (Reynolds et al., 2020) (Fig. 3b). Also, field-based measurements of stalk bending strength using lodging instrument - DARLING - have provided a reliable estimation of stalk lodging incidence in maize (Sekhon et al., 2020) (Fig. 3a). Kashiwagi and Ishimaru (2004) tested a lodging instrument - prostrate tester - capable of characterizing the pushing resistance of the lower part of rice culms cultivars.

5.4 Remote sensing for lodging

Several studies attempted to use RS techniques to measure lodging using different traits such as lodging score (natural lodging), the chemical composition of biomass, and root traits. Singh et al. (2019) provided a proof of concept application based on phenomics using UAVs that can be a high-throughput alternative for lodging score screenings. This application can be scalable to tens-of-thousands of plots in breeding and genetic studies in wheat and possibly to other crops. Hames et al. (2003) described outdoors spectrometers capable of collecting data from visible and NIR regions of the electromagnetic spectrum from 400 to 2500 nm that can be used to collect compositional information for cellulose, hemicellulose, protein, and lignin. Recently, ground-penetrating radar (GPR) has been used to predict root biomass (below-ground biomass) and root diameter with sensible precision in winter wheat and energy cane (Liu et al., 2018) and cassava (Delgado et al., 2017). The estimation of these root parameters using GPR can help breeders in the crop improvement decision-making (Liu et al., 2018) including lodging resistance enhancement. Thus, the use of GPR would greatly facilitate root phenomics, something that is commonly achieved by excavating and evaluating field-grown plants for basic root traits using or commonly known as shovelomics (Trachsel et al., 2011; Colombi et al., 2015) techniques.

6 Selecting for disease resistance

The conventional selection for disease resistance is prone to subjectivity. Even if well-trained eyes can identify and score for different levels of pathogen infection in a quite standardized manner, it is very difficult to unify criteria or avoid biases when evaluating large populations. Therefore, the prospects of having tools that overcome this subjectivity, by ensuring higher accuracy and repeatability, has been a primary motivation for the development of sensor-based approaches for the detection and quantification of diseases in breeding plots.

Several studies have demonstrated the feasibility of using sensors for the detection of changes in the optical properties of the vegetation caused by pathogen infections (Camargo and Smith, 2009; Bauriegel et al., 2011; Hillnhütter et al., 2011; Zhang et al., 2012; Behmann et al., 2015). The spatial resolution and spectral information are two important characteristics of the data that contribute to the ability to detect, identify, and localize pathogen infections in plants. Unfortunately, these two attributes do not come necessarily together, meaning that sensors providing more complete spectral information usually lack high resolution and vice versa. For instance, digital RGB cameras compensate their poor spectral information with very high spatial resolution. This resolution - together with their information in the red, green, and blue channels - has facilitated the use of different image analysis techniques ranging from basic analyses of color and texture patterns (Bock et al., 2010) to complex DL algorithms for the identification and quantification of visual symptoms (Camargo and Smith, 2009; Neumann et al., 2014). On the other hand, multi- and hyperspectral imaging sensors usually have a poor resolution compared to digital RGB cameras. However, they are capable of detecting specific changes in the spectral reflectance caused by pathogenesis, making them a promising tool for detailed assessment of the host-pathogen interaction. Depending on their spectral range, they can be suitable to detect pathogen-induced changes in the pigment composition using the visible part of the spectrum (i.e. 400–700 nm) or changes in water content or structural alteration when measuring the NIR (i.e. 700–1000 nm). These alterations in the spectrum can be captured by calculating reflectance indices (Huang et al., 2007; Ashourloo et al., 2014) or sophisticated statistical analysis that accounts for changes in the whole spectrum (Hamid Muhammed and Larsolle, 2003; Muhammed, 2005). Thomas et al. (2018) successfully used hyperspectral images and machine learning techniques to quantify disease severity in small plots under controlled conditions. Similarly, Yu et al. (2018) used hyperspectral data for the development of quantitative metrics of *Septoria tritici* blotch infection in wheat.

Pathogen infections also cause alteration to metabolic processes that can be reflected in changes in temperature or the emission of chlorophyll fluorescence. For example, thermography has been used to detect changes in transpiration due to early disease infection from organ to canopy level (Oerke et al., 2006; Mahlein, 2016). Similarly, the spatial distribution of alterations in photosynthetic rates induced by disease infection has been detected using active chlorophyll fluorescence imaging (Bürling et al., 2011). Although promising, both thermography and imaging fluorometry have been used mainly under controlled conditions and still present limitations, such as limited resolution, for their implementation as high-throughput approaches under field conditions.

Notwithstanding the advances in digital phenotyping for disease resistance of crops, the technology is still far from being at a developmental stage that can be adopted by breeding programs. The vast majority of work conducted on this topic is focused on specific host-pathogen interactions and consequently the models or parameters cannot be generalized (Mahlein et al., 2019). Secondly, pathogen identification and quantification with sensors require highly accurate and good quality data, which currently is achieved mainly under controlled environments and in low throughput. Transferring these approaches to field conditions and to platforms that would enable higher throughput will need to overcome the confounding factors such as unstable environmental conditions (especially illumination) and the presence of different biotic and abiotic stresses that may cause similar lesions in the vegetation (Zhang et al., 2012). Finally, the consolidation of image-based approaches for the assessment of disease resistance would require increasing the amount of training datasets for the development of robust models using machine learning (Mahlein et al., 2019).

7 How might trait selection look in the future

It only takes one glance at the evolution of digital imaging in the past two decades to realize that technology has advanced at a vertiginous pace and that at this point our imagination might not have limits to conceive how plant phenotyping may look in the future. The flourishment of artificial intelligence and computer vision, together with an ongoing increase in computing capacity and interconnectivity, has resulted in phenotyping applications that were unimaginable even a decade ago, and they are leading us to a steady expansion of our possibilities for more complex phenotyping.

It should not be surprising if, in the future, phenotyping will rely on multiple platforms that relate to each other by up- or downscaling models of traits and their interaction with the environment. The interplay between a plant genome and the environment results in phenotypic expressions that differ in complexity depending on their spatiotemporal scale. A good example of this is photosynthesis, which depending on the time scale or whether we are assessing it at a leaf or at canopy level, may be affected by different limiting factors (Rascher and Pieruschka, 2008). Consequently, integrative solutions for phenotyping at different levels may consider the use of terrestrial, aerial, and space platforms combined and complemented by detailed environmental information from the microplot to the field - or even regional - scale.

Terrestrial and aerial platforms have demonstrated their value for plant phenotyping. Therefore, we should expect further developments and uses for both for phenotyping in breeding programs. The biggest advantages of terrestrial vehicles are their capacity for capturing extremely high-resolution data, as well as for simultaneous measurement of a number of dynamic traits, and consequently they will probably be the *de facto* platforms for the development of new and more complex DL algorithms. Terrestrial carts have the possibility to measure under controlled illumination conditions using lamps and shades or chambers to cut the sunlight, which, together with automatic driving modes, make them the main candidates for continuous monitoring platforms in the future. On the other hand, UAVs will be equipped with smaller sensors, enabling the simultaneous recording of different kinds of data. Further, the use of DL algorithms on UAV images will be more common as the resolution of the images increases. However, the potential of UAVs for phenotyping extensive breeding sites will only be achieved if the flying autonomy and camera speed are improved.

Another important aspect to consider for the future use of UAVs is the increasing number of restrictions being imposed by many countries for safety concerns, which are limiting the speed, altitude, and the possibility of unsupervised UAV operations. In this context, the use of manned aircraft for phenotyping small plots may become a good alternative to UAVs as higher resolution and faster sensors are developed. Such platforms can cover extensive experimental sites in a matter of minutes, drastically reducing the occurrence of environmental confounding factors. Nevertheless, the operational costs of manned aircraft may still be restrictive for many breeding programs in the near future.

The use of satellites for high-throughput phenotyping of breeding plots has started to become a possibility with the recent advent of novel sensors capable of recording images at resolutions of less than a meter. Although the spatial resolution may be a limitation for some time, satellites would allow multisite observations, facilitating the phenotyping in sites that may not have easy access to other platforms and improving the characterization of the GxE interaction. The possibility of phenotyping from space may look like the ultimate solution for multisite breeding programs. However, before that happens, satellites would need to improve their revisit period and land coverage.

Current tendencies indicate that the spatial resolution of imaging sensors will continue increasing. The higher resolution has advantages as it should facilitate the identification and localization of different plant organs from aerial platforms and would provide more information for training DL algorithms. Nevertheless, higher resolution also implies the capture of more heterogeneity in illumination and texture within the canopy, which can impose some additional challenges when interpreting spectral or thermal images.

Additionally, higher resolution implies larger volumes of data, and therefore it should be accompanied by proper developments of processing and storing capacity. In spectroscopy, we should also see improvements in the spectral resolution and the signal to noise ratio, making these systems more sensitive to subtle changes or differences in the optical properties of the vegetation, and consequently improving the training datasets for ML algorithms.

All the potential advances for high-throughput phenotyping mentioned above imply the management and processing of enormous datasets, which can only be possible with high performing computers and adequate algorithms. There seems to be no ceiling for informatics, and therefore we can expect that phenotypic information will flow almost in real-time from the moment of acquisition to the breeders for decision-making, all supported by high onboard computing capacity, fast wireless data transmission, and most importantly, thanks to efficient models able to accurately detect features from high spatial and spectral resolution data. In fact, ML and DL algorithms are quickly evolving and solving extremely complex tasks in a very short time. However, an expected bottleneck will be the training process of these models as a considerable amount of training and validation data points is needed to obtain generalizable models applicable to most conditions. That would also mean that it will be almost mandatory to have a complete characterization of the environment even at the level of microclimate within each plot to reduce all possible confounding factors. For such a level of detail, it will be necessary to build robust networks of environmental sensors distributed across trials and synchronized with all the other measuring platforms.

What the discipline of digital phenotyping is experiencing right now is a phase of discovery of what technology can do. Advances so far are impressive, but it is also true that there is still a way to go for their consolidation and final integration in breeding pipelines. This will require a multidisciplinary work combining public and private efforts, and a proper strategy for transferring these new technologies to breeders and farmers. Furthermore, the wide adoption of digital phenotyping would only succeed if these High-Tec approaches showed a simple face to their final users.

8 Where to look for further information

The International Plant Phenotyping Network (IPPN) is an association that represents major plant phenotyping centers and aims to increase visibility and impact of the disciplne (https://www.plant-phenotyping.org/IPPN_home). The IPPN - and other regional phenotyping networks listed in the IPPN website - enable cooperation and sharing of ideas and techniques among stakeholders in academia and industry. A number of articles cited herein can provide more information, context and speculation about the role of phenotyping now and in the future (Araus et al., 2018; Fahlgren et al., 2015; Furbank et al., 2019).

9 References

Abbate, P. E., Pontaroli, A. C., Lázaro, L. and Gutheim, F. (2013). A method of screening for spike fertility in wheat, *The Journal of Agricultural Science* 151(3), 322–330. doi: 10.1017/S0021859612000068.

Acreche, M. M. and Slafer, G. A. (2011). Lodging yield penalties as affected by breeding in Mediterranean wheats, *Field Crops Research* 122(1), 40–48. doi: 10.1016/j.fcr.2011.02.004.

Agrios, G. (2005). *Plant Pathology*. London: Academic Press.

Aisawi, K. A. B., Reynolds, M. P., Singh, R. P. and Foulkes, M. J. (2015). The physiological basis of the genetic progress in yield potential of CIMMYT spring wheat cultivars from 1966 to 2009, *Crop Science* 55(4), 1749–1764. doi: 10.2135/cropsci2014.09.0601.

Alonso, M. P., Mirabella, N. E., Panelo, J. S., Cendoya, M. G. and Pontaroli, A. C. (2018). Selection for high spike fertility index increases genetic progress in grain yield and stability in bread wheat, *Euphytica* 214(7), 112. doi: 10.1007/s10681-018-2193-4.

Ammar, K., Mergoum, M. and Rajaram, S. (2004). The history and evolution of Triticale. In: Mergoum, M. and Gomez-Macpherson, H. (Eds) *Triticale Improvement and Production*. Rome, Italy: Food and Agriculture Organization, pp. 1-9.

Araus, J. L. and Cairns, J. E. (2014). Field high-throughput phenotyping: The new crop breeding frontier, *Trends in Plant Science* 19(1), 52–61. doi: 10.1016/j.tplants.2013.09.008.

Araus, J. L., Casadesus, J. and Bort, J. (2001). Recent tools for the screening of physiological traits determining yield. In: Reynolds, M. P., Ortiz-Monasterio, J. I. and McNab, A. (Eds) *Application of Physiology in Wheat Breeding*. Mexico D.F.: CIMMYT, pp. 59–77.

Araus, J. L. and Kefauver, S. C. (2018). Breeding to adapt agriculture to climate change: Affordable phenotyping solutions, *Current Opinion in Plant Biology* 45(B), 237–247. doi: 10.1016/j.pbi.2018.05.003.

Araus, J. L., Kefauver, S. C., Zaman-Allah, M., Olsen, M. S. and Cairns, J. E. (2018). Translating high-throughput phenotyping into genetic gain, *Trends in Plant Science* 23(5), 451–466. doi: 10.1016/j.tplants.2018.02.001.

Ashourloo, D., Mobasheri, M. R. and Huete, A. (2014). Evaluating the effect of different wheat rust disease symptoms on vegetation indices using hyperspectral measurements, *Remote Sensing* 6(6), 5107–5123. doi: 10.3390/rs6065107.

Atzberger, C. (2013). Advances in remote sensing of agriculture: Context description, existing operational monitoring systems and major information needs, *Remote Sensing* 5(2), 949–981. doi: 10.3390/rs5020949.

Babar, M. A., Reynolds, M. P., van Ginkel, M., Klatt, A. R., Raun, W. R. and Stone, M. L. (2006). Spectral reflectance to estimate genetic variation for in-season biomass, leaf chlorophyll, and canopy temperature in wheat, *Crop Science* 46(3), 1046–1057. doi: 10.2135/cropsci2005.0211.

Baker, C. J., Berry, P. M., Spink, J. H., Sylvester-Bradley, R., Griffin, J. M., Scott, R. K. and Clare, R. W. (1998). A method for the assessment of the risk of wheat lodging, *Journal of Theoretical Biology* 194(4), 587–603. doi: 10.1006/jtbi.1998.0778.

Baker, C. J., Sterling, M. and Berry, P. (2014). A generalised model of crop lodging, *Journal of Theoretical Biology* 363, 1–12. doi: 10.1016/j.jtbi.2014.07.032.

Baret, F. and Buis, S. (2008). Estimating canopy characteristics from remote sensing observations: Review of methods and associated problems. In: Liang, S. (Ed) *Advances in Land Remote Sensing*. Springer, Dordrecht . doi: 10.1007/978-1-4020-6450-0_7.

Bauriegel, E., Giebel, A., Geyer, M., Schmidt, U. and Herppich, W. B. (2011). Early detection of Fusarium infection in wheat using hyper-spectral imaging, *Computers and Electronics in Agriculture* 75(2), 304–312. doi: 10.1016/j.compag.2010.12.006.

Behmann, J., Mahlein, A., Rumpf, T., Römer, C. and Plümer, L. (2015). A review of advanced machine learning methods for the detection of biotic stress in precision crop protection, *Precision Agriculture* 16(3), 239–260. doi: 10.1007/s11119-014-9372-7.

Berger, K., Atzberger, C., Danner, M., D'Urso, G., Mauser, W., Vuolo, F. and Hank, T. (2018). Evaluation of the PROSAIL model capabilities for future hyperspectral model environments: A review study, *Remote Sensing* 10(2). doi: 10.3390/rs10010085.

Bernardo, R. (2013). Genomewide selection when major genes are known, *Crop Science* 54(1), 68–75. doi: 10.2135/cropsci2013.05.0315.

Berry, P. M. (1998). Predicting lodging in wheat. Ph. D. Thesis. UK: The University of Nottingham, 210 pp.

Berry, P. M., Baker, C. J., Hatley, D., Dong, R., Wang, X., Blackburn, G. A., Miao, Y., Sterling, M. and Whyatt, J. D. (2021). Development and application of a model for calculating the risk of stem and root lodging in maize, *Field Crops Research* 262. doi: 10.1016/j.fcr.2020.108037.

Berry, P. M., Griffin, J. M., Sylvester-Bradley, R., Scott, R. K., Spink, J. H., Baker, C. J. and Clare, R. W. (2000). Controlling plant form through husbandry to minimise lodging in wheat, *Field Crops Research* 67(1), 59–81. doi: 10.1016/S0378-4290(00)00084-8.

Berry, P. M. and Spink, J. (2012). Predicting yield losses caused by lodging in wheat, *Field Crops Research* 137, 19–26. doi: 10.1016/j.fcr.2012.07.019.

Berry, P. M., Spink, J., Sterling, M. and Pickett, A. A. (2003a). Methods for rapidly measuring the lodging resistance of wheat cultivars, *Journal of Agronomy and Crop Science* 189(6), 390–401. doi: 10.1046/j.0931-2250.2003.00062.x.

Berry, P. M., Sterling, M., Baker, C. J., Spink, J. and Sparkes, D. L. (2003b). A calibrated model of wheat lodging compared with field measurements, *Agricultural and Forest Meteorology* 119(3–4), 167–180. doi: 10.1016/S0168-1923(03)00139-4.

Berry, P. M., Sterling, M. and Mooney, S. J. (2006). Development of a model of lodging for barley, *Journal of Agronomy and Crop Science* 192(2), 151–158.

Berry, P. M., Sterling, M., Spink, J. H., Baker, C. J., Sylvester-Bradley, R., Mooney, S. J., Iams, A. R. and Ennos, A. R. (2004). Understanding and reducing lodging in cereals, *Advances in Agronomy* 217–271. doi: 10.1016/S0065-2113(04)84005-7.

Berry, P. M., Sylvester-Bradley, R. and Berry, S. (2007). Ideotype design for lodging-resistant wheat, *Euphytica* 154(1–2), 165–179. doi: 10.1007/s10681-006-9284-3.

Bidinger, F., Musgrave, R. B. and Fischer, R. A. (1977). Contribution of stored pre-anthesis assimilate to grain yield in wheat and barley, *Nature* 270(5636), 431–433. doi: 10.1038/270431a0.

Blackburn, G. A. (2007). Hyperspectral remote sensing of plant pigments, *Journal of Experimental Botany* 58(4), 855–867. doi: 10.1093/jxb/erl123.

Blum, A. (1998). Improving wheat grain filling under stress by stem reserve mobilisation, *Euphytica* 100(1–3), 77–83. doi: 10.1023/A:1018303922482.

Bock, C. H., Poole, G. H., Parker, P. E. and Gottwald, T. R. (2010). Plant disease severity estimated visually, by digital photography and image analysis and by hyperspectral imaging, *Critical Reviews in Plant Sciences* 29(2), 59–107. doi: 10.1080/07352681003617285.

Bosch, M., Mayer, C. D., Cookson, A. and Donnison, I. S. (2011). Identification of genes involved in cell wall biogenesis in grasses by differential gene expression profiling of elongating and non-elongating maize internodes, *Journal of Experimental Botany* 62(10), 3545–3561. doi: 10.1093/jxb/err045.

Britannica, T. E. (2021). *George Harrison Shull, Encyclopedia Britannica*. Available at: https://www.britannica.com/biography/George-Harrison-Shull (accessed 26 May 2021).

Bürling, K., Hunsche, M. and Noga, G. (2011). Use of blue-green and chlorophyll fluorescence measurements for differentiation between nitrogen deficiency and pathogen infection in winter wheat, *Journal of Plant Physiology* 168(14), 1641–1648. doi: 10.1016/j.jplph.2011.03.016.

Camargo, A. and Smith, J. S. (2009). Image pattern classification for the identification of disease causing agents in plants, *Computers and Electronics in Agriculture* 66(2), 121–125. doi: 10.1016/j.compag.2009.01.003.

Campoy, J., Campos, I., Plaza, C., Calera, M., Bodas, V. and Calera, A. (2020). Estimation of harvest index in wheat crops using a remote sensing-based approach, *Field Crops Research* 256, 107910. doi: 10.1016/j.fcr.2020.107910.

CCRC. (2007). *Plant Cell Walls*. Athens GA: Complex Carbohydrate Research Center, BioEnergy Science Center, University of Georgia.

Cen, H., Wan, L., Zhu, J., Li, Y., Li, X., Zhu, Y., Weng, H., Wu, W., Yin, W., Xu, C., Bao, Y., Feng, L., Shou, J. and He, Y. (2019). Dynamic monitoring of biomass of rice under different nitrogen treatments using a lightweight UAV with dual image-frame snapshot cameras, *Plant Methods* 15(1), 32. doi: 10.1186/s13007-019-0418-8.

Cendrero-Mateo, M. P., Bennertz, S., Burkart, A., Julitta, T., Cogliati, S., Scharr, H., Rademske, P., Alonso, L., Pinto, F. and Rascher, U. (2018). Sun induced fluorescence calibration and validation for field phenotyping. In: *International Geoscience and Remote Sensing Symposium (IGARSS)*. doi: 10.1109/IGARSS.2018.8519174.

Ching, A., Dhugga, K. S., Appenzeller, L., Meeley, R., Bourett, T. M., Howard, R. J. and Rafalski, A. (2006). Brittle stalk 2 encodes a putative glycosylphosphatidylinositol-anchored protein that affects mechanical strength of maize tissues by altering the composition and structure of secondary cell walls, *Planta* 224(5), 1174–1184. doi: 10.1007/s00425-006-0299-8.

Chlingaryan, A., Sukkarieh, S., and Whelan, B. (2018). Machine learning approaches for crop yield prediction and nitrogen status estimation in precision agriculture: A review, *Computers and Electronics in Agriculture* 151, 61–69. doi: 10.1016/j.compag.2018.05.012.

Colombi, T., Kirchgessner, N., Le Marié, C. A., York, L. M., Lynch, J. P. and Hund, A. (2015). Next generation shovelomics: Set up a tent and REST, *Plant and Soil* 388(1–2), 1–20. doi: 10.1007/s11104-015-2379-7.

Conaty, W. C., Mahan, J. R., Neilsen, J. E., Tan, D. K. Y., Yeates, S. J. and Sutton, B. G. (2015). The relationship between cotton canopy temperature and yield, fibre quality and water-use efficiency, *Field Crops Research* 183, 329–341. doi: 10.1016/j.fcr.2015.08.010.

Costa, C., Schurr, U., Loreto, F., Menesatti, P. and Carpentier, S. (2019). Plant phenotyping research trends, a science mapping approach, *Frontiers in Plant Science* 9(January), 1–11. doi: 10.3389/fpls.2018.01933.

Curran, P. J. (1989). Remote sensing of foliar chemistry, *Remote Sensing of Environment* 30(3), 271–278. doi: 10.1016/0034-4257(89)90069-2.

Delgado, A., Hays, D. B., Bruton, R. K., Ceballos, H., Novo, A., Boi, E. and Selvaraj, M. G. (2017). Ground penetrating radar: A case study for estimating root bulking rate in cassava (*Manihot esculenta Crantz*), *Plant Methods* 13, 65. doi: 10.1186/s13007-017-0216-0.

Denison, R. F., Kiers, E. T. and West, S. A. (2003). Darwinian agriculture: When can humans find solutions beyond the reach of natural selection? *The Quarterly Review of Biology* 78(2), 145–168.

Dong, Y., Liu, J., Zhang, Y., Geng, H., Rasheed, A., Xiao, Y., Cao, S., Fu, L., Yan, J., Wen, W., Zhang, Y., Jing, R., Xia, X. and He, Z. (2016). Genome-wide association of stem water soluble carbohydrates in bread wheat, *PLoS ONE* 11(11), e0164293. doi: 10.1371/journal.pone.0164293.

Dreccer, M. F., Barnes, L. R. and Meder, R. (2014). Quantitative dynamics of stem water soluble carbohydrates in wheat can be monitored in the field using hyperspectral reflectance, *Field Crops Research* 159, 70–80. doi: 10.1016/j.fcr.2014.01.001.

Dreccer, M. F., Condon, A. G., Macdonald, B., Rebetzke, G. J., Awasi, M., Borgognone, M. G., Peake, A., Piñera-Chavez, F. J., Hundt, A., Jackway, P. and McIntyre, C. L. (2020). Genotypic variation for lodging tolerance in spring wheat: Wider and deeper root plates, a feature of low lodging, high yielding germplasm, *Field Crops Research* 258(February). doi: 10.1016/j.fcr.2020.107942.

Dreccer, M. F., Molero, G., Rivera-Amado, C., John-Bejai, C. and Wilson, Z. (2019). Yielding to the image: How phenotyping reproductive growth can assist crop improvement and production, *Plant Science* 282, 73–82. doi: 10.1016/j.plantsci.2018.06.008.

Easson, D. L., White, E. M. and Pickles, S. J. (1993). The effects of weather, seed rate and cultivar on lodging and yield in winter wheat, *The Journal of Agricultural Science* 121(2), 145–156. doi: 10.1017/S0021859600077005.

El-Hendawy, S., Al-Suhaibani, N., Dewir, Y., Elsayed, S., Alotaibi, M., Hassan, W., Refay, Y. and Tahir, M. (2019). Ability of modified spectral reflectance indices for estimating growth and photosynthetic efficiency of wheat under saline field conditions, *Agronomy* 9(1), 35. doi: 10.3390/agronomy9010035.

Fahlgren, N., Gehan, M. A. and Baxter, I. (2015). Lights, camera, action: High-throughput plant phenotyping is ready for a close-up, *Current Opinion in Plant Biology* 24, 93–99. doi: 10.1016/j.pbi.2015.02.006.

Fava, F., Colombo, R., Bocchi, S., Meroni, M., Sitzia, M., Fois, N. and Zucca, C. (2009). Identification of hyperspectral vegetation indices for Mediterranean pasture characterization, *International Journal of Applied Earth Observation and Geoinformation* 11(4), 233–243. doi: 10.1016/j.jag.2009.02.003.

Fernandez-Gallego, J. A., Kefauver, S. C., Gutiérrez, N. A., Nieto-Taladriz, M. T. and Araus, J. L. (2018). Wheat ear counting in-field conditions: High throughput and low-cost approach using RGB images, *Plant Methods* 14(1), 22. doi: 10.1186/s13007-018-0289-4.

Fischer, K. S., Edmeades, G. O. and Johnson, E. C. (1989). Selection for the improvement of maize yield under moisture deficits, *Field Crops Research* 22(3), 227–243. doi: 10.1016/0378-4290(89)90094-4.

Fischer, R. A. (2011). Wheat physiology: A review of recent developments, *Crop and Pasture Science* 62(2), 95–114.

Fischer, R. A., Byerlee, D. and Edmeades, G. O. (2014). *Crop yields and global food security: will yield increase continue to feed the world?*. ACIAR Monograph No. 158. Australian Centre for International Agricultural Research: Canberra, pp. xxii + 634.

Fischer, R. A. and Stapper, M. (1987). Lodging effects on high-yielding crops of irrigated semidwarf wheat, *Field Crops Research* 17(3-4), 245–258.

Fisher, R. A. and Mackenzie, W. A. (1923). Studies in crop variation. II. The manurial response of different potato varieties, *The Journal of Agricultural Science* 13(3), 311–320. doi: 10.1017/S0021859600003592.

Foulkes, M. J., Scott, R. K. and Sylvester-Bradley, R. (2002). The ability of wheat cultivars to withstand drought in UK conditions: Formation of grain yield, *The Journal of Agricultural Science* 138(2), 153–169. doi: 10.1017/S0021859601001836.

Foulkes, M. J., Slafer, G. A., Davies, W. J., Berry, P. M., Sylvester-Bradley, R., Martre, P., Calderini, D. F., Griffiths, S. and Reynolds, M. P. (2011). Raising yield potential of wheat. III. Optimizing partitioning to grain while maintaining lodging resistance, *Journal of Experimental Botany* 62(2), 469–486. doi: 10.1093/jxb/erq300.

Fu, Y., Yang, G., Wang, J., Song, X. and Feng, H. (2014). Winter wheat biomass estimation based on spectral indices, band depth analysis and partial least squares regression using hyperspectral measurements, *Computers and Electronics in Agriculture* 100, 51–59. doi: 10.1016/j.compag.2013.10.010.

Furbank, R. T., Jimenez-Berni, J. A., George-Jaeggli, B., Potgieter, A. B. and Deery, D. M. (2019). Field crop phenomics: Enabling breeding for radiation use efficiency and biomass in cereal crops, *New Phytologist* 223(4), 1714–1727. doi: 10.1111/nph.15817.

Gebbing, T., Schnyder, H. and KUhbauch, W. (1999). The utilization of pre-anthesis reserves in grain filling of wheat. Assessment by steady-state 13 CO_2/12CO_2 labelling, *Plant, Cell and Environment* 22(7), 851–858. doi: 10.1046/j.1365-3040.1999.00436.x.

Ghosal, S., Zheng, B., Chapman, S. C., Potgieter, A. B., Jordan, D. R., Wang, X., Singh, A. K., Singh, A., Hirafuji, M., Ninomiya, S., Ganapathysubramanian, B., Sarkar, S. and Guo, W. (2019). A weakly supervised deep learning framework for sorghum head detection and counting, *Plant Phenomics* 2019, 1525874. doi: 10.34133/2019/1525874.

Godwin, I. D., Rutkoski, J., Varshney, R. K. and Hickey, L. T. (2019). Technological perspectives for plant breeding, *TAG. Theoretical and Applied Genetics. Theoretische und Angewandte Genetik* 132(3), 555–557. doi: 10.1007/s00122-019-03321-4.

Govind, A., Guyon, D., Roujean, J., Yauschew-Raguenes, N., Kumari, J., Pisek, J. and Wigneron, J. (2013). Effects of canopy architectural parameterizations on the modeling of radiative transfer mechanism, *Ecological Modelling* 251, 114–126. doi: 10.1016/j.ecolmodel.2012.11.014.

Gutierrez, M., Reynolds, M. P. and Klatt, A. R. (2010a). Association of water spectral indices with plant and soil water relations in contrasting wheat genotypes, *Journal of Experimental Botany* 61(12), 3291–3303. doi: 10.1093/jxb/erq156.

Gutierrez, M., Reynolds, M. P. and Klatt, A. R. (2015). Effect of leaf and spike morphological traits on the relationship between spectral reflectance indices and yield in wheat, *International Journal of Remote Sensing* 36(3), 701–718. doi: 10.1080/01431161.2014.999878.

Gutierrez, M., Reynolds, M. P., Raun, W. R., Stone, M. L. and Klatt, A. R. (2010b). Spectral water indices for assessing yield in elite bread wheat genotypes under well-irrigated, water-stressed, and high-temperature conditions, *Crop Science* 50(1), 197–214. doi: 10.2135/cropsci2009.07.0381.

Guzmán, C., Autrique, E., Mondal, S., Huerta-Espino, J., Singh, R. P., Vargas, M., Crossa, J., Amaya, A. and Peña, R. J. (2017). Genetic improvement of grain quality traits for CIMMYT semi-dwarf spring bread wheat varieties developed during 1965-2015: 50 years of breeding, *Field Crops Research* 210(March), 192–196. doi: 10.1016/j.fcr.2017.06.002.

Hames, B. R., Thomas, S. R., Sluiter, A. D., Roth, C. J. and Templeton, D. W. (2003). Rapid biomass analysis: New tools for compositional analysis of corn stover feedstocks and process intermediates from ethanol production. *Applied Biochemistry and Biotechnology* 105(108), 5-16. doi: 10.1385/ABAB:105:1-3:5.

Hamid Muhammed, H. and Larsolle, A. (2003). Feature vector based analysis of hyperspectral crop reflectance data for discrimination and quantification of fungal disease severity in wheat, *Biosystems Engineering* 86(2), 125-134. doi: 10.1016/S1537-5110(03)00090-4.

Han, L., Yang, G., Dai, H., Xu, B., Yang, H., Feng, H., Li, Z. and Yang, X. (2019). Modeling maize above-ground biomass based on machine learning approaches using UAV remote-sensing data, *Plant Methods* 15(1), 10. doi: 10.1186/s13007-019-0394-z.

Harlan, J. R. (1972). Genetics of disaster, *Journal of Environmental Quality* 1(3), 212-215. doi: 10.2134/jeq1972.00472425000100030002x.

Hay, R. K. M. (1995). Harvest index: A review of its use in plant breeding and crop physiology, *Annals of Applied Biology* 126(1), 197-216. doi: 10.1111/j.1744-7348.1995.tb05015.x.

Hillnhütter, C., Mahlein, A.-K., Sikora, R. A. and Oerke, E.-C. (2011). Remote sensing to detect plant stress induced by *Heterodera schachtii* and *Rhizoctonia solani* in sugar beet fields, *Field Crops Research* 122(1), 70-77. doi: 10.1016/j.fcr.2011.02.007.

Houborg, R., Fisher, J. B. and Skidmore, A. K. (2015). Advances in remote sensing of vegetation function and traits, *International Journal of Applied Earth Observation and Geoinformation* 43, 1-6. doi: 10.1016/j.jag.2015.06.001.

Hu, H., Meng, Y., Wang, H., Liu, H. and Chen, S. (2012). Identifying quantitative trait loci and determining closely related stalk traits for rind penetrometer resistance in a high-oil maize population, *TAG. Theoretical and Applied Genetics. Theoretische und Angewandte Genetik* 124(8), 1439-1447. doi: 10.1007/s00122-012-1799-5.

Hu, H., Scheben, A. and Edwards, D. (2018). Advances in integrating genomics and bioinformatics in the plant breeding pipeline, *Agriculture* 8(6), 1-18. doi: 10.3390/agriculture8060075.

Huang, W., Lamb, D. W., Niu, Z., Zhang, Y., Liu, L. and Wang, J. (2007). Identification of yellow rust in wheat using in-situ spectral reflectance measurements and airborne hyperspectral imaging, *Precision Agriculture* 8(4-5), 187-197. doi: 10.1007/s11119-007-9038-9.

Jensen, J. R. (2000). *Remote Sensing of the Environment: An Earth Resource Perspective, Prentice Hall Series in Geographic Information Science*. Upper Saddle River, NJ: Prentice Hall.

Jin, H. and Eklundh, L. (2014). A physically based vegetation index for improved monitoring of plant phenology, *Remote Sensing of Environment* 152, 512-525. doi: 10.1016/j.rse.2014.07.010.

Joseph, G. M. D., Mohammadi, M., Sterling, M., Baker, C. J., Gillmeier, S. G., Soper, D., Jesson, M., Blackburn, G. A., Whyatt, J. D., Gullick, D., Murray, J., Berry, P., Hartley, D. and Finnan, J. (2020). Determination of crop dynamic and aerodynamic parameters for lodging prediction, *Journal of Wind Engineering and Industrial Aerodynamics* 202(January), 104169. doi: 10.1016/j.jweia.2020.104169.

Kashiwagi, T. and Ishimaru, K. (2004). Identification and functional analysis of a locus for improvement of lodging resistance in rice, *Plant Physiology* 134(2), 676-683. doi: 10.1104/pp.103.029355.

Keller, B., Vass, I., Matsubara, S., Paul, K., Jedmowski, C., Pieruschka, R., Nedbal, L., Rascher, U. and Muller, O. (2019). Maximum fluorescence and electron transport kinetics determined by light-induced fluorescence transients (LIFT) for photosynthesis phenotyping, *Photosynthesis Research* 140(2), 221–233. doi: 10.1007/s11120-018-0594-9.

Kemanian, A. R., Stöckle, C. O., Huggins, D. R. and Viega, L. M. (2007). A simple method to estimate harvest index in grain crops, *Field Crops Research* 103(3), 208–216. doi: 10.1016/j.fcr.2007.06.007.

Kendall, S. L., Holmes, H., White, C. A., Clarke, S. M. and Berry, P. M. (2017). Quantifying lodging-induced yield losses in oilseed rape, *Field Crops Research* 211, 106–113. doi: 10.1016/j.fcr.2017.06.013.

Knapp, J. S., Harms, C. L. and Volenec, J. J. (1987). Growth regulator effects on wheat culm nonstructural and structural carbohydrates and lignin 1, *Crop Science* 27(6), 1201–1205.

Knipling, E. B. (1970). Physical and physiological basis for the reflectance of visible and near-infrared radiation from vegetation, *Remote Sensing of Environment* 1(3), 155–159. doi: 10.1016/S0034-4257(70)80021-9.

Kobayashi, H. and Iwabuchi, H. (2008). A coupled 1-D atmosphere and 3-D canopy radiative transfer model for canopy reflectance, light environment, and photosynthesis simulation in a heterogeneous landscape, *Remote Sensing of Environment* 112(1), 173–185. doi: 10.1016/j.rse.2007.04.010.

Komatsu, S. and Yanagawa, Y. (2013). Cell wall proteomics of crops, *Frontiers in Plant Science* 4, 17. doi: 10.3389/fpls.2013.00017.

Kong, E., Liu, D., Guo, X., Yang, W., Sun, J., Li, X., Zhan, K., Cui, D., Lin, J. and Zhang, A. (2013). Anatomical and chemical characteristics associated with lodging resistance in wheat, *The Crop Journal* 1(1), 43–49. doi: 10.1016/j.cj.2013.07.012.

Korohou, T., Okinda, C., Li, H., Cao, Y., Nyalala, I., Huo, L., Potcho, M., Li, X. and Ding, Q. (2020). Wheat grain yield estimation based on image morphological properties and wheat biomass, *Journal of Sensors* 2020, 1–11. doi: 10.1155/2020/1571936.

Kromdijk, J., Głowacka, K., Leonelli, L., Gabilly, S. T., Iwai, M., Niyogi, K. K. and Long, S. P. (2016). Improving photosynthesis and crop productivity by accelerating recovery from photoprotection, *Science* 354(6314), 857–861. doi: 10.1126/science.aai8878.

Li, W., Zhang, B., Li, R., Chang, X. and Jing, R. (2015). Favorable alleles for stem water-soluble carbohydrates identified by association analysis contribute to grain weight under drought stress conditions in wheat, *PLoS ONE* 10(3), e0119438. doi: 10.1371/journal.pone.0119438.

Liu, L., Ye, X. P., Womac, A. R. and Sokhansanj, S. (2010). Variability of biomass chemical composition and rapid analysis using FT-NIR techniques, *Carbohydrate Polymers* 81(4), 820–829. doi: 10.1016/j.carbpol.2010.03.058.

Liu, S., Baret, F., Andrieu, B., Burger, P. and Hemmerlé, M. (2017). Estimation of wheat plant density at early stages using high resolution imagery, *Frontiers in Plant Science* 8, 739. doi: 10.3389/fpls.2017.00739.

Liu, X., Dong, X., Xue, Q., Leskovar, D. I., Jifon, J., Butnor, J. R. and Marek, T. (2018). Ground penetrating radar (GPR) detects fine roots of agricultural crops in the field, *Plant and Soil* 423(1–2), 517–531.

Lobell, D. B., Cassman, K. G. and Field, C. B. (2009). Crop yield gaps: Their importance, magnitudes, and causes, *Annual Review of Environment and Resources* 34(1), 179–204. doi: 10.1146/annurev.environ.041008.093740.

Long, S. P., Marshall-Colon, A. and Zhu, X. G. (2015). Meeting the global food demand of the future by engineering crop photosynthesis and yield potential, *Cell* 161(1), 56–66. doi: 10.1016/j.cell.2015.03.019.

Lopes, M. S. and Reynolds, M. P. (2010). Partitioning of assimilates to deeper roots is associated with cooler canopies and increased yield under drought in wheat, *Functional Plant Biology* 37(2), 147. doi: 10.1071/FP09121.

Lopes, M. S. and Reynolds, M. P. (2012). Stay-green in spring wheat can be determined by spectral reflectance measurements (normalized difference vegetation index) independently from phenology, *Journal of Experimental Botany* 63(10), 3789–3798. doi: 10.1093/jxb/ers071.

Lu, N., Zhou, J., Han, Z., Li, D., Cao, Q., Yao, X., Tian, Y., Zhu, Y., Cao, W. and Cheng, T. (2019). Improved estimation of aboveground biomass in wheat from RGB imagery and point cloud data acquired with a low-cost unmanned aerial vehicle system, *Plant Methods* 15(1), 17. doi: 10.1186/s13007-019-0402-3.

Ma, Q. H. (2009). The expression of caffeic acid 3-O-methyltransferase in two wheat genotypes differing in lodging resistance, *Journal of Experimental Botany* 60(9), 2763–2771. doi: 10.1093/jxb/erp132.

Mahlein, A. K. (2016). Plant disease detection by imaging sensors – Parallels and specific demands for precision agriculture and plant phenotyping, *Plant Disease* 100(2), 241–251. doi: 10.1094/PDIS-03-15-0340-FE.

Mahlein, A. K., Kuska, M. T., Thomas, S., Wahabzada, M., Behmann, J., Rascher, U. and Kersting, K. (2019). Quantitative and qualitative phenotyping of disease resistance of crops by hyperspectral sensors: Seamless interlocking of phytopathology, sensors, and machine learning is needed!, *Current Opinion in Plant Biology* 50, 156–162. doi: 10.1016/j.pbi.2019.06.007.

Marcelis, L. F. M. (1996). Sink strength as a determinant of dry matter partitioning in the whole plant, *Journal of Experimental Botany* 47(Spec No), 1281–1291. doi: 10.1093/jxb/47.Special_Issue.1281.

Maxwell, K. and Johnson, G. N. (2000). Chlorophyll fluorescence–A practical guide, *Journal of Experimental Botany* 51(345), 659–668. doi: 10.1093/jxb/51.345.659.

McAusland, L., Vialet-Chabrand, S., Jauregui, I., Burridge, A., Hubbart-Edwards, S., Fryer, M. J., King, I. P., King, J., Pyke, K., Edwards, K. J., Carmo-Silva, E., Lawson, T. and Murchie, E. H. (2020). Variation in key leaf photosynthetic traits across wheat wild relatives is accession dependent not species dependent, *New Phytologist* 228(6), 1767–1780. doi: 10.1111/nph.16832.

McIntyre, C. L., Seung, D., Casu, R. E., Rebetzke, G. J., Shorter, R. and Xue, G. P. (2012). Genotypic variation in the accumulation of water soluble carbohydrates in wheat, *Functional Plant Biology* 39(7), 560–568. doi: 10.1071/FP12077.

Mergoum, M. and Gómez-Macpherson, H. (Eds). (2004). *Triticale Improvement and Production (FAO Plant Production and Protection Paper 179)*. Rome: Food and Agriculture Organization of the United Nations.

Metje-Sprink, J., Menz, J., Modrzejewski, D. and Sprink, T. (2019). DNA-free genome editing: Past, present and future, *Frontiers in Plant Science* 9. doi: 10.3389/fpls.2018.01957.

Milla, R. and Osborne, C. P. (2021). Crop origins explain variation in global agricultural relevance, *Nature Plants* 7(5), 598–607. doi: 10.1038/s41477-021-00905-1.

Molero, G., Joynson, R., Pinera-Chavez, F. J., Gardiner, L. J., Rivera-Amado, C., Hall, A. and Reynolds, M. P. (2019). Elucidating the genetic basis of biomass accumulation and radiation use efficiency in spring wheat and its role in yield potential, *Plant Biotechnology Journal* 17(7), 1276–1288. doi: 10.1111/pbi.13052.

Muhammed, H. H. (2005). Hyperspectral crop reflectance data for characterising and estimating fungal disease severity in wheat, *Biosystems Engineering* 91(1), 9–20. doi: 10.1016/j.biosystemseng.2005.02.007.

Murchie, E. H. and Lawson, T. (2013). Chlorophyll fluorescence analysis: A guide to good practice and understanding some new applications, *Journal of Experimental Botany* 64(13), 3983–3998. doi: 10.1093/jxb/ert208.

Murchie, E. H. and Niyogi, K. K. (2011). Manipulation of photoprotection to improve plant photosynthesis, *Plant Physiology* 155(1), 86–92. doi: 10.1104/pp.110.168831.

Neumann, M., Hallau, L., Klatt, B., Kersting, K. and Bauckhage, C. (2014). Erosion band features for cell phone image based plant disease classification. In: *Proceeding of the 22nd International Conference on Pattern Recognition (ICPR)*, Stockholm, Sweden, 24–28 August 2014, pp. 3315–3320. doi: 10.1109/ICPR.2014.571.

Niu, L., Feng, S., Ru, Z., Li, G., Zhang, Z. and Wang, Z. (2012). Rapid determination of single-stalk and population lodging resistance strengths and an assessment of the stem lodging wind speeds for winter wheat, *Field Crops Research* 139, 1–8. doi: 10.1016/j.fcr.2012.10.014.

Niu, Y., Zhang, L., Zhang, H., Han, W. and Peng, X. (2019). Estimating above-ground biomass of maize using features derived from UAV-based RGB imagery, *Remote Sensing* 11(11), 1261. doi: 10.3390/rs11111261.

Oerke, E. C., Steiner, U., Dehne, H. W. and Lindenthal, M. (2006). Thermal imaging of cucumber leaves affected by downy mildew and environmental conditions, *Journal of Experimental Botany* 57(9), 2121–2132. doi: 10.1093/jxb/erj170.

Pantazi, X. E., Moshou, D., Alexandridis, T., Whetton, R. L. and Mouazen, A. M. (2016). Wheat yield prediction using machine learning and advanced sensing techniques, *Computers and Electronics in Agriculture* 121, 57–65. doi: 10.1016/j.compag.2015.11.018.

Pantuwan, G., Fukai, S., Cooper, M., Rajatasereekul, S. and O'Toole, J. C. (2002). Yield response of rice (*Oryza sativa* L.) genotypes to drought under rainfed lowland, *Field Crops Research* 73(2–3), 181–200. doi: 10.1016/S0378-4290(01)00194-0.

Peake, A. S., Huth, N. I., Carberry, P. S., Raine, S. R. and Smith, R. J. (2014). Quantifying potential yield and lodging-related yield gaps for irrigated spring wheat in sub-tropical Australia, *Field Crops Research* 158, 1–14. doi: 10.1016/j.fcr.2013.12.001.

Pellegrini, P., Cossani, C. M., Bella, C. M. D., Piñeiro, G., Sadras, V. O. and Oesterheld, M. (2020). Simple regression models to estimate light interception in wheat crops with Sentinel-2 and a handheld sensor, *Crop Science* 60(3), 1607–1616. doi: 10.1002/csc2.20129.

Peng, D., Chen, X., Yin, Y., Lu, K., Yang, W., Tang, Y. and Wang, Z. (2014). Lodging resistance of winter wheat (*Triticum aestivum* L.): Lignin accumulation and its related enzymes activities due to the application of paclobutrazol or gibberellin acid, *Field Crops Research* 157, 1–7. doi: 10.1016/j.fcr.2013.11.015.

Piñera-Chavez, F. J., Berry, P. M., Foulkes, M. J., Jesson, M. A. and Reynolds, M. P. (2016). Avoiding lodging in irrigated spring wheat. I. Stem and root structural requirements, *Field Crops Research* 196, 325–336. doi: 10.1016/j.fcr.2016.06.009.

Pinthus, M. J. (1974). Lodging in wheat, barley, and oats: The phenomenon, its causes, and preventive measures, *Advances in Agronomy* 25, 209–263.

Pinto, R. S. and Reynolds, M. P. (2015). Common genetic basis for canopy temperature depression under heat and drought stress associated with optimized root distribution in bread wheat, *TAG. Theoretical and Applied Genetics. Theoretische und Angewandte Genetik* 128(4), 575–585. doi: 10.1007/s00122-015-2453-9.

Pothula, A. K., Igathinathane, C., Kronberg, S. and Hendrickson, J. (2014). Digital image processing based identification of nodes and internodes of chopped biomass stems, *Computers and Electronics in Agriculture* 105, 54–65. doi: 10.1016/j.compag.2014.04.006.

Rascher, U. and Pieruschka, R. (2008). Spatio-temporal variations of photosynthesis: Understand and scale light use efficiency and stresses of plant ecosystems, *Precision Agriculture* 9(6), 355–366. doi: 10.1007/s11119-008-9074-0.

Rasti, S., Bleakley, C. J., Silvestre, G. C. M., Holden, N. M., Langton, D. and O'Hare, G. M. P. (2020). Crop growth stage estimation prior to canopy closure using deep learning algorithms, *Neural Computing and Applications* 33(5), 1733–1743. doi: 10.1007/s00521-020-05064-6.

Rebetzke, G. J., van Herwaarden, A. F., Jenkins, C., Weiss, M., Lewis, D., Ruuska, S., Tabe, L., Fettell, N. A. and Richards, R. A. (2008). Quantitative trait loci for water-soluble carbohydrates and associations with agronomic traits in wheat, *Australian Journal of Agricultural Research* 59(10), 891–905. doi: 10.1071/AR08067.

Reynolds, M., Chapman, S., Crespo-Herrera, L., Molero, G., Mondal, S., Pequeno, D. N. L., Pinto, F., Pinera-Chavez, F. J., Poland, J., Rivera-Amado, C., Saint Pierre, C. and Sukumaran, S. (2020). Breeder friendly phenotyping, *Plant Science* 295, 110396. doi: 10.1016/j.plantsci.2019.110396.

Reynolds, M. and Langridge, P. (2016). Physiological breeding, *Current Opinion in Plant Biology* 31, 162–171. doi: 10.1016/j.pbi.2016.04.005.

Reynolds, M. and Pinto, F. (2019). Theory and application of phenotyping in wheat for different target environments. In: Ordon, F. and Friedt, W. (Eds) *Advances in crop breeding techniques*. Cambridge: Burleigh Dodds Science Publishing, pp. 293–326.

Reynolds, M. P. (2010). *Climate Change and Crop Production*. Wallingford: CABI Publishing. doi: 10.1079/9781845936334.0000.

Reynolds, M. P. and Borlaug, N. E. (2006). Impacts of breeding on international collaborative wheat improvement, *The Journal of Agricultural Science* 144(1), 3–17. doi: 10.1017/S0021859606005867.

Reynolds, M. P., Pask, A. J. D., Hoppitt, W. J. E., Sonder, K., Sukumaran, S., Molero, G., Pierre, C. S., Payne, T., Singh, R. P., Braun, H. J., Gonzalez, F. G., Terrile, I. I., Barma, N. C. D., Hakim, A., He, Z., Fan, Z., Novoselovic, D., Maghraby, M., Gad, K. I. M., Galal, E. G., Hagras, A., Mohamed, M. M., Morad, A. F. A., Kumar, U., Singh, G. P., Naik, R., Kalappanavar, I. K., Biradar, S., Sai Prasad, S. V., Chatrath, R., Sharma, I., Panchabhai, K., Sohu, V. S., Mavi, G. S., Mishra, V. K., Balasubramaniam, A., Jalal-Kamali, M. R., Khodarahmi, M., Dastfal, M., Tabib-Ghaffari, S. M., Jafarby, J., Nikzad, A. R., Moghaddam, H. A., Ghojogh, H., Mehraban, A., Solís-Moya, E., Camacho-Casas, M. A., Figueroa-López, P., Ireta-Moreno, J., Alvarado-Padilla, J. I., Borbón-Gracia, A., Torres, A., Quiche, Y. N., Upadhyay, S. R., Pandey, D., Imtiaz, M., Rehman, M. U., Hussain, M., Hussain, M., Ud-Din, R., Qamar, M., Sohail, M., Mujahid, M. Y., Ahmad, G., Khan, A. J., Sial, M. A., Mustatea, P., von Well, E., Ncala, M., de Groot, S., Hussein, A. H. A., Tahir, I. S. A., Idris, A. A. M., Elamein, H. M. M., Manes, Y. and Joshi, A. K. (2017). Strategic crossing of biomass and harvest index—Source and sink—Achieves genetic gains in wheat, *Euphytica* 213(11), 23. doi: 10.1007/s10681-017-2040-z.

Richards, R. A. (2006). Physiological traits used in the breeding of new cultivars for water-scarce environments, *Agricultural Water Management* 80(1-3), 197–211. doi: 10.1016/j.agwat.2005.07.013.

Richards, R. and Townley-Smith, T. (1987). Variation in leaf area development and its effect on water use, yield and harvest index of droughted wheat, *Australian Journal of Agricultural Research* 38(6), 983. doi: 10.1071/AR9870983.

Rivera-Amado, C., Trujillo-Negrellos, E., Molero, G., Reynolds, M. P., Sylvester-Bradley, R. and Foulkes, M. J. (2019). Optimizing dry-matter partitioning for increased spike growth, grain number and harvest index in spring wheat, *Field Crops Research* 240, 154–167. doi: 10.1016/j.fcr.2019.04.016.

Robles-Zazueta, C. A., Molero, G., Pinto, F., Foulkes, M. J., Reynolds, M. P. and Murchie, E. H. (2021). Field based remote sensing models predict radiation use efficiency in wheat, *Journal of Experimental Botany* 72(10), 3756–3773. doi: 10.1093/jxb/erab115.

Roitsch, T., Cabrera-Bosquet, L., Fournier, A., Ghamkhar, K., Jiménez-Berni, J., Pinto, F. and Ober, E. S. (2019). Review: New sensors and data-driven approaches—A path to next generation phenomics, *Plant Science* 282, 2–10. doi: 10.1016/j.plantsci.2019.01.011.

Rouse, J. W., Rouse Jr., J. W., Haas, R. H., Schell, J. A. and Deering, D. W. (1974). Monitoring vegetation systems in the Great Plains with ERTS. In: *Proceedings of the Third Earth Resources Technology Satellite- 1 Symposium*. Greenbelt: NASA SP 351, 325–333.

Rutkoski, J., Poland, J., Mondal, S., Autrique, E., Pérez, L. G., Crossa, J., Reynolds, M. and Singh, R. (2016). Canopy temperature and vegetation indices from high-throughput phenotyping improve accuracy of pedigree and genomic selection for grain yield in wheat, *Genetics Society of America* 6(9), 2799–2808. doi: 10.1534/g3.116.032888.

Ruuska, S. A., Rebetzke, G. J., van Herwaarden, A. F., Richards, R. A., Fettell, N. A., Tabe, L. and Jenkins, C. L. D. (2006). Genotypic variation in water-soluble carbohydrate accumulation in wheat, *Functional Plant Biology* 33(9), 799–809. doi: 10.1071/FP06062.

Sadeghi-Tehran, P., Virlet, N., Ampe, E. M., Reyns, P. and Hawkesford, M. J. (2019). DeepCount: In-fieldautomatic quantification of wheat spikes using simple linear iterative clustering and deep convolutional neural networks, *Frontiers in Plant Science* 10, 1176. doi: 10.3389/fpls.2019.01176.

Sadras, V. O. and Connor, D. J. (1991). Physiological basis of the response of harvest index to the fraction of water transpired after anthesis: A simple model to estimate harvest index for determinate species, *Field Crops Research* 26(3–4), 227–239. doi: 10.1016/0378-4290(91)90001-C.

Sadras, V. O. and Richards, R. A. (2014). Improvement of crop yield in dry environments: Benchmarks, levels of organisation and the role of nitrogen, *Journal of Experimental Botany* 65(8), 1981–1995. doi: 10.1093/jxb/eru061.

Salvi, S., Porfiri, O. and Ceccarelli, S. (2013). Nazareno Strampelli, the "Prophet" of the green revolution, *The Journal of Agricultural Science* 151(1), 1–5. doi: 10.1017/S0021859612000214.

Schnyder, H. (1993). The role of carbohydrate storage and redistribution in the source-sink relations of wheat and barley during grain filling – A review, *New Phytologist* 123(2), 233–245. doi: 10.1111/j.1469-8137.1993.tb03731.x.

Sekhon, R. S., Joyner, C. N., Ackerman, A. J., McMahan, C. S., Cook, D. D. and Robertson, D. J. (2020). Stalk bending strength is strongly associated with maize stalk lodging incidence across multiple environments, *Field Crops Research* 249(February), 107737. doi: 10.1016/j.fcr.2020.107737.

Serbin, S. P., Dillaway, D. N., Kruger, E. L. and Townsend, P. A. (2012). Leaf optical properties reflect variation in photosynthetic metabolism and its sensitivity to temperature, *Journal of Experimental Botany* 63(1), 489–502. doi: 10.1093/jxb/err294.

Shearman, V. J., Scott, R. K. and Foulkes, M. J. (2005). Crop physiology & metabolism physiological processes associated with wheat yield progress in the UK, *Crop Sci.* 45, 175–185.

Sierra-Gonzalez, A., Molero, G., Rivera-Amado, C., Babar, M. A., Reynolds, M. P. and Foulkes, M. J. (2021). Exploring genetic diversity for grain partitioning traits to enhance yield in a high biomass spring wheat panel, *Field Crops Research* 260, 107979. doi: 10.1016/j.fcr.2020.107979.

Silva-Perez, V., Molero, G., Serbin, S. P., Condon, A. G., Reynolds, M. P., Furbank, R. T. and Evans, J. R. (2018). Hyperspectral reflectance as a tool to measure biochemical and physiological traits in wheat, *Journal of Experimental Botany* 69(3), 483–496. doi: 10.1093/jxb/erx421.

Singh, D., Wang, X., Kumar, U., Gao, L., Noor, M., Imtiaz, M., Singh, R. P. and Poland, J. (2019). High-throughput phenotyping enabled genetic dissection of crop lodging in wheat, *Frontiers in Plant Science* 10(April), 394. doi: 10.3389/fpls.2019.00394.

Singh, R. P., Chintagunta, A. D., Agarwal, D. K., Kureel, R. S. and Kumar, S. P. J. (2020). Varietal replacement rate: Prospects and challenges for global food security, *Global Food Security* 25, 100324. doi: 10.1016/j.gfs.2019.100324.

Singh, R. P., Hodson, D. P., Jin, Y., Lagudah, E. S., Ayliffe, M. A., Bhavani, S., Rouse, M. N., Pretorius, Z. A., Szabo, L. J., Huerta-Espino, J., Basnet, B. R., Lan, C. and Hovmøller, M. S. (2015). Emergence and spread of new races of wheat stem rust fungus: Continued threat to food security and prospects of genetic control, *Phytopathology* 105(7), 872–884. doi: 10.1094/PHYTO-01-15-0030-FI.

Slafer, G. A., Calderini, D. F. and Miralles, D. J. (1996). Yield components and compensation in wheat: Opportunities for further increasing yield potential. In: *Increasing Yield Potential in Wheat: Breaking the Barriers*. Mexico, DF: CIMMYT, pp. 101–133.

Slafer, G. A., Elia, M., Savin, R., García, G. A., Terrile, I. I., Ferrante, A., Miralles, D. J. and González, F. G. (2015). Fruiting efficiency: An alternative trait to further rise wheat yield, *Food and Energy Security* 4(2), 92–109. doi: 10.1002/fes3.59.

Smith, C., Karunaratne, S., Badenhorst, P., Cogan, N., Spangenberg, G. and Smith, K. (2020). Machine learning algorithms to predict forage nutritive value of in situ perennial ryegrass plants using hyperspectral canopy reflectance data, *Remote Sensing* 12(6), 928. doi: 10.3390/rs12060928.

Taiz, L. (1984). Plant cell wall expansion: Regulation of cell wall mechanical properties, *Annual Review of Plant Physiology* 35(1), 585 657.

Taiz, L. and Zeiger, E. (2015). Plant physiology (6th edn). Sunderland: Sinauer Associates Inc.

Tattaris, M., Reynolds, M. P. and Chapman, S. C. (2016). A direct comparison of remote sensing approaches for high-throughput phenotyping in plant breeding, *Frontiers in Plant Science* 7, 1131. doi: 10.3389/fpls.2016.01131.

ten Harkel, J., Bartholomeus, H. and Kooistra, L. (2019). Biomass and crop height estimation of different crops using UAV-based lidar, *Remote Sensing* 12(1), 17. doi: 10.3390/rs12010017.

Thomas, S., Behmann, J., Steier, A., Kraska, T., Muller, O., Rascher, U. and Mahlein, A. K. (2018). Quantitative assessment of disease severity and rating of barley cultivars based on hyperspectral imaging in a non-invasive, automated phenotyping platform, *Plant Methods* 14(1), 45. doi: 10.1186/s13007-018-0313-8.

Thorp, K. R., Gore, M. A., Andrade-Sanchez, P., Carmo-Silva, A. E., Welch, S. M., White, J. W. and French, A. N. (2015). Proximal hyperspectral sensing and data analysis approaches for field-based plant phenomics, *Computers and Electronics in Agriculture* 118, 225–236. doi: 10.1016/j.compag.2015.09.005.

Trachsel, S., Kaeppler, S. M., Brown, K. M. and Lynch, J. P. (2011). Shovelomics: High throughput phenotyping of maize (*Zea mays* L.) root architecture in the field, *Plant and Soil* 341(1–2), 75–87. doi: 10.1007/s11104-010-0623-8.

Tripathi, S. C., Sayre, K. D. and Kaul, J. N. (2005). Planting systems on lodging behavior, yield components and yield of irrigated spring bread wheat, *Crop Science* 45(4), 1448–1455. doi: 10.2135/cropsci2003-714.

UKRI. (2021). *United Kingdom Research and Innovation. Understanding and Reducing Lodging in Maize and Rice*.

van Herwaarden, A. F., Angus, J. F., Richards, R. A. and Farquhar, G. D. (1998). The negative grain yield response of dryland wheat to nitrogen fertiliser II. Carbohydrate and protein dynamics, *Australian Journal of Agricultural Research* 49(7), 1083–1094. doi: 10.1071/A97040.

Velumani, K., Oude Elberink, S., Yang, M. Y. and Baret, F. (2017). Wheat ear detection in plots by segmenting mobile laser scanner data, In: *ISPRS Ann. Photogramm. Remote Sens. Spatial Inf. Sci.* IV-2/W4, 149–156. doi: 10.5194/isprs-annals-IV-2-W4-149-2017.

Wang, J., Zhu, J., Huang, R. and Yang, Y. (2012). Investigation of cell wall composition related to stem lodging resistance in wheat (*Triticum aestivum* L.) by FTIR spectroscopy, *Plant Signaling and Behavior* 7(7), 856–863.

Wang, X., Zhao, C., Guo, N., Li, Y., Jian, S. and Yu, K. (2015). Determining the canopy water stress for spring wheat using canopy hyperspectral reflectance data in loess plateau semiarid regions, *Spectroscopy Letters* 48(7), 492–498. doi: 10.1080/00387010.2014.909495.

Wang, Z., Liu, X., Li, R., Chang, X. and Jing, R. (2011). Development of near-infrared reflectance spectroscopy models for quantitative determination of water-soluble carbohydrate content in wheat stem and glume, *Analytical Letters* 44(15), 2478–2490. doi: 10.1080/00032719.2011.551859.

Warren, J. (2015). *The Nature of Crops: How We Came to Eat the Plants We Do*. Wales, UK: Aberystwyth University.

Weibel, R. O. and Pendleton, J. W. (1964). Effect of artificial lodging on winter wheat grain yield and quality 1, *Agronomy Journal* 56(5), 487–488.

Wiersma, J. J., Dai, J. and Durgan, B. R. (2011). Optimum timing and rate of trinexapac-ethyl to reduce lodging in spring wheat, *Agronomy Journal* 103(3), 864–870. doi: 10.2134/agronj2010.0398.

Xiang, L., Tang, L., Gai, J. and Wang, L. (2020). PhenoStereo: A high-throughput stereo vision system for field-based plant phenotyping – With an application in sorghum stem diameter estimation. In: *ASABE 2020 annual international meeting*. doi: 10.13031/aim.202001190.

Yamamoto, K., Guo, W. and Ninomiya, S. (2016). Node detection and internode length estimation of tomato seedlings based on image analysis and machine learning, *Sensors* 16(7). doi: 10.3390/s16071044.

Yasrab, R., Atkinson, J. A., Wells, D. M., French, A. P., Pridmore, T. P. and Pound, M. P. (2019). RootNav 2.0: Deep learning for automatic navigation of complex plant root architectures, *GigaScience* 8(11), 1–16. doi: 10.1093/gigascience/giz123.

Yemm, E. W. and Willis, A. J. (1954). The estimation of carbohydrates in plant extracts by anthrone, *The Biochemical Journal* 57(3), 508–514.

Yu, F. H., Xu, T. Y., Du, W., Ma, H., Zhang, G. S. and Chen, C. L. (2017). Radiative transfer models (RTMs) for field phenotyping inversion of rice based on UAV hyperspectral remote sensing, *International Journal of Agricultural and Biological Engineering* 10(4), 150–157. doi: 10.25165/j.ijabe.20171004.3076.

Yu, K., Anderegg, J., Mikaberidze, A., Karisto, P., Mascher, F., McDonald, B. A., Walter, A. and Hund, A. (2018). Hyperspectral canopy sensing of wheat Septoria tritici blotch disease, *Frontiers in Plant Science* 9, 1195. doi: 10.3389/fpls.2018.01195.

Yue, J., Feng, H., Jin, X., Yuan, H., Li, Z., Zhou, C., Yang, G. and Tian, Q. (2018). A comparison of crop parameters estimation using images from UAV-mounted snapshot hyperspectral sensor and high-definition digital camera, *Remote Sensing* 10(7), 1138. doi: 10.3390/rs10071138.

Zhang, C., Filella, I., Garbulsky, M. and Peñuelas, J. (2016). Affecting factors and recent improvements of the photochemical reflectance index (PRI) for remotely sensing foliar, canopy and ecosystemic radiation-use efficiencies, *Remote Sensing* 8(9). doi: 10.3390/rs8090677.

Zhang, J., Huang, Y., Pu, R., Gonzalez-Moreno, P., Yuan, L., Wu, K. and Huang, W. (2019). Monitoring plant diseases and pests through remote sensing technology: A review, *Computers and Electronics in Agriculture* 165, 104943. doi: 10.1016/j.compag.2019.104943.

Zhang, J., Pu, R., Huang, W., Yuan, L., Luo, J. and Wang, J. (2012). Using in-situ hyperspectral data for detecting and discriminating yellow rust disease from nutrient stresses, *Field Crops Research* 134, 165–174. doi: 10.1016/j.fcr.2012.05.011.

Zhang, Z., Masjedi, A., Zhao, J. and Crawford, J. J. (2017). Prediction of sorghum biomass based on image based features derived from time series of UAV images. In: *2017 IEEE International Geoscience and Remote Sensing Symposium (IGARSS)*, pp. 6154–6157. doi: 10.1109/IGARSS.2017.8128413.

Zhang, Z., Kayacan, E., Thompson, B. and Chowdhary, G. (2020). High precision control and deep learning-based corn stand counting algorithms for agricultural robot, *Autonomous Robots* 44(7), 1289–1302. doi: 10.1007/s10514-020-09915-y.

Zheng, H., Cheng, T., Yao, X., Deng, X., Tian, Y., Cao, W. and Zhu, Y. (2016). Detection of rice phenology through time series analysis of ground-based spectral index data, *Field Crops Research* 198, 131–139. doi: 10.1016/j.fcr.2016.08.027.

Part 2

Sensor types

Chapter 3

Advances in optical analysis for crop phenotyping

Jian Jin and Tanzeel U. Rehman, Purdue University, USA; and Qin Zhang, Washington State University, USA

1 Introduction

With the human population expected to reach 9.7 billion by the end of 2050, global crop production needs to double to meet rising demand (Ehrlich and Harte, 2015; Ray et al., 2013). We will need higher-yielding plants that are more nutritious, use water and nutrients more efficiently and can tolerate variation in the environment (Conceição and Mendoza, 2009). Genomic tools are already commonly used in crop breeding programs (Mullet et al., 2014). A major bottleneck is a lack of high-quality and highly automatic phenotyping methods, which makes breeding and plant science research efforts slower and more expensive. Many programs still rely on traditional phenotyping methods that are costly and have low throughput. To select high-yield plants faster and more efficiently, modern phenotyping is key.

Plant phenotyping technology is critical to support breeding and decision-making in crop production. Accurate measurement, timely analysis and decision tools are key to increased crop production. Recent years have seen rapid growth in remote sensing applications in the field of agriculture (Gracia-Romero et al., 2019; Li et al., 2014; Wang et al., 2016). Plant canopy reflectance equals the ratio of the amount of light leaving the canopy to the amount of incoming light (Kitchen and Goulding, 2001). Current remote sensing techniques rely heavily on this ratio to obtain information on plant physiological features (Meyer and

http://dx.doi.org/10.19103/AS.2022.0102.03

Neto, 2008). Advances in low-weight, low-cost imaging platforms have resulted in improved agricultural data collection. Through various sensors such as red-green-blue (RGB), hyperspectral and 3D cameras carried by these platforms, plant phenotypic properties are captured in the images that largely facilitate the process of crop analyses, such as accessing plant biomass, nutrient levels, disease stresses and so on (Li et al., 2014; Ma et al., 2019; Wang et al., 2020b; Zhang et al., 2019b).

The literature regarding optical sensing for plant phenotyping continues to develop. In 2017, Perez-Sanz et al. classified the image acquisition phenotyping systems into seven groups, namely mono-RGB vision, stereo vision, multi- and hyperspectral cameras, time of flight (ToF) cameras, light detection and ranging (LiDAR) technology, thermography and fluorescence imaging and tomography imaging (Perez-Sanz et al., 2017). One or more of these sensors are integrated into ground, aerial, and/or satellite platforms for field crops phenotyping. Many indoor phenotyping facilities have been constructed in both industry and academia over the last few decades. In the most typical setting, plants grown in the greenhouse and growth chambers are transferred by conveyor systems to various imaging stations, where they are imaged individually and then transferred back to the growing environment once the imaging is complete (Ge et al., 2016; Gehan et al., 2017; Golzarian et al., 2011). More recently, other indoor phenotyping systems have moved the sensors to the plants instead, with the aim of minimizing disturbance to plant growth. From all these high-throughput fields and greenhouse phenotyping systems, large amounts of images can be collected in a short amount of time. The general image-processing algorithms are similar between these systems; almost all conduct segmentation of the green canopy objects in the image, then calculate color or morphological features of these objects and use these features to predict more advanced plant physiological features such as nutrient stress, diseases, chemical impacts and so on with correlation-based machine learning tools. Based on practical phenotyping applications with different objectives and quality requirements, the machine learning tools can vary from linear multivariate models such as partial linear squares, nonlinear models such as support vector machine (SVM), and the increasingly popular deep learning models.

Optical plant sensing is complicated due to the complexity of the plant object and the imaging environment. Various challenges exist, which often result in unsuccessful phenotyping projects. Plant canopy reflectance actively changes over time, especially in the field, which introduces a great deal of variation in plant phenotyping results (Beneduzzi et al., 2017; Maji et al., 2015; Padilla et al., 2019). The variability in canopy reflection could be highly significant due to the diversity in the canopy size, shape, composition and arrangement of cells, leaves and stems (Ollinger, 2011). In addition, the plants' chemical constituents change during growth in line with their environment

(Gamon et al., 2015; Ishihara et al., 2015; Jackson et al., 1979; Ranson et al., 1985). Environmental factors such as air temperature, solar zenith angle, solar azimuth angle, shadows, humidity and wind speed were found to correlate strongly with changes in chemical constituents in plants (An et al., 2017; Gamon et al., 2015; Oliveira and Scharf, 2014; Zhao et al., 2018; Zhou et al., 2017). Plant physiological properties usually follow periodical behavior, which is commonly referred to as circadian rhythmicity (Greenham et al., 2015; Tindall et al., 2015). These circadian rhythms have been observed in a diverse range of plant species including corn, potato, rice, wheat, barley and soybean over the past few decades (Dunford et al., 2002; Kloosterman et al., 2013; Preuss et al., 2012; Tindall et al., 2015; Turner, 2007), raising the question of when to image and to what extent the different imaging time matters.

In this chapter, instead of re-enumerating the different types of sensors, the most popularly used optical phenotyping sensors such as RGB, multi- and hyperspectral cameras and their carrier platforms are reviewed. In order to help users select the right sensor for the right phenotyping application, this chapter will focus on comparing the advantages and disadvantages of the different sensor options, including the bottleneck challenges around different optical sensors. Some potential solutions to these challenges are provided in the case studies.

2 Popular optical sensors

2.1 RGB cameras

Red-green-blue (RGB) cameras (also known as regular digital or color cameras) have been used since the earliest plant phenotyping studies, either as a standalone sensor or in combination with other sensing devices (Harris et al., 2011; Humplík et al., 2015). Projected leaf area, biomass, height, size, morphology, nitrogen and yield have commonly been predicted/estimated using RGB imaging with subsequent image-processing algorithms (Bendig et al., 2013, 2014; Hoyos-Villegas et al., 2014; Holman et al., 2016; Jannoura et al., 2015; Li et al., 2016; Schirrmann et al., 2016; Swinfield et al., 2019). For plant phenotyping applications, RGB images are generally used to extract color (color/vegetation indices), geometrical/morphological and/or textural features (Chang and Rehman, 2017), which can later be used with a wide variety of machine learning algorithms to predict the relevant phenotypic feature (Rehman et al., 2019a).

Color features are generally extracted from the original RGB images or by changing their color model (e.g. HSV, HSI, XYZ, L*a*b, etc.) and their combinations. Wang et al. (2014) extracted 13 color features using 3 color models including RGB, HSV and L*a*b to predict the leaf nitrogen concentration (LNC) in 6 rice cultivars. Results indicated a significant correlation between SPAD and LNC and, among the color features, *b index represented the best possible relationship with both SPAD and LNC. Li et al. (2015) found that the dark green color index, based on the HSB color model from aerial images, was able to predict the canopy nitrogen concentration with an R^2 of 0.672. In addition to nitrogen concentration, Schirrmann et al. (2016) predicted the biophysical parameters (fresh biomass, dry biomass, leaf area index and plant height) of wheat crops using color features (RGB ratios), crop coverage and plant height. Biophysical parameters were predicted with R^2 in the range of 0.70–0.97, while the R^2 for the nitrogen concentration was in the range of 0.24–0.64. For a list of commonly used color indices for agronomic images, readers can refer to Xue and Su (2017).

The gray-level co-occurrence matrices (GLCM) are most commonly used for extracting textural features (Armi and Fekri-Ershad, 2019). These matrices quantify the texture by statistically estimating the probability of spatial distribution of an image pixel and its neighboring tonal variations at different orientations (Haralick et al., 1973). Yuan et al. (2019) predicted soybean end-season traits using color- and GLCM-based textural features extracted from early-season RGB canopy images. The study concluded that yield, maturity and seed size were all soybean traits that could be predicted early. Details of these matrices and an explanation of their different textural features can be seen in the work of Rehman et al. (2018, 2019b).

The morphological/geometrical features extracted from the RGB images are commonly used to predict the biomass, projected leaf area and plant height. A 3D crop surface model (CSM or a similar elevation model) is usually generated by a series of overlapping RGB images via structure from motion (SfM) algorithms. Bendig et al. (2014) also estimated the fresh and dry biomass of 18 cultivars of summer barley treated with 2 nitrogen levels using the plant height. The CSM was developed by extracting the 3D structure from the RGB images using the SfM technique. Results indicated that plant heights extracted from the CSM have an R^2 of 0.81 and 0.82 for fresh and dry biomass, respectively. Li et al. (2016) used a similar technique and estimated the canopy height and above-ground biomass with an R^2 of 0.88 and 0.78, respectively. Geipel et al. (2014) combined the vegetation indices and 3D CSM extracted from the geo-referenced RGB images to better predict the corn grain yield. The results demonstrated that the developed strategy can achieve a coefficient of determination (R^2) of up to 0.74 with the root mean square error ranging from 0.67 t/ha to 1.28 t/ha. Other common studies that have used SfM for modeling biomass, height and leaf area include Mathews and Jensen (2013), Holman et al. (2016), Brocks and Bareth (2018) and Swinfield et al. (2019).

Although very high spatial resolution RGB cameras are commercially available at a relatively low cost, their major drawback is that they can only provide the information at three wavebands (red, green and blue). Moreover, continuous spectral information is not available.

2.2 Near-infrared-enabled RGB cameras

Although RGB cameras are widely used in remote sensing due to their low cost, light weight and ease of use, they are not customized for crop phenotyping. For example, most charged-coupled devices or complementary metal oxide semiconductor RGB sensors' spectral sensitivity ranges from visible to near-infrared (NIR) wavelengths up to 1000 nm. The NIR sensitivity is of critical importance to plant phenotyping as the reflective ratio between NIR and visible wavelengths is the most characteristic measure of the leaf's chlorophyll content and health. Unfortunately, in order to better mimic human vision, most commercial RGB camera producers apply a high pass filter in front of the sensor chip to remove the NIR signal from the images. This significantly reduces the RGB camera's capability in plant phenotyping. Many phenotyping researchers have tried to remove the filter to improve plant phenotyping quality, but this is not recommended by the authors due to the safety issues regarding charged capacitors in the camera. The RGB camera with the NIR filter removed is known as an RGBN (NIR-sensitive RGB) camera. However, the red band of the RGBN camera is still a mixture of red and NIR, making it difficult to calculate the ratio for chlorophyll content such as normalized difference vegetative index (NDVI).

Many studies have explored the feasibility of using an RGBN camera to measure NDVI. A two-camera system containing a standard RGB camera and an RGBN camera is reported to calculate NDVI (Sakamoto et al., 2012). The standard RGB camera captures the red reflectance, while the RGBN camera captures the NIR reflectance. However, the two-camera system creates new challenges, such as pixel-to-pixel matching and cross-calibration. An RGBN camera with a filter-switching unit is reported to calculate NDVI without the limitations of the two-camera system (Petach et al., 2014). With the filter-switching unit, the visible image and NIR image are captured respectively by switching the NIR rejection filter. Subsequently, the NDVI is calculated with both images. The limitation is that the position switching of the filter-switching unit is time-consuming, which creates issues with pixel-to-pixel matching when the camera is mounted on a moving platform such as an unmanned aerial vehicle (UAV). An RGBN camera with a special filter to help obtain pure NIR has been studied in recent years (Rabatel et al., 2014; Nijland et al., 2014). The new filter is usually either a long-pass filter or a two-band-pass filter. The long-pass filter cuts off the visible light from a particular wavelength, while the two-band-pass filter allows two narrow bands of light to pass through. Therefore, the RGBN camera records visible and NIR light in a specified channel(s) depending on the filter applied. Most earlier studies calculated NDVI using the formula (NIR – red) / (NIR + red) by using the corresponding channel values obtained from RGBN camera as red and NIR, respectively (Sakamoto et al., 2011; Netto et al., 2019; Zheng et al., 2018; Stow et al., 2000; Dworak et al.,

2013). Vega et al. (2015) used an RGBN camera of this type to assess the NDVI on four different days and predicted grain yield, aerial biomass and nitrogen content using linear regression. The NDVI was found to correlate significantly with all crop parameters at 99% confidence level, except during the early growth stages. In 2020, Wang et al. designed a new device called Ncam, composed of an RGBN camera, a filter and a microcontroller at a total material cost of just $70–85. This new NDVI estimation solution compares to a high-end hyperspectral camera in an experiment with corn plants under different nitrogen and water treatments. The results showed that the Ncam with two-band-pass filter achieved high performance ($R^2 = 0.99$, root mean square error (RMSE) = 0.0093) to estimate NDVI with the machine learning model (Wang et al., 2020b).

2.3 Multispectral and hyperspectral cameras

Multispectral and hyperspectral imaging has become one of the most important phenotyping technologies in recent decades (Li et al., 2014a). Multispectral cameras generally provide more useful information compared to ordinary RGB sensors by packing between 3 and 40 carefully selected wavebands in both visible and NIR regions. Svensgaard et al. (2014) used a nine-waveband multispectral camera to acquire images of four different wheat varieties treated with two different nitrogen levels and repeated on two different soil types at four different time points. The vegetation coverage and NDVI enabled them to differentiate between the varieties, especially at the early growth stage. Xu et al. (2019) used five-band multispectral images for flower detection, canopy cover and plant height in cotton fields. The mean canopy NDVI and canopy cover were found to correlate with R^2 in the range of 0.33–0.56. The spectral difference of the canopy, flower, ground and ground shade was used with SVM for flower detection with an accuracy in the range of 56–100%. Veys et al. (2019) used a 36 narrow-band multispectral camera covering the 365–960 nm spectral range for the early detection of light leaf spot on oilseed rape. The study concluded that the proposed method was able to detect light leaf spot 13 days before the appearance of visible symptoms with an accuracy of 92%. Navarro et al. (2021) used a 37-band camera in the range of 400–1000 nm with an ad hoc deep learning architecture containing 3D filter banks to classify five seedless table grape varieties. The authors reported that the proposed method was able to correctly classify 100% of the images. Although multispectral imaging provides multiple wavebands, these are not continuous and therefore unable to provide the continuous spectrum.

Compared to other imaging technologies such as grayscale, RGB and multispectral imaging, hyperspectral imaging captures hundreds of bands, enabling researchers and scientists to measure many physiological features of plants efficiently and accurately, including plant biomass, leaf moisture content and chlorophyll content (Clevers et al., 2008; Di Gennaro et al., 2018; Li et al., 2014). Ge et al. (2016) studied the temporal dynamics of plant growth, water use and leaf water content of two maize genotypes subjected to two water treatments. The water content was successfully predicted for both genotypes (R^2 of 0.81 and 0.92) using the plant reflectance extracted from the visible near-infrared (VNIR) hyperspectral camera. Rehman et al. (2020) compared the performance of partial least squares regression (PLSR), support vector regression (SVR) and deep learning regression for predicting the water content of corn using the mean spectral reflectance extracted from the VNIR camera. The deep learning architecture containing inception-like 1D filter banks outperformed both linear (PLSR) and nonlinear (SVR) approaches. Zhao et al. (2020) used a hyperspectral camera in the range of 900–1700 nm to assess the relative water content and water thickness for five cultivars of tomatoes grown in plant factories. Results showed that the best results for water content were obtained by using a two-band index from the spectral reflectance pre-processed with the first-order derivative.

Zhang et al. (2019a) optimized the imaging angle of a swing hyperspectral imaging system for the plant-level and pixel-level water and nitrogen content. Their results showed that 15° and 75° were the most optimized angles to distinguish the different nitrogen and water treatments among the plants. Ma et al. (2020) used a nonlinear kernel partial least squares regression (KPLSR) to predict the biomass of maize plants using the spectral reflectance from the hyperspectral images. In contrast to conventional linear models, the R^2 of KPLSR with radial basis function was better (0.924 vs. 0.896). In addition to VNIR, several researchers have also explored the complete or earlier short wave infrared (SWIR) region with regard to its application in plant phenotyping. Pandey et al. (2017) investigated the chemical properties (including leaf water content and concentrations of macronutrients) of maize and soybean plants using hyperspectral imaging in the range of 550–1700 nm. The authors found that the water content was predicted with the highest R^2, while the sodium and boron were not predicted accurately. Readers interested in the application of SWIR can refer to Banerjee et al. (2020), Bruning et al. (2019), Camino et al. (2018), Kicherer et al. (2017) and Mertens et al. (2021).

2.4 Other optical sensors such as 3D LiDAR and thermal cameras

The 3D sensors provide the dense point clouds that can be used to provide information regarding the 3D structure/architecture of plant canopies (Qiu et al., 2019) or to extract the morphological traits of individual plants (Panjvani et al., 2019). Qiu et al. (2019) used a mobile robot-mounted LiDAR unit for measuring plant height and row spacing in experimental maize fields. The R^2 and RMSE between LiDAR extracted and manually measured row spacing were found to be 0.237 m and 0.0916 m, respectively. Panjvani et al. (2019) developed a low-cost LiDAR platform for extracting leaf traits such as length, width, area and leaf annotations in indoor settings. The study reported an R^2 in the range of 0.504–0.681, 0.29–0.56 and 0.336-0.595 for leaf length, width and area, respectively. Christiansen et al. (2017) used a LiDAR unit mounted on a UAV for developing a correlation between plant height and the amount of nitrogen applied to an experimental winter wheat field. The authors found that the application of nitrogen beyond 200 kg/ha did not increase crop height estimates during the current growth stage. Madec et al. (2017) made a comparison between plant heights of well-watered and water-stressed wheat genotypes extracted from the LiDAR scanner mounted on a ground vehicle and from an RGB camera carried via UAV. The RGB images were processed using the SfM technique to generate the CSM. The results found that LiDAR and RGB height estimates were consistent, although the SfM showed slight underestimation due to the relatively coarser spatial resolution of RGB images.

Li et al. (2020) combined the canopy structural features extracted from the LiDAR and the vegetative indices extracted from the red-edge camera for predicting the biomass. The authors reported that canopy height, canopy perimeter and vegetative indices yielded a robust biomass model with a maximum correlation coefficient of 0.90. Guo et al. (2017) developed a multi-sensor system containing LiDAR, RGB, hyperspectral and thermal cameras that can extract plant height, plant width, leaf width, leaf length, leaf area, leaf inclination angle and other similar traits for study. Ravi et al. (2019) integrated LiDAR, RGB, VNIR and SWIR to predict the end-of-season biomass of sorghum. The LiDAR data was used to extract the 30th, 50th, 75th, 90th, 95th, 99th and 100th percentile height values for each plot of the sorghum calibration panel. Those spectral and geometric features were used to predict the end-of-season biomass. Similarly, Masjedi et al. (2020) studied the potential of predicting the sorghum biomass using multi-sensor (LiDAR, VNIR, SWIR) data acquired by mounting these sensors on a UAV. The geometrical features from the LiDAR data, including height percentiles, canopy volume and canopy cover, were combined with the average spectral reflectance and vegetation indices for predicting the biomass. The study indicated the R^2 ranged between 0.64 and 0.89 for end-of-season biomass. In addition to LiDAR, ToF and stereo cameras have been commonly employed to extract 3D information for plant phenotyping. Some studies using these techniques can be seen in the work of Song et al. (2011), Azzari et al. (2013), Paulus et al. (2014) and McCormick et al. (2016).

Plant phenotyping researchers have also exploited the use of thermal cameras to assess canopy temperature dynamics (Messina and Modica, 2020).

Ludovisi et al. (2017) studied the response of well-watered and moderate drought on the population of different tree genotypes using thermal infrared images acquired with the help of UAV. Results found that thermal imaging could effectively assess the genotype variability under drought stress conditions. Deery et al. (2016) used a radiometrically calibrated camera with a manned helicopter for monitoring the canopy temperature. Readers interested in the latest applications of thermal imaging in plant phenotyping should refer to Messina and Modica (2020).

3 Major challenges in optical sensing

While the optical sensors with their carrier platforms and data infrastructure have rapidly evolved, plant measurement quality still suffers from several major challenges regarding both the complexity of the canopies and changing environmental conditions. Each of these challenges is discussed in this section.

3.1 Variance in pigments, nutrients and stresses across the canopy

Chlorophyll, nitrogen content, water content and stresses from diseases are all non-uniformly distributed across the plant canopy. If we take the corn plant as an example, the SPAD 502 Plus Chlorophyll Meter's measurement results of different locations on the same V9 stage corn plant can be viewed in Fig. 1. The chlorophyll measurements range from 3 to 46, which have >15 times (1500%) variance among them. The difference exists not only between different leaves (especially between the lower senescent leaves and newer leaves above) but also between different locations on the same leaf. The chlorophyll tends to have a higher accumulation density toward the leaf tip direction compared with the base part of the leaf, with a roughly linear ascending trend. However, the ascending curve varies between different leaves. Besides chlorophyll content, other analyzed items such as metabolome and transcriptome analysis, protein measurements as well as dry weight determination all showed continuous gradients across the canopy (Pick et al., 2011; Sun et al., 2015).

The phenotyping quality from both remote sensing cameras and single-location proximal sensors such as spectrometers and SPAD meters suffers from such significant variances. As mentioned above, the common image-processing algorithms calculate the averaged color intensities or spectrum across all the green canopy pixels. The mean color or spectra cannot represent the whole highly un-uniform canopy, especially when many leaves are partially occluded by others and there is no control of the percentages of the visible area of each leaf to the camera.

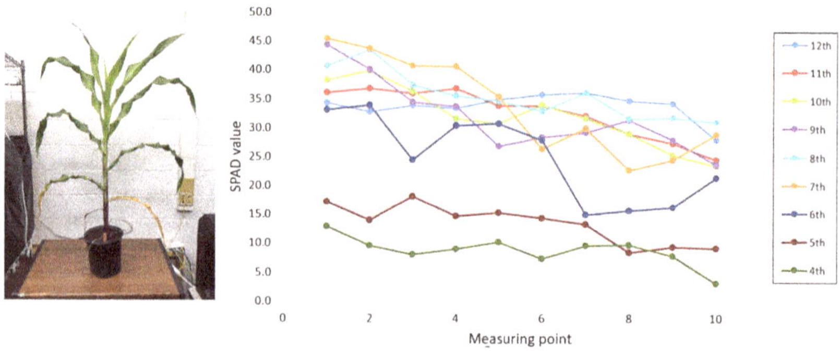

Figure 1 SPAD 502 Plus Chlorophyll Meter's measurement results of different locations on the same V8 stage corn plant in 2017 at Purdue University. Each curve contains ten measurements from each leaf, and the ten measurement locations are uniformly distributed on each leaf in the middle between the midrib and leaf edge. The bottom four leaves were skipped as they either completely withered or were too short for the measurements.

In order to reduce the color variance's impact on the proximal sensors' measurement quality, standard protocols have been introduced to guide the selection of sampling locations. For example, the top matured corn leaf should be selected and the measurement should be in the middle of the leaf, avoiding the midrib. These protocols are more or less the same as those used for leaf sampling for crop nutrient assessment purposes since both face the same variance issue (Spectrum Analytic Inc.: https://www.spectrumanalytic.com/services/analysis/plantguide.pdf; accessed 11 April 2021). However, these protocols cannot completely remove the variance noise because significant variance still exists even at a small local area of the leaf (Fig. 2). In some cases, an offset of just 1-2 cm on the corn leaf can result in >20% of the color measurement differences.

3.2 Impact from the leaf surface slopes

In addition to the variance in color across the canopy, the slope of the leaf surface is also a significant factor affecting the imaging result. A test at Purdue university in 2020, as shown in Fig. 3, highlighted the effects of rotating the leaf (with the same camera and lighting setup and the same imaging distance) from 0° (normal to the camera lens) to 80°, and how the reflected spectrum received by the camera changed significantly. The changed spectrum resulted in a significant variation in plant feature prediction values. For example, the NDVI varied by 18% for corn and 12% for soybean (Fig. 3). The result also indicates a certain pattern of the color change by leaf slope, which could potentially be

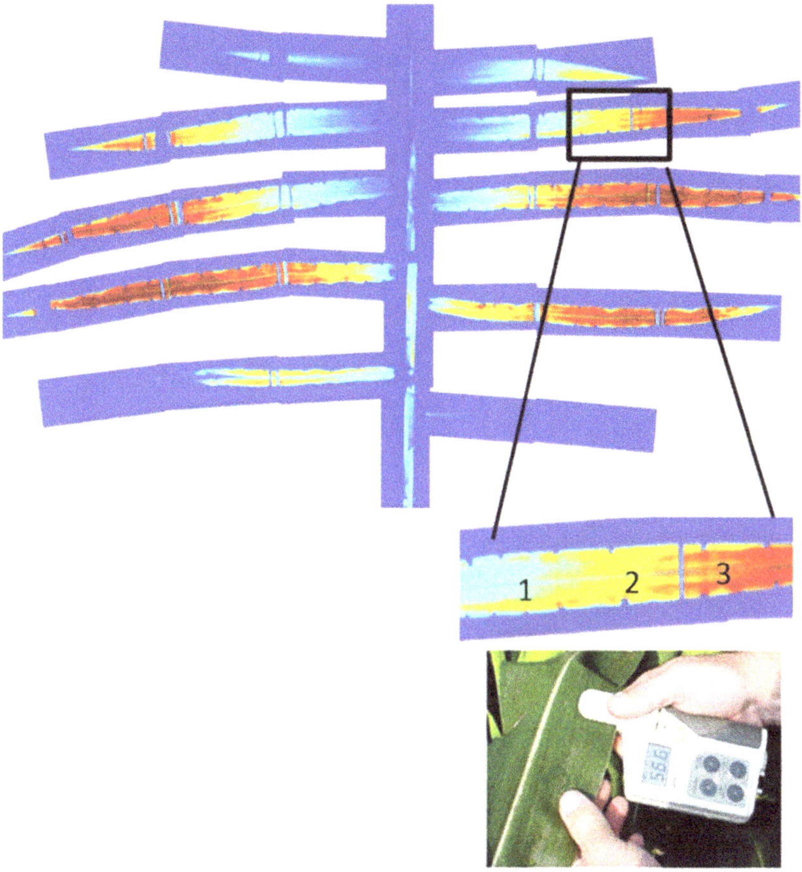

Figure 2 NDVI heatmap of a V9 corn plant. The image is composed of an NDVI heatmap of each individual leaf. Each leaf was taken off the plant and laid flat on a horizontal surface when imaged by a hyperspectral camera. The locations labeled on the leaf are within 2 cm of the closest neighbor but showed >20% difference in NDVI values.

calibrated. The PROSAIL model (Jacquemoud and Baret, 1990; Jacquemoud et al., 2009) was introduced and widely used to estimate the relationship between canopy structural variables such as the leaf slope angle. PROSAIL is easy to use (Salas and Henebry, 2013) but only predicts the general changing trend, which cannot be used to precisely calibrate the spectrum drift in each specific imaging scenario (Berger et al., 2018). Recalibrating for each imaging scenario is usually needed, but it is a time-consuming and complicated task that requires pixel-to-pixel matching between 3D vision (for calculating the slope at each pixel) and RGB/multispectral/hyperspectral cameras (Zhang et al., 2020).

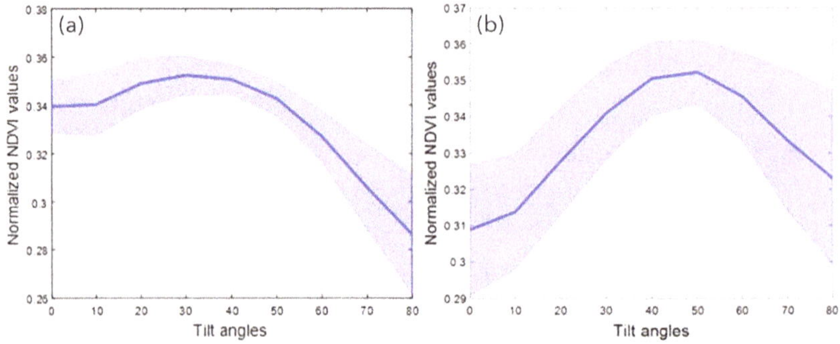

Figure 3 Change in NDVI from the same leaf spot caused by the change in the tilt angle of the leaf surface from 0° to 80° for (a) corn and (b) soybean.

3.3 Plants' diurnal changes

The time at which to image matters. Like human beings, most plants change throughout the day. The diurnal impact on plants' reflectance characteristics is complicated. For example, the reflectance characteristics of plants at noon can be very different from those in the afternoon. One cause of these variabilities could be the interaction between the camera's sensitivity and view angle, the canopy geometry, solar zenith angle, solar azimuth angle and shadows (Danner et al., 2019; Ishihara et al., 2015; Jackson et al., 1979; Ranson et al., 1985). Meanwhile, the plant itself is endogenously sensitive to the environmental conditions, with complicated interactions between genetic backgrounds, the external environment and the treatments (G(Genotype) × E(Environm ent) × T(Treatment)). All these affect the overall reflectance characteristics of plants throughout the day, causing diurnal variabilities. Even in the controlled environment of a greenhouse, the imaging result of the same plant can be quite different over 24 h. Figure 4 shows the change in the averaged canopy spectrum of a corn plant in Purdue's Lilly 13-4 automatic phenotyping greenhouse, which collects hyperspectral images of each plant continuously over a 24 h period (Ma et al., 2019).

The diurnal variance on phenotyping data has been reported as an unignorable impact in many plant studies. Gardener concluded that diurnal variance was a major unresolved noise issue when using reflectance measurements for estimating plant features such as leaf area, plant biomass and phenology (Gardner, 1985). In the last few decades, diurnal variability has been documented in various plant phenotyping studies for corn, soybean and wheat with both passive and active optical sensors (Oliveira and Scharf, 2014; Beneduzzi et al., 2017). These variances are retained in the captured images, weakening the signal power of the data. However, for most current remote sensing studies, the impacts of diurnal variabilities are rarely considered. For

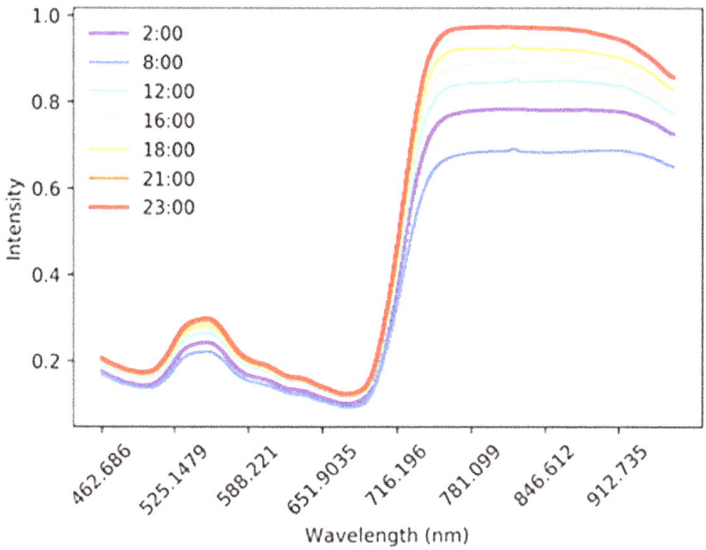

Figure 4 The spectra of the same maize plant at different imaging time points during 1 day (2:00 am, 8:00 am, 12:00 pm, 4:00 pm, 6:00 pm, 09:00 pm and 11:00 pm on 14 June 2017).

example, the NDVI values had over 10% difference at different times of the day on the same plant from the raw remote sensing measurements (Beneduzzi et al., 2017; Sticksel et al., 2004). Similar diurnal variances are also found in many other plant features such as plant temperature, spectra features, chlorophyll content and so on (Gamon et al., 2015; Zhao et al., 2018).

To reduce the impacts of diurnal variability, airborne remote sensing systems usually have strict rules regarding the sampling time window (around noon) and weather condition requirements (Barbedo, 2019; Di Gennaro et al., 2018; Gracia-Romero et al., 2019).

3.4 Other environmental impacts

In addition to diurnal variances, changing environmental conditions have a significant impact on the imaging result (Gamon et al., 2015). Fans, cooling walls and temperature management systems create non-uniform microclimates that make plants in the greenhouse look different from one another (Aaslyng et al., 2007). These non-uniformities will cause unwanted noise in plant phenotyping. The environmental impacts are more severe in the field. Plant imaging data is greatly affected by changing daylight and other environmental factors such as cloud coverage, wind speed and temperature (Gamon et al., 2015). The noise caused by the changing environment could be much stronger than the signals to be measured in phenotyping projects, resulting in poor phenotyping quality or even measurement failure. It is important to discern between plant signal and environmental noise in plant imaging.

4 Case studies

4.1 LeafSpec

As discussed above, modern phenotyping systems still suffer from various noise factors. One way of improving phenotyping quality is to move the sensor closer to the plants. With proximal sensors, imaging can take place inside a closed chamber to remove ambient lighting impact. The leaf can be positioned in the chamber at a fixed angle and distance from the camera and artificial lighting source. The leaf imaging quality is no longer impacted by the soil background. Leaf images can be taken at a much higher resolution, so that the nutrients and stress distribution can be analyzed at the tissue/pixel level. Purdue University's Davidson Prize winner 'LeafSpec' leaf imager (patents applied in 2018) represents this 'proximal sensing' mode (Wang et al., 2020a). It takes 3–5 s for LeafSpec to scan a leaf and get a hyperspectral image with 0.5 mm/pixel resolution. The software on the device processes the images immediately with an onboard microprocessor to predict the physiological features of the plant, such as leaf moisture, nutrient levels, disease stresses and so on. Each measurement can be geo-referenced and uploaded to the web-based geographic information system, which provides various map viewing functions (Fig. 5).

As mentioned above, the LeafSpec device design removes many types of ambient noise that arise in current greenhouse and field hyperspectral imaging systems. Due to its touch-based sensing mode, the LeafSpec camera is much closer to the object for high-resolution images with even leaf veins clearly visible (Fig. 5b). With the unprecedented resolution of hyperspectral leaf images, advanced algorithms can analyze the distribution of nutrition and stress on the leaf surface. Compared with traditional image-processing

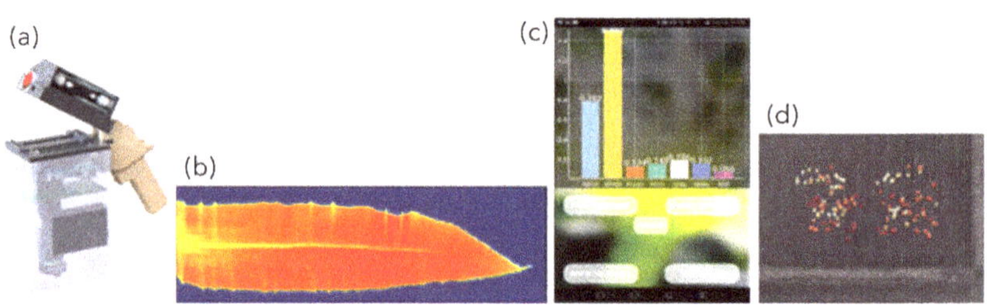

Figure 5 (a) The handheld plant sensor device LeafSpec; (b) scanned hyperspectral image; (c) plant health features predictions on a smartphone APP; (d) GIS map.

methods that simply average all leaf pixels, the new algorithms enable more precise phenotyping measurements. Figure 6 shows preliminary data in which the new leaf stress distribution analysis algorithm separates leaves with different nitrogen content levels more clearly than traditional algorithms. Combining the hardware design of the touch-based hyperspectral imager with the more advanced image-processing software, LeafSpec achieves higher signal-over-noise for leaf measurement.

4.2 Diurnal variance and environmental calibration

As mentioned above, an as-yet under-explored challenge in agricultural plant remote sensing is that the time of the day and environmental conditions such as sun angle, cloud coverage and temperature can significantly alter plant appearance and thus affect the imaging sensor's accuracy upon plant feature measurements. These image alterations result from the complicated interaction between real-time environments and plants. Analysis of these impacts requires continuous monitoring of the changes through various environmental conditions, which is difficult with current airborne remote sensing systems.

In order to collect sufficient time-series data to analyze the correlation between images and environmental conditions, in 2019, Purdue deployed a 150-feet-long, 24-feet-high field hyperspectral imaging gantry (Fig. 7a) in its Agronomy Center for Research and Education field. The fully automated imaging gantry works as a 'mock drone' system carrying a weather-proof VNIR hyperspectral camera with 1.4 nm spectral resolution to continuously scan the crops in the 3-acre field. The facility also includes a local weather station (Fig. 7b) and other environmental sensors that record comprehensive environmental condition data such as temperature, radiation, wind speed and so on during the imaging process. It is also the first fixed field imaging system to not apply any shadow on the crops, due to the special design created by Purdue's engineers.

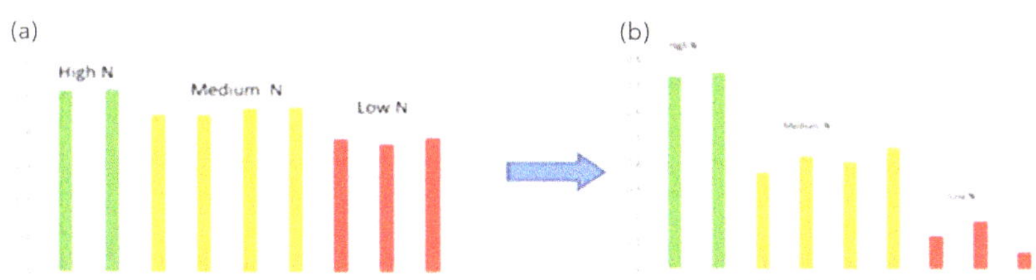

Figure 6 The leaf nutrition and stress distribution analysis algorithm showed improved measurement quality. (a) The averaged DNVI across the leaf's surface showed numerical differences between leaves with high, medium and low nitrogen levels, but differences were not significant; (b) the leaf greenness index calculated by LeafSpec's new distribution analysis algorithm is able to separate nitrogen levels more clearly.

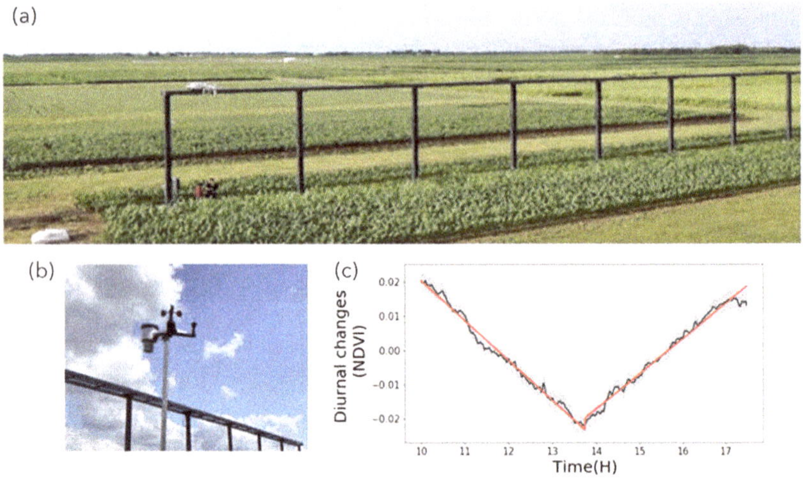

Figure 7 (a) The field imaging gantry; (b) the weather station by the gantry; (c) the 2019 time-series images showed a stable 'V'-shaped daily changing pattern from 10:00 am to 18:00 pm.

During the summer of 2019, over 8000 repetitive images of the same corn field were taken at 2.5 min intervals from sunrise to sunset, every day, for 31 days. The corn had been planted between ~30 and 60 days during this imaging period. Time-tagged local environment data including solar zenith angle, solar irradiation, temperature, wind speed and so on were also recorded during the imaging time.

The 2019 summer results showed consistent diurnal changing patterns among the images. For instance, the popular NDVI changed by 15–30% throughout each day. The result also confirmed the hypothesis that these changes follow certain patterns so they can be precisely modeled and calibrated. For example, Fig. 7c shows the 'V'-shaped NDVI diurnal changing trend, which has been validated across genotypes, nitrogen and water treatments, between 2019 and 2020. It was also observed with time-series UAV imaging results in 2020.

Although the results above suggested that the diurnal impacts could be calibrated, the causes of this significant diurnal variance in crop imagery remain unclear. The Purdue researchers originally suspected that changing sunlight conditions were the major contribution to the variance. However, the team observed a similar diurnal changing pattern in a corn assay in Purdue's greenhouse imaging facility, which scanned each individual plant every hour continuously for 9 days (Ma et al., 2019). Since the greenhouse imaging took place in an enclosed chamber with only fixed artificial lights, the diurnal variance could only be due to changes in the plants themselves. Still, two possible causes for the diurnal variance remain: (1) the intrinsic change in the leaf color over the hours; or (2) the impact from the plants' morphological changes. It is well known that leaf motion such as the leaf angle changes occur in response to an environmental stimulus such as light (Guo et al., 2015). At the same time, corn leaf rolling occurs as a reaction to water stress development during the day (Baret et al., 2018). Both of these plant morphological changes cause variations in the leaf surface slope (to the camera), which in turn could significantly alter the plant spectrum in the images (Zhang et al., 2020). In 2020, the Purdue team collected time-series LeafSpec (mentioned in Section 4.1) images to determine whether they could confirm or rule out the plant's morphological changes as the major cause of the diurnal variance, as the LeafSpec images do not suffer from the leaf slope change as much. They expect to publish their discoveries following the experiment.

The time-series data collected from the field imaging gantry was also used to investigate the correlation between the changes due to other environmental conditions and variance in the images. At first, the single-factor correlation analysis was conducted to figure out which environmental conditions have the most significant impacts on field hyperspectral imaging. The results showed that the measured changes of radiation, diurnal time, zenith angle, temperature and wind speed all correlated with the environment-induced variation in NDVI, except for humidity and its derivatives. This confirmed the findings from the previous literature that plant-sensing data were strongly impacted by air temperature, solar radiation, sun zenith angle and diurnal time (de Souza et al., 2010; Ishihara et al., 2015; Jones et al., 2009), while humidity was rarely reported to demonstrate a similar impact. Thus, the Purdue research team removed humidity and its derivatives in the correlation analysis.

An artificial neural network (ANN) was then successfully built to model the relationship between the phenotyping data variation and environmental changes. The ANN model was able to predict precisely the environmental effects in remote sensing results and thus could be used to effectively eliminate the environment-induced variation in the phenotyping features. Testing the NDVI calculated from the hyperspectral images showed that variance in NDVI was significantly reduced by 79% (Fig. 8). In addition to NDVI, similar performances of the ANN modeling method were confirmed with other phenotyping features such as relative water content predictions. Therefore, this modeling method showed great potential to be applied in airborne remote sensing applications in agriculture to significantly improve the imaging quality.

5 Summary and future trends

Various optical sensors and their platforms have been developed and applied in crop phenotyping. Based on the distance between the sensor and crops,

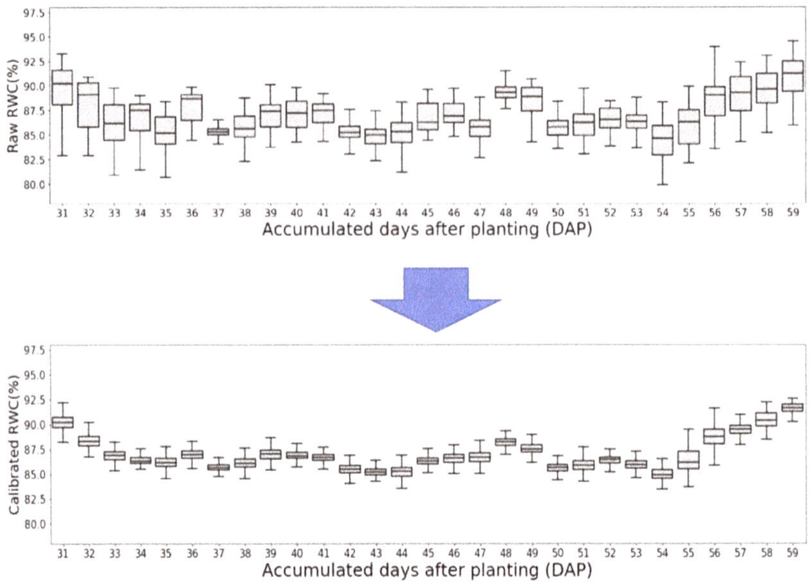

Figure 8 Box plots for the five-fold cross-validated correction result of the sample data set. The raw NDVI showed huge daily variances, but the ANN model-corrected NDVI had much more condensed boxes.

they can be categorized into remote sensing systems such as satellite and airborne imagers and proximal sensing systems such as the SPAD meter and spectrometer. While the performance of an optical phenotyping system is often evaluated by its signal quality and throughput, both are directly related to the imaging distance: the further away from the crops, the higher the throughput but lower the resolution or signal-over-noise ratio (SNR) it can have (Fig. 9).

In order to improve the signal quality of remote sensing platforms, the bottleneck noise issues need to be addressed. Impacts from changing environmental conditions and atmosphere need to be further investigated in order to effectively calibrate the imaging result.

Most traditional proximal sensors such as SPAD or spectrometer only measure a small spot on the leaf. Due to the great variance in color/stress/nutrient distribution on the leaf, these probe-like single-spot optical sensors have not taken good advantage of the proximal sensing distance. The newly developed award-winning handheld hyperspectral imager LeafSpec described above is able to capture very high-resolution hyperspectral leaf images with unprecedented SNR. However, there is great potential to further develop advanced color/stress/nutrient distribution analysis algorithms to take full advantage of the high-quality hyperspectral leaf images from LeafSpec.

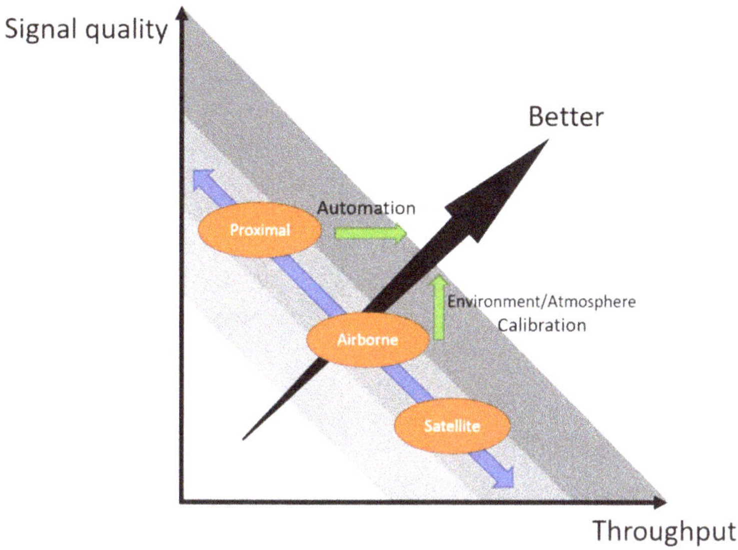

Figure 9 The performance of different types of optical sensors in signal quality and throughput, and their future development trend.

In order to improve the throughput of proximal sensors, robotic technologies could be the solution. In 2018, Bao et al. developed a robotic 3D plant perception and leaf probing system with collision-free motion planning for automated indoor plant phenotyping. In 2021, Chen et al. successfully demonstrated a robotic system to automatically detect target leaves with deep learning and control a robotic arm to scan the leaves with Purdue's LeafSpec hyperspectral leaf imager in the field. Meanwhile, these robotic systems for proximal leaf sensors need to be further developed in order to better meet the throughput and reliability requirements in practical uses.

6 Where to look for further information

6.1 Further reading

The papers listed below provide a good review of current aspects and future advances in the field of plant phenotyping:

- Mulla, D. J. (2013). Twenty-five years of remote sensing in precision agriculture: Key advances and remaining knowledge gaps. *Biosystems Engineering* 114(4), 358–371.
- Roitsch, T., Cabrera-Bosquet, L., Fournier, A., Ghamkhar, K., Jiménez-Berni, J., Pinto, F., and Ober, E. S. (2019). New sensors and data-driven

approaches - A path to next generation phenomics. *Plant Science* 282, 2-10.

- Liu, H., Bruning, B., Garnett, T., and Berger, B. (2020). Hyperspectral imaging and 3D technologies for plant phenotyping: From satellite to close-range sensing. *Computers and Electronics in Agriculture* 175, 105621.
- Xie, C., and Yang, C. (2020). A review on plant high-throughput phenotyping traits using UAV-based sensors. *Computers and Electronics in Agriculture* 178, 105731.
- Zhang, C., Marzougui, A., and Sankaran, S. (2020). High-resolution satellite imagery applications in crop phenotyping: An overview. *Computers and Electronics in Agriculture* 175, 105584.

6.2 Key journals/conferences

- *Computers and Electronics in Agriculture.*
- *Frontiers in Plant Science.*
- *Remote Sensing.*
- *Biosystems Engineering.*
- *Plant Phenomics.*
- *Plant Methods.*
- *Sensors.*
- Transactions of the ASABE.
- *Precision Agriculture.*
- American Society of Agricultural and Biological Engineers (ASABE) Annual International Meeting.
- International Plant Phenotyping Network (IPPN) International Symposium and Phenome conference. Also several regional sub-bodies.

7 References

Aaslyng, J. M., Andreassen, A. U., Allee, H. and Holst, N. (2007). Microclimate prediction for dynamic greenhouse climate control. *HortScience* 42, 272-279.

An, N., Welch, S. M., Markelz, R. J. C., Baker, R. L., Palmer, C. M., Ta, J., Maloof, J. N. and Weinig, C. (2017). Quantifying time-series of leaf morphology using 2D and 3d photogrammetry methods for high-throughput plant phenotyping. *Computers and Electronics in Agriculture* 135, 222-232. doi:10.1016/j.compag.2017.02.001.

Armi, L. and Fekri-Ershad, S. (2019). *Texture Image Analysis and Texture Classification Methods-A Review.* arXiv Preprint ArXiv:1904.06554.

Azzari, G., Goulden, M. L. and Rusu, R. B. (2013). Rapid characterization of vegetation structure with a Microsoft Kinect sensor. *Sensors* 13(2), 2384-2398. doi:10.3390/s130202384.

Banerjee, B. P., Joshi, S., Thoday-Kennedy, E., Pasam, R. K., Tibbits, J., Hayden, M., Spangenberg, G. and Kant, S. (2020). High-throughput phenotyping using digital and hyperspectral imaging-derived biomarkers for genotypic nitrogen response. *Journal of Experimental Botany* 71(15), 4604–4615. doi:10.1093/jxb/eraa143.

Bao, Y., Shah, D. S. and Tang, L. (2018). 3D perception-based collision-free robotic leaf probing for automated indoor plant phenotyping. *Transactions of the ASABE* 61(3), 859–872. doi:10.13031/trans.12653.

Barbedo, J. G. A. (2019). A review on the use of unmanned aerial vehicles and imaging sensors for monitoring and assessing plant stresses. *Drones* 3(2), 40. doi:10.3390/drones3020040.

Baret, F., Madec, S., Irfan, K., Lopez, J., Comar, A., Hemmerlé, M., Dutartre, D., Praud, S. and Tixier, M. H. (2018). Leaf-rolling in maize crops: From leaf scoring to canopy-level measurements for phenotyping. *Journal of Experimental Botany* 69(10), 2705–2716. doi:10.1093/jxb/ery071.

Bendig, J., Bolten, A. and Bareth, G. (2013). Monitoring des Pflanzenwachstums mit Hilfe multitemporaler und HOCH AUFLÖSENDER Oberflächenmodelle von getreidebeständen AUF Basis VON bildern AUS UAV-Befliegungen. *Photogrammetrie - Fernerkundung - Geoinformation* 2013(6), 551–562. doi:10.1127/1432-8364/2013/0200.

Bendig, J., Bolten, A., Bennertz, S., Broscheit, J., Eichfuss, S. and Bareth, G. (2014). Estimating biomass of barley using crop surface models (CSMS) derived from UAV-based RGB imaging. *Remote Sensing* 6(11), 10395–10412. doi:10.3390/rs61110395.

Beneduzzi, H. M., Souza, E. G., Bazzi, C. L. and Schenatto, K. (2017). Temporal variability in active reflectance sensor-measured NDVI in soybean and wheat crops. *Engenharia Agrícola* 37(4), 771–781. doi:10.1590/1809-4430-eng.agric.v37n4p771-781/2017.

Berger, K., Atzberger, C., Danner, M., D'Urso, G., Mauser, W., Vuolo, F., Hank, T. (2018). Evaluation of the PROSAIL model capabilities for future hyperspectral model environments: A review study. *Remote Sensing* 10. doi:10.3390/rs10010085.

Brocks, S. and Bareth, G. (2018). Estimating barley biomass with CROP surface models from oblique RGB imagery. *Remote Sensing* 10(2), 268. doi:10.3390/rs10020268.

Bruning, B., Liu, H., Brien, C., Berger, B., Lewis, M. and Garnett, T. (2019). The development of Hyperspectral distribution maps to predict the content and distribution of nitrogen and water in Wheat (*Triticum aestivum*). *Frontiers in Plant Science* 10, 1380. doi:10.3389/fpls.2019.01380.

Camino, C., González-Dugo, V., Hernández, P., Sillero, J. C. and Zarco-Tejada, P. J. (2018). Improved nitrogen retrievals with airborne-derived fluorescence and plant traits quantified From VNIR-SWIR hyperspectral imagery in the context of precision agriculture. *International Journal of Applied Earth Observation and Geoinformation* 70, 105–117. doi:10.1016/j.jag.2018.04.013.

Chang, Y. K. and Rehman, T. U. (2017). Current and future applications of cost-effective smart cameras in agriculture. *Robotics and Mechatronics for Agriculture*, 83–128. doi:10.1201/9781315203638-4.

Chen, Z., Wang, J., Wang, T., Song, Z., Li, Y., Huang, Y., Wang, L. and Jin, J. (2021). Automated in-field leaf-level hyperspectral imaging of corn plants using a Cartesian robotic platform. *Computers and Electronics in Agriculture* 183, 105996. doi:10.1016/j.compag.2021.105996.

Christiansen, M. P., Laursen, M. S., Jørgensen, R. N., Skovsen, S. and Gislum, R. (2017). Designing and testing a UAV mapping system for agricultural field surveying. *Sensors* 17(12), 2703. doi:10.3390/s17122703.

Clevers, J. G. P. W., Kooistra, L., and Schaepman, M. E. (2008). Using spectral information from the NIR water absorption features for the retrieval of canopy water content. *International Journal of Applied Earth Observation and Geoinformation* 10, 388–397. doi:10.1016/j.jag.2008.03.003.

Conceição, P. and Mendoza, R. U. (2009). Anatomy of the global food crisis. *Third World Quarterly* 30(6), 1159–1182. doi:10.1080/01436590903037473.

Danner, M., Berger, K., Wocher, M., Mauser, W. and Hank, T. (2019). Fitted PROSAIL parameterization of leaf inclinations,water content and brown pigment content for winter wheat and maize canopies. *Remote Sensing* 11(10). doi:10.3390/rs11101150.

de Souza, E. G., Scharf, P. C. and Sudduth, K. A. (2010). Sun position and cloud effects on reflectance and vegetation indices of corn. *Agronomy Journal* 102(2), 734–744. doi:10.2134/agronj2009.0206.

Deery, D. M., Rebetzke, G. J., Jimenez-Berni, J. A., James, R. A., Condon, A. G., Bovill, W. D., Hutchinson, P., Scarrow, J., Davy, R. and Furbank, R. T. (2016). Methodology for high-throughput field phenotyping of canopy temperature using airborne thermography. *Frontiers in Plant Science* 7, 1808. doi:10.3389/fpls.2016.01808.

Di Gennaro, S. F., Rizza, F., Badeck, F. W., Berton, A., Delbono, S., Gioli, B., Toscano, P., Zaldei, A. and Matese, A. (2018). UAV-based high-throughput phenotyping to discriminate barley vigour with visible and near-infrared vegetation indices. *International Journal of Remote Sensing* 39(15–16), 5330–5344. doi:10.1080/01431161.2017.1395974.

Dunford, R. P., Yano, M., Kurata, N., Sasaki, T., Huestis, G., Rocheford, T. and Laurie, D. A. (2002). Comparative mapping of the barley Ppd-H1 photoperiod response gene region, which lies close to a junction between two rice linkage segments. *Genetics* 161(2), 825–834.

Dworak, V., Selbeck, J., Dammer, K. H., Hoffmann, M., Zarezadeh, A. A. and Bobda, C. (2013). Strategy for the development of a smart ndvi camera system for outdoor plant detection and agricultural embedded systems. *Sensors* 13(2), 1523–1538. doi:10.3390/s130201523.

Ehrlich, P. R. and Harte, J. (2015). Opinion: To feed the world in 2050 will require a global revolution. *Proceedings of the National Academy of Sciences of the United States of America* 112(48), 14743–14744. doi:10.1073/pnas.1519841112.

Gamon, J. A., Kovalchuck, O., Wong, C. Y. S., Harris, A. and Garrity, S. R. (2015). Monitoring seasonal and diurnal changes in photosynthetic pigments with automated PRI and NDVI sensors. *Biogeosciences* 12(13), 4149–4159. doi:10.5194/bg-12-4149-2015.

Gardner, B. R. (1985). *Techniques for Remotely Monitoring Canopy Development and Estimating Grain Yield of Moisture Stressed Corn (Landsat, Sensed).*

Ge, Y., Bai, G., Stoerger, V. and Schnable, J. C. (2016). Temporal dynamics of maize plant growth, water use and leaf water content using automated high throughput RGB and hyperspectral imaging. *Computers and Electronics in Agriculture* 127, 625–632. doi:10.1016/j.compag.2016.07.028.

Gehan, M. A., Fahlgren, N., Abbasi, A., Berry, J. C., Callen, S. T., Chavez, L., Doust, A. N., Feldman, M. J., Gilbert, K. B., Hodge, J. G., Hoyer, J. S., Lin, A., Liu, S., Lizárraga, C., Lorence, A., Miller, M., Platon, E., Tessman, M. and Sax, T. (2017). Plantcv v2: Image analysis software for high-throughput plant phenotyping. *PeerJ* 5, e4088. doi:10.7717/peerj.4088.

Geipel, J., Link, J. and Claupein, W. (2014). Combined spectral and spatial modeling of corn yield based on aerial images and crop surface models acquired with an unmanned aircraft system. Remote sensing 6(11), 10335–10355.

Golzarian, M. R., Frick, R. A., Rajendran, K., Berger, B., Roy, S., Tester, M. and Lun, D. S. (2011). *Accurate Inference of Shoot Biomass from High-Throughput Images of Cereal Plants 1-11*.

Gracia-Romero, A., Kefauver, S. C., Fernandez-Gallego, J. A., Vergara-Díaz, O., Nieto-Taladriz, M. T. and Araus, J. L. (2019). UAV and ground image-based phenotyping: A proof of concept With durum wheat. *Remote Sensing* 11(10), 1244. doi:10.3390/rs11101244.

Greenham, K., Lou, P., Remsen, S. E., Farid, H. and McClung, C. R. (2015). TRiP: Tracking rhythms in plants, an automated leaf movement analysis program for circadian period estimation. *Plant Methods* 11(1), 1-11. doi:10.1186/s13007-015-0075-5.

Guo, Q., Dai, E., Han, X., Xie, S., Chao, E. and Chen, Z. (2015). Fast nastic motion of plants and bioinspired structures. *Journal of the Royal Society Interface/The Royal Society* 12(110), 1-14.

Guo, Q., Wu, F., Pang, S., Zhao, X., Chen, L., Liu, J., Xue, B., Xu, G., Li, L., Jing, H. and Chu, C. (2017). Crop 3D–A lidar based platform for 3D high-throughput crop phenotyping. *Science in China (Life Sciences)* 61(3), 328-339. doi:10.1007/s11427-017-9056-0.

Haralick, R. M., Shanmugam, K. and Dinstein, I. H. (1973). Textural features for image classification. *IEEE Transactions on Systems, Man* and *Cybernetics SMC* 3(6), 610-621.

Harris, G. J., Krishnaswamy, A. and Byer, S. (2011). U.S. Patent No. 7,970,206. Washington, DC: U.S. Patent and Trademark Office.

Holman, F., Riche, A., Michalski, A., Castle, M., Wooster, M. and Hawkesford, M. (2016). High throughput field phenotyping of wheat plant height and growth rate in field plot trials using UAV based remote sensing. *Remote Sensing* 8(12), 1031. doi:10.3390/rs8121031.

Hoyos-Villegas, V., Houx, J. H., Singh, S. K. and Fritschi, F. B. (2014). Ground-based digital imaging as a tool to assess soybean growth and yield. *Crop Science* 54(4), 1756-1768. doi:10.2135/cropsci2013.08.0540.

Humplík, J. F., Lazár, D., Husičková, A. and Spíchal, L. (2015). Automated phenotyping of plant shoots using imaging methods for analysis of plant stress responses - A review. *Plant Methods* 11(1), 29. doi:10.1186/s13007-015-0072-8.

Ishihara, M., Inoue, Y., Ono, K., Shimizu, M. and Matsuura, S. (2015). The impact of Sunlight conditions on the consistency of vegetation indices in croplands–Effective usage of vegetation indices from continuous ground-based spectral measurements. *Remote Sensing* 7(10), 14079-14098. doi:10.3390/rs71014079.

Jackson, R. D., Pinter, P. J., Idso, S. B. and Reginato, R. J. (1979). Wheat spectral reflectance: Interactions between crop configuration, sun elevation and azimuth angle. *Applied Optics* 18(22), 3730-3732. doi:10.1364/AO.18.003730.

Jacquemoud, S. and Baret, F. (1990). Prospect: A model of leaf optical properties spectra. *Remote Sensing of Environment* 34(2), 75-91. doi:10.1016/0034-4257(90)90100-Z.

Jacquemoud, S., Verhoef, W., Baret, F., Bacour, C., Zarco-Tejada, P. J., Asner, G. P., François, C. and Ustin, S. L. (2009). PROSPECT + SAIL models: A review of use for vegetation characterization. *Remote Sensing of Environment* 113, S56-S66. doi:10.1016/j.rse.2008.01.026.

Jannoura, R., Brinkmann, K., Uteau, D., Bruns, C. and Joergensen, R. G. (2015). Monitoring of crop biomass using true colour aerial photographs taken from a remote controlled hexacopter. *Biosystems Engineering* 129, 341-351. doi:10.1016/j.biosystemseng.2014.11.007.

Jones, H. G., Serraj, R., Loveys, B. R., Xiong, L., Wheaton, A. and Price, A. H. (2009). Thermal infrared imaging of crop canopies for the remote diagnosis and quantification of plant responses to water stress in the field. *Functional Plant Biology* 36(11), 978–989.

Kicherer, A., Herzog, K., Bendel, N., Klück, H. C., Backhaus, A., Wieland, M., Rose, J. C., Klingbeil, L., Läbe, T., Hohl, C., Petry, W., Kuhlmann, H., Seiffert, U. and Töpfer, R. (2017). Phenoliner: A new FIELD phenotyping platform for Grapevine research. *Sensors* 17(7), 1625. doi:10.3390/s17071625.

Kitchen, N. and Goulding, K. (2001). On-farm technologies and practices to improve nitrogen use efficiency. In: *Nitrogen in the Environment: Sources, Problems and Management* (pp. 335–369). doi:10.1016/b978-044450486-9/50015-7.

Kloosterman, B., Abelenda, J. A., Gomez, Mdel M., Oortwijn, M., de Boer, J. M., Kowitwanich, K., Horvath, B. M., van Eck, H. J., Smaczniak, C., Prat, S., Visser, R. G. and Bachem, C. W. (2013). Naturally occurring allele diversity allows potato cultivation in northern latitudes. *Nature* 495(7440), 246–250. doi:10.1038/nature11912.

Li, F., Piasecki, C., Millwood, R. J., Wolfe, B., Mazarei, M. and Stewart, C. N. (2020). High-throughput switchgrass phenotyping and biomass modeling by UAV. *Frontiers in Plant Science* 11, 574073. doi:10.3389/fpls.2020.574073.

Li, L., Zhang, Q. and Huang, D. (2014). A review of imaging techniques for plant phenotyping. *Sensors* 14(11), 20078–20111. doi:10.3390/s141120078.

Li, J., Zhang, F., Qian, X., Zhu, Y. and Shen, G. (2015). Quantification of rice canopy nitrogen balance index with digital imagery from unmanned aerial vehicle. *Remote Sensing Letters* 6(3), 183–189. doi:10.1080/2150704X.2015.1021934.

Li, W., Niu, Z., Chen, H., Li, D., Wu, M. and Zhao, W. (2016). Remote estimation of canopy height and aboveground biomass of maize using high-resolution stereo images from a low-cost unmanned aerial vehicle system. *Ecological Indicators* 67, 637–648. doi:10.1016/j.ecolind.2016.03.036.

Ludovisi, R., Tauro, F., Salvati, R., Khoury, S., Mugnozza Scarascia, G. and Harfouche, A. (2017). UAV-based thermal imaging for high-throughput field phenotyping of black poplar response to drought. *Frontiers in Plant Science* 8, 1681. doi:10.3389/fpls.2017.01681.

Ma, D., Carpenter, N., Amatya, S., Maki, H., Wang, L., Zhang, L., Neeno, S., Tuinstra, M. R. and Jin, J. (2019). Removal of greenhouse microclimate heterogeneity with conveyor system for indoor phenotyping. *Computers and Electronics in Agriculture* 166 , 104979. doi:10.1016/j.compag.2019.104979.

Ma, D., Maki, H., Neeno, S., Zhang, L., Wang, L. and Jin, J. (2020). Application of non-linear partial least squares analysis on prediction of biomass of maize plants using hyperspectral images. *Biosystems Engineering* 200, 40–54. doi:10.1016/j.biosystemseng.2020.09.002.

Madec, S., Baret, F., De Solan, B., Thomas, S., Dutartre, D., Jezequel, S., Hemmerlé, M., Colombeau, G. and Comar, A. (2017). High-throughput phenotyping of plant height: Comparing unmanned aerial vehicles and ground lidar estimates. *Frontiers in Plant Science* 8, 2002. doi:10.3389/fpls.2017.02002.

Maji, S., Chakraborty, P., Bhowmick, M., Dutta, S. K., Nath, R. and Chakraborty, P. K. (2015). Diurnal variation in spectral properties of potato under different dates of planting and N-doses. *Environment & Ecology* 33(1), 478–483.

Masjedi, A., Crawford, M. M., Carpenter, N. R. and Tuinstra, M. R. (2020). Multi-temporal predictive modelling of Sorghum biomass using UAV-based hyperspectral and lidar data. *Remote Sensing* 12(21), 3587. doi:10.3390/rs12213587.

Mathews, A. and Jensen, J. (2013). Visualizing and quantifying vineyard canopy LAI using an unmanned aerial vehicle (UAV) collected high density structure from motion point cloud. *Remote Sensing* 5(5), 2164–2183. doi:10.3390/rs5052164.

McCormick, R. F., Truong, S. K. and Mullet, J. E. (2016). 3D sorghum reconstructions from depth images identify QTL regulating shoot architecture. *Plant Physiology* 172(2), 823–834. doi:10.1104/pp.16.00948.

Mertens, S., Verbraeken, L., Sprenger, H., Demuynck, K., Maleux, K., Cannoot, B., De Block, J., Maere, S., Nelissen, H., Bonaventure, G., Crafts-Brandner, S. J., Vogel, J. T., Bruce, W., Inzé, D. and Wuyts, N. (2021). Proximal hyperspectral imaging detects diurnal and drought-induced changes in maize physiology. *Frontiers in Plant Science* 12, 640914. doi:10.3389/fpls.2021.640914.

Messina, G. and Modica, G. (2020). Applications of UAV thermal imagery in precision agriculture: State of the art and future research outlook. *Remote Sensing* 12(9), 1491. doi:10.3390/rs12091491.

Meyer, G. E. and Neto, J. C. (2008). Verification of color vegetation indices for automated crop imaging applications. *Computers and Electronics in Agriculture* 63(2), 282–293. doi:10.1016/j.compag.2008.03.009.

Mullet, J., Morishige, D., McCormick, R., Truong, S., Hilley, J., McKinley, B., Anderson, R., Olson, S. N. and Rooney, W. (2014). Energy sorghum—A genetic model for the design of C4 grass bioenergy crops. *Journal of Experimental Botany* 65(13), 3479-3489. doi:10.1093/jxb/eru229.

Navarro, P. J., Miller, L., Gila-Navarro, A., Díaz-Galián, M. V., Aguila, D. J. and Egea-Cortines, M. (2021). 3Deepm: An ad hoc architecture based on deep learning methods for multispectral image classification. *Remote Sensing* 13(4), 729. doi:10.3390/rs13040729.

Netto, A. F. A., Martins, R. N., De Souza, G. S. A., Dos Santos, F. F. L. and Rosas, J. T. F. (2019). Evaluation of a low-cost camera for agricultural applications. *Journal of Experimental Agriculture International*, 1-9. doi:10.9734/jeai/2019/v32i530117.

Nijland, W., De Jong, R., De Jong, S. M., Wulder, M. A., Bater, C. W. and Coops, N. C. (2014). Monitoring plant condition and phenology using infrared sensitive consumer grade digital cameras. *Agricultural and Forest Meteorology* 184, 98–106. doi:10.1016/j.agrformet.2013.09.007.

Oliveira, L. F. and Scharf, P. C. (2014). Diurnal variability in reflectance measurements from cotton. *Crop Science* 54(4), 1769–1781. doi:10.2135/cropsci2013.04.0217.

Ollinger, S. V. (2011). Sources of variability in canopy reflectance and the convergent properties of plants. *New Phytologist* 189(2), 375-394. doi:10.1111/j.1469-8137.2010.03536.x.

Padilla, F. M., De Souza, R., Peña-Fleitas, M. T., Grasso, R., Gallardo, M. and Thompson, R. B. (2019). Influence of time of day on measurement with chlorophyll meters and canopy reflectance sensors of different crop N status. *Precision Agriculture* 20(6), 1087–1106. doi:10.1007/s11119-019-09641-1.

Pandey, P., Ge, Y., Stoerger, V. and Schnable, J. C. (2017). High throughput in vivo analysis of plant leaf chemical properties using hyperspectral imaging. *Frontiers in Plant Science* 8, 1348. doi:10.3389/fpls.2017.01348.

Panjvani, K., Dinh, A. V. and Wahid, K. A. (2019). LiDARPheno - A low-cost LiDAR-based 3D scanning system for leaf morphological trait extraction. *Frontiers in Plant Science* 10, 147. doi:10.3389/fpls.2019.00147.

Paulus, S., Behmann, J., Mahlein, A. K., Plümer, L. and Kuhlmann, H. (2014). Low-cost 3D systems: Suitable tools for plant phenotyping. *Sensors* 14(2), 3001-3018. doi:10.3390/s140203001.

Perez-Sanz, F., Navarro, P. J. and Egea-Cortines, M. (2017). Plant phenomics: An overview of image acquisition technologies and image data analysis algorithms. *GigaScience* 6(11), 1-18. doi:10.1093/gigascience/gix092.

Petach, A. R., Toomey, M., Aubrecht, D. M. and Richardson, A. D. (2014). Monitoring vegetation phenology using an infrared-enabled security camera. *Agricultural and Forest Meteorology* 195-196, 143-151. doi:10.1016/j.agrformet.2014.05.008.

Pick, T. R., Bräutigam, A., Schlüter, U., Denton, A. K., Colmsee, C., Scholz, U., Fahnenstich, H., Pieruschka, R., Rascher, U., Sonnewald, U. and Weber, A. P. (2011). Systems analysis of a maize leaf developmental gradient redefines the current C4 model and provides candidates for regulation. *The Plant Cell* 23(12), 4208-4220. doi:10.1105/tpc.111.090324.

Preuss, S. B., Meister, R., Xu, Q., Urwin, C. P., Tripodi, F. A., Screen, S. E., Anil, V. S., Zhu, S., Morrell, J. A., Liu, G., Ratcliffe, O. J., Reuber, T. L., Khanna, R., Goldman, B. S., Bell, E., Ziegler, T. E., McClerren, A. L., Ruff, T. G. and Petracek, M. E. (2012). Expression of the Arabidopsis thaliana BBX32 gene in soybean increases grain yield. *PLoS ONE* 7(2), e30717. doi:10.1371/journal.pone.0030717.

Qiu, Q., Sun, N., Bai, H., Wang, N., Fan, Z., Wang, Y., Meng, Z., Li, B. and Cong, Y. (2019). Field-based high-throughput phenotyping for maize plant using 3D lidar point cloud generated with a "phenomobile". *Frontiers in Plant Science* 10, 554. doi:10.3389/fpls.2019.00554.

Rabatel, G., Gorretta, N. and Labbé, S. (2014). Getting simultaneous red and near-infrared band data from a single digital camera for plant monitoring applications: Theoretical and practical study. *Biosystems Engineering* 117, 2-14. doi:10.1016/j.biosystemseng.2013.06.008.

Ranson, K. J., Daughtry, C. S. T., Biehl, L. L. and Bauer, M. E. (1985). Sun-view angle effects on reflectance factors of corn canopies. *Remote Sensing of Environment* 18(2), 147-161. doi:10.1016/0034-4257(85)90045-8.

Ravi, R., Hasheminasab, S. M., Zhou, T., Masjedi, A., Quijano, K., Flatt, J. E., Crawford, M. and Habib, A. (2019). UAV-based multi-sensor multi-platform integration for high throughput phenotyping. In: *Autonomous Air and Ground Sensing Systems for Agricultural Optimization and Phenotyping* IV, 11008. doi:10.1117/12.2519190.

Ray, D. K., Mueller, N. D., West, P. C., and Foley, J. A. (2013). Yield trends are insufficient to double global crop production by 2050. *PLoS ONE* 8. doi:10.1371/journal.pone.0066428.

Rehman, T. U., Ma, D., Wang, L., Zhang, L. and Jin, J. (2020). Predictive spectral analysis using an end-to-end deep model from hyperspectral images for high-throughput plant phenotyping. *Computers and Electronics in Agriculture* 177, 105713. doi:10.1016/j.compag.2020.105713.

Rehman, T. U., Mahmud, M. S., Chang, Y. K., Jin, J. and Shin, J. (2019a). Current and future applications of statistical machine learning algorithms for agricultural machine vision systems. *Computers and Electronics in Agriculture* 156, 585-605. doi:10.1016/j.compag.2018.12.006.

Rehman, T. U., Zaman, Q. U., Chang, Y. K., Schumann, A. W. and Corscadden, K. W. (2019b). Development and field evaluation of a machine vision based in-season weed detection system for wild blueberry. *Computers and Electronics in Agriculture* 162, 1–13. doi:10.1016/j.compag.2019.03.023.

Rehman, T. U., Zaman, Q. U., Chang, Y. K., Schumann, A. W., Corscadden, K. W. and Esau, T. J. (2018). Optimising the parameters influencing performance and weed (goldenrod) identification accuracy of colour co-occurrence matrices. *Biosystems Engineering* 170, 85–95. doi:10.1016/j.biosystemseng.2018.04.002.That

Sakamoto, T., Gitelson, A. A., Nguy-Robertson, A. L., Arkebauer, T. J., Wardlow, B. D., Suyker, A. E., Verma, S. B. and Shibayama, M. (2012). An alternative method using digital cameras for continuous monitoring of crop status. *Agricultural and Forest Meteorology* 154–155, 113–126. doi:10.1016/j.agrformet.2011.10.014.

Sakamoto, T., Shibayama, M., Kimura, A. and Takada, E. (2011). Assessment of digital camera-derived vegetation indices in quantitative monitoring of seasonal rice growth. *ISPRS Journal of Photogrammetry and Remote Sensing* 66(6), 872–882. doi:10.1016/j.isprsjprs.2011.08.005.

Salas, E. and Henebry, G. (2013). A new approach for the analysis of hyperspectral data: theory and sensitivity analysis of the moment distance method. *Remote Sensing* 6(1), 20–41. doi:10.3390/rs6010020.

Schirrmann, M., Giebel, A., Gleiniger, F., Pflanz, M., Lentschke, J. and Dammer, K. (2016). Monitoring agronomic parameters of winter wheat crops with low-cost uav imagery. *Remote Sensing* 8(9), 706. doi:10.3390/rs8090706.

Song, Y., Glasbey, C. A., Van der Heijden, G. W. A. M., Polder, G. and Dieleman, J. A. (2011). Combining stereo and time-of-flight images with application to automatic plant phenotyping. In: *Lecture Notes in Computer Science* (pp. 467–478). Berlin, Heidelberg: Springer. doi:10.1007/978-3-642-21227-7_44.

Spectrum Analytic Inc (n.d.). Illustrated guide to sampling for plant analysis. Retrieved April 11, 2021. Available at: https://www.spectrumanalytic.com/services/analysis/plantguide.pdf.

Sticksel, E., Schächtl, J., Huber, G., Liebler, J. and Maidl, F. (2004). Diurnal variation in hyperspectral vegetation indices related to winter wheat biomass formation. *Precision Agriculture* 5(5), 509–520. doi:10.1007/s11119-004-5322-0.

Stow, D., Hope, A., Richardson, D., Chen, D., Garrison, C. and Service, D. (2000). Potential of colour-infrared digital camera imagery for inventory and mapping of alien Plant invasions in South African shrublands. *International Journal of Remote Sensing* 21(15), 2965–2970. doi:10.1080/01431160050121384.

Sun, Y., Wang, H., Sheng, H., Liu, X., Yao, Y. and Gong, C. (2015). Variations in internal water distribution and leaf anatomical structure in maize under persistently reduced soil water content and growth recovery after re-watering. *Acta Physiologiae Plantarum* 37(12), 263rd ser. doi:10.1007/s11738-015-2006-5.

Svensgaard, J., Roitsch, T. and Christensen, S. (2014). Development of a mobile multispectral imaging platform for precise field phenotyping. *Agronomy* 4(3), 322–336. doi:10.3390/agronomy4030322.

Swinfield, T., Lindsell, J. A., Williams, J. V., Harrison, R. D., Agustiono, H., Gemita, E., Schönlieb, C. B. and Coomes, D. A. (2019). Accurate measurement of tropical forest canopy heights and aboveground carbon using structure from motion. *Remote Sensing* 11(8), 928. doi:10.3390/rs11080928.

Tindall, A. J., Waller, J., Greenwood, M., Gould, P. D., Hartwell, J. and Hall, A. (2015). A comparison of high-throughput techniques for assaying circadian rhythms in plants. *Plant Methods* 11(1), 32. doi:10.1186/s13007-015-0071-9.

Turner, A.(2007). The pseudo-response regulator Ppd-H1 provides adaptation to photoperiod in barley. *Science* 310(5750), 1031–1034. doi:10.1126/science.1117619.

Vega, F. A., Ramírez, F. C., Saiz, M. P. and Rosúa, F. O. (2015). Multi-temporal imaging using an unmanned aerial vehicle for monitoring a sunflower crop. *Biosystems Engineering* 132, 19–27. doi:10.1016/j.biosystemseng.2015.01.008.

Veys, C., Chatziavgerinos, F., AlSuwaidi, A., Hibbert, J., Hansen, M., Bernotas, G., Smith, M., Yin, H., Rolfe, S. and Grieve, B. (2019). Multispectral imaging for presymptomatic analysis of light leaf spot in oilseed rape. *Plant Methods* 15(1), 4. doi:10.1186/s13007-019-0389-9.

Wang, L., Duan, Y., Zhang, L., Rehman, T., Ma, D. and Jin, J. (2020a). Precise estimation of NDVI with a simple nir sensitive RGB camera and machine learning methods for corn plants. *Sensors* 20(11), 3208. doi:10.3390/s20113208.

Wang, L., Jin, J., Song, Z., Wang, J., Zhang, L., Rehman, T. U., Ma, D., Carpenter, N. R. and Tuinstra, M. R. (2020b). Leafspec: An accurate and portable hyperspectral corn leaf imager. *Computers and Electronics in Agriculture* 169, 105209. doi:10.1016/j.compag.2019.105209.

Wang, R., Cherkauer, K. and Bowling, L. (2016). Corn response to climate stress detected with satellite-based NDVI time series. *Remote Sensing* 8(4), 269. doi:10.3390/rs8040269.

Wang, Y., Wang, D., Shi, P. and Omasa, K. (2014). Estimating rice chlorophyll content and leaf nitrogen concentration with a digital still color camera under natural light. *Plant Methods* 10(1), 36. doi:10.1186/1746-4811-10-36.

Xu, R., Li, C. and Paterson, A. H. (2019). Multispectral imaging and unmanned aerial systems for cotton plant phenotyping. *PLoS ONE* 14(2), e0205083. doi:10.1371/journal.pone.0205083.

Xue, J. and Su, B. (2017). Significant remote sensing vegetation indices: A review of developments and applications. Journal of sensors 2017, 1353691.

Yuan, W., Wijewardane, N. K., Jenkins, S., Bai, G., Ge, Y. and Graef, G. L. (2019). Early prediction of soybean traits through color and texture features of canopy rgb imagery. *Scientific Reports* 9(1), 14089. doi:10.1038/s41598-019-50480-x.

Zhang, L., Jin, J., Wang, L., Huang, P. and Ma, D. (2020). A 3D white referencing method for soybean leaves based on fusion of hyperspectral images and 3D point clouds. *Precision Agriculture* 21(6), 1173–1186. doi:10.1007/s11119-020-09713-7.

Zhang, L., Maki, H., Ma, D., Sánchez-Gallego, J. A., Mickelbart, M. V., Wang, L., Rehman, T. U. and Jin, J. (2019a). Optimized angles of the swing hyperspectral imaging system for single corn plant. *Computers and Electronics in Agriculture* 156, 349–359. doi:10.1016/j.compag.2018.11.030.

Zhang, L., Wang, L., Wang, J., Song, Z., Rehman, T. U., Bureetes, T., Ma, D., Chen, Z., Neeno, S. and Jin, J. (2019b). Leaf scanner: A portable and low-cost multispectral corn leaf scanning device for precise phenotyping. *Computers and Electronics in Agriculture* 167, 105069. doi:10.1016/j.compag.2019.105069.

Zhao, L., Liu, Z., Xu, S., He, X., Ni, Z., Zhao, H. and Ren, S. (2018). Retrieving the diurnal FPAR of a maize canopy from the jointing stage to the tasseling stage with vegetation indices under different water stresses and light conditions. *Sensors* 18(11), 3965. doi:10.3390/s18113965.

Zhao, T., Nakano, A., Iwaski, Y. and Umeda, H. (2020). Application of hyperspectral imaging for assessment of tomato leaf water status in plant factories. *Applied Sciences* 10(13), 4665. doi:10.3390/app10134665.

Zheng, H., Cheng, T., Li, D., Zhou, X., Yao, X., Tian, Y., Cao, W. and Zhu, Y. (2018). Evaluation of RGB, color-infrared and multispectral images acquired from unmanned aerial systems for the estimation of nitrogen accumulation in rice. *Remote Sensing* 10(6), 824. doi:10.3390/rs10060824.

Zhou, X., Xu, Y. and Zhang, F. (2017). Evaluation of effect of diurnal ambient temperature range on solar chimney power plant performance. *International Journal of Heat and Mass Transfer* 115, 398–405. doi:10.1016/j.ijheatmasstransfer.2017.07.051.

Chapter 4

Advances in the use of thermography in crop phenotyping

David M. Deery, CSIRO Agriculture and Food, Australia

1 Introduction

The quantitative dependency between the plant canopy temperature (CT) and the transpiration occurring at the interface of the atmosphere and plant surfaces is foundational to using thermal sensing in plant phenotyping. The energy required for the phase transition of water to vapour, that occurs at the plant surface, manifestly reduces the temperature there. Thus, reduced transpiration rate lowers the latent heat flux at the canopy surface, concomitantly increasing the sensible heat and the resultant CT. However, the relationship between the CT and actual transpiration rate is highly influenced by the meteorological conditions and the radiative properties of the crop surface. These challenges, although highlighted as early as the 1960s by Fuchs and colleagues (Fuchs et al. 1967; Fuchs and Tanner 1966), are still not resolved in sufficient detail to turn CT analyses into a readily usable tool for crop phenotyping and breeding. Studies are required that reveal more of the details about how to acquire and interpret thermography data in the context of field experiments under often rapidly varying irradiation regimes. This chapter will firstly introduce the principles of thermography, as applied to plant phenotyping. Next, the available technologies and measurement techniques are discussed. Finally,

http://dx.doi.org/10.19103/AS.2022.0102.05

future trends are discussed within the context of exemplar case studies from the literature.

2 Foundational theory of thermography

Thermography enables contactless temperature measurement of an object of interest. This is because, according to Planck's law (Planck 1900), all materials above the absolute zero of temperature (0 K or −273.15°C) emit thermal radiation, with the amount emitted increasing with temperature. Planck's law describes the energy emitted per unit wavelength, E_λ (by a unit projected area, into a unit solid angle), from a true blackbody - an idealized object that is a perfect absorber and emitter of radiation. Planck's law, Equation 1, is a function of wavelength (λ) and surface temperature (T), where T is the absolute temperature in Kelvin, h is Planck's constant, k is Boltzmann's constant and c is the speed of light.

$$E_\lambda = \frac{2hc^2}{\lambda^5} \frac{1}{e^{[hc/(\lambda kT)]} - 1} \tag{1}$$

Examples of the radiated spectrum predicted by Planck's law are shown in Fig. 1 for a blackbody at 300 K, 800 K and 5800 K. The temperature of the sun is approximately 5800 K, while 300 K is a temperature typical

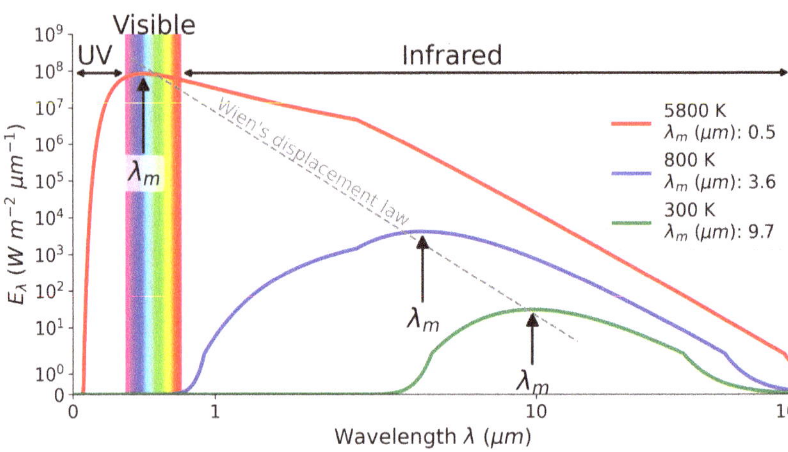

Figure 1 Planck's law (Equation 1) energy intensity, E_λ, of the thermal radiation emitted from a blackbody, as a function of wavelength, λ, for temperatures of 5800 K, 800 K and 300 K. The temperature of the sun is approximately 5800 K, while 300 K is typical of the earth's surface temperature. For 800 K and 300 K, all the thermal radiation is in the infrared spectrum. In contrast, the peak emissions at 5800 K occur in the visible spectrum. Note the considerable shift in the peak wavelengths, denoted λ_m, between the three curves as approximated by Wien's law (Equation 2).

of the earth's surface. For blackbodies of 300 and 800 K, all the thermal radiation is in the infrared spectrum. In contrast, the peak emissions from the sun, at 5800 K, occur in the visible spectrum. Figure 1 also shows the considerable shift in the maximum wavelength, denoted by λ_m, between the three blackbody temperature curves. These peak wavelengths where the maximum energy is emitted, λ_m (μm), from the Planck distribution describing blackbody radiation curves, are inversely related to temperature, T. This relationship is approximated by Wien's displacement law (Wien 1896) (Equation 2).

$$\lambda_m = \frac{2897.7}{T} \tag{2}$$

A blackbody can be defined as a hypothetical object that perfectly absorbs and radiates the maximum thermal radiation possible for a given temperature. The total amount of thermal radiation emitted per unit area from a blackbody is obtained by integrating Planck's law over all wavelengths. The resulting equation is the Stefan-Boltzman law (Boltzmann 1884), Equation 3:

$$P = \varepsilon\sigma T^4 \tag{3}$$

where the power, P (W m^{-2}), radiated per unit area of a blackbody is proportional to the fourth power of its absolute temperature, T (in K), σ is the Stefan-Boltzman constant that has a value of 5.67×10^{-8} W m^{-2} K^{-2} and e is the emissivity of the surface.

The emissivity of a surface is the ratio of the amount of radiation actually emitted to that emitted by a blackbody at the same temperature (see Kirchoff's law of radiation (Kirchhoff 1860)). Thus, the emissivity of a blackbody is unity at all wavelengths. In practice, however, the emissivity is always less than 1 and dependent upon the surface characteristics and wavelength. In the case of plant canopies and their components, published emissivity values in the wavelength region of 8–14 μm range from 0.96 to 0.98 (refer page 60 Table 3.5 of Jones and Vaughan 2010). The emissivity of a dense plant canopy will tend to be greater than that of individual leaves within the canopy. This is because the radiation emitted from a canopy will comprise a greater proportion of scattered radiation than that emitted from a sample leaf within the canopy. Furthermore, the emissivity of bare soil can vary rather more than that of plant canopies, with values of 0.88 reported for sand (Idso et al. 1969). The precise surface emissivity will likely vary between plant species, angle of incidence, canopy structure and degree of canopy ground cover; it is therefore a source of measurement uncertainty. However, for dense canopies, the emissivity is particularly high (Huband and Monteith 1986; Sobrino et al. 2001, 2002), thereby reducing the potential measurement error.

3 Principles of thermography measurement

Thermography sensors measure a specific part of the infrared region of the electromagnetic spectrum emitted by an object of interest and relate it to the object's temperature according to the theory discussed above (Ballico 2004). The infrared spectrum encapsulates a wide range of wavelengths nominally extending from the visible red edge at 0.75 μm to the far infrared at 1000 μm. This part of the spectrum is often conveniently divided into sub-regions as follows: near-infrared (NIR), 0.7–1.4 μm; short-wave infrared (SWIR), 1.4–3.0 μm; medium-wave infrared (MWIR), 3.0–5.0 μm; long-wave infrared (LWIR), 8.0–14.0 μm and far-infrared (FIR), 14.0–1000 μm. The LWIR is considered optimal for thermography of vegetation outdoors, as it is defined by a high proportion of atmospheric transmission (Fig. 2).

Further, the radiated spectral energy density predicted by Planck's law (Equation 1) and Wien's displacement law (Wien 1896) peaks at 10 μm for temperatures typical of vegetation (ca 300 K) (Fig. 1). In addition, the SWIR and MWIR regions are generally considered not suitable for daytime measurements outdoors, owing to interference from reflected sunlight (George et al. 2019; Peckham et al. 2015).

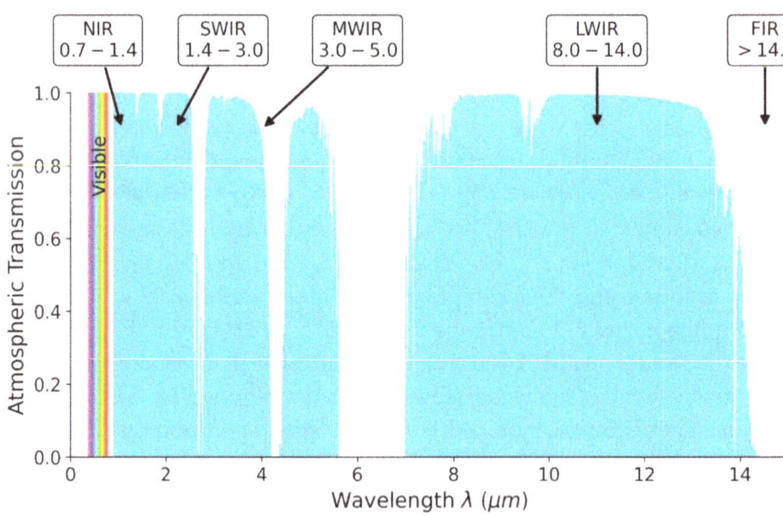

Figure 2 Atmospheric transmission of infrared radiation as a function of wavelength, λ. The shaded areas represent the highest infrared transmission, these are sometimes referred to as 'atmospheric windows.' The sub-regions of the infrared spectrum are indicated, as well as the visible spectrum. The LWIR is considered optimal for thermography of vegetation outdoors, as it is defined by a high proportion of atmospheric transmission. Based on data obtained at Mauna Kea by the Gemini Observatory and processed using the ATRAN modelling software (Lord 1992).

4 Technologies available and thermography methods

LWIR thermography sensors typically use uncooled microbolometers to measure temperature. The microbolometer sensor is heated by the incoming radiation (LWIR) and the change in temperature is quantified by the change in resistance of the microbolometer. This sensing technology is currently available in three main formats, as described below: (a) hand-held infrared thermometers (HHIRTs), (b) ground-based static sensors (GBSSs) and (c) thermal cameras.

4.1 Hand-held infrared thermometers

HHIRTs, as demonstrated in Fig. 3, have been used extensively over the past few decades to phenotype CT in a range of crops and research applications (for example, a selection of more highly cited papers include Ayeneh et al. 2002; Hatfield et al. 1987; Lopes and Reynolds 2010; Olivares-Villegas et al. 2007; Pinter et al. 1990; Rashid et al. 1999; Rattey et al. 2011; Reynolds et al. 1994,

Figure 3 Measurement of CT using the hand-held infrared thermometer method. In this example, the following hand-held infrared thermometer is used: Mikron 1600 (Mikron Infrared Instrument Co., Inc., Oakland, NJ, USA). Photo reproduced with permission from Rebetzke et al. (2013a).

2007a). In particular, there are several examples where HHIRT sensing has been used to phenotype for stomatal conductance or adaptation to water limitation (e.g. Amani et al. 1996; Blum et al. 1989; Fischer et al. 1998; Giunta et al. 2008; Rebetzke et al. 2013b). Although HHIRTs have proved effective in the past, they are laborious and not particularly suitable for large experiments comprising hundreds of experimental plots. This is because the CT measurements can be confounded by weather changes during the time required to manually measure all plots. However, HHIRTs may be suitable for experiments comprising a small number of plots, where the operator(s) can measure CT on all plots without large changes in weather conditions. Documentation describing the use of HHIRTs for CT phenotyping is readily available (e.g. Pask et al. 2012; Rebetzke et al. 2013a). The method and key considerations are summarised as follows:

- **Cloud and wind**. Measurements should be taken when weather conditions are sunny and clear and the winds are light (≤20 km/h). This is because transpiration at the leaf surface is strongly impacted by the prevailing micrometeorological conditions (Maes and Steppe 2012). For example, slightly increased wind speed can lead to higher transpiration by the destruction of moist boundary layers on leaf surfaces.
- **Canopy cover**. Complete canopy closure is preferable to avoid the influence of soil. If the canopy cover is incomplete, care should be taken to avoid the background soil, which is often warmer than the crop.
- **Time of day**. It is generally recommended that CT measurements be taken in the time window of 1–2 h before and after solar noon.
- **Sun position**. The sun should be behind the operator, while ensuring that the operator does not shade the plot.
- **Crop growth stage and canopy height**. Plant reproductive organs, including spikes and inflorescences, typically have different temperatures than the canopy. Because of this, minimal variation in the plant growth stage (i.e. Zadoks growth stage in cereals (Zadoks et al. 1974)) is recommended. Measurements during grain filling should occur before leaf senescence of the upper canopy. Canopy height should also be measured, due to the potential for negative correlations between canopy height and CT (e.g. Giunta et al. 2008; Olivares-Villegas et al. 2007; Rattey et al. 2011; Rebetzke et al. 2013b; Sadras and Lawson 2011). This, again, is caused by differences in micrometeorological conditions: plants that are a few cm higher than their canopy neighbours are typically exposed to higher wind speed. Therefore, moist boundary layers around the leaves are decreased, vapour pressure deficit (VPD) is altered and transpiration increases.
- **Measurement**. To minimize capturing soil in the infrared thermometer's field-of-view (FOV), the instrument is held obliquely to each plot and scanned across the canopy at an angle of *ca.* 20° (to the horizontal) for

6-8 s to derive an average CT value for each plot (Fig. 3). The operator distance from the plot is determined by the infrared thermometer's FOV to ensure that only the plant canopy is measured.

4.2 Ground-based static sensors

GBSSs (Fig. 4) provide continuous measurement of the plant CT (Jones et al. 2018; Mahan et al. 2010; O'Shaughnessy et al. 2011a,b), often with the capacity to wirelessly transmit their data to the internet (Salehi et al. 2015). GBSSs typically comprise a single-pixel microbolometer sensor, whereby the output represents an aggregate statistical mean of the temperature distribution within the sensor FOV. Their sensing FOV is typically narrow (e.g. 10°), designed to enable accurate CT measurement of row crops without the influence of the background soil. GBSSs are used in both research and production agriculture settings, principally as a quantitative measure of plant stress (Coast et al. 2020; Conaty et al. 2014). In cotton, they are used to optimize the scheduling of irrigation (Conaty et al. 2012) as a means of maximizing the productive output from the crop (Conaty et al. 2015). The capacity of GBSSs to continuously record the CT offers significant advantages in phenotyping, particularly for irrigation scheduling.

Figure 4 Ground-based static sensors (GBSSs) in a cotton crop. These sensors provide continuous measurement of the plant canopy temperature and wirelessly transmit their data to the internet.

In terms of sensor deployment, GBSSs are typically height adjustable to maintain a consistent height above the crop canopy throughout the crop growing season. In row crops where significant amounts of bare soil are evident between individual rows, care must be taken to ensure each individual sensor is installed to view the canopy at an angle that minimizes the chance of viewing background soil. Additionally, to obtain the most reliable measurement of CT in the Southern Hemisphere, it is recommended that sensors face approximately north to avoid viewing the sunlit side of the canopy. Sunlit canopies will tend to have a far greater range in temperature variation than shaded canopies (Jones et al. 2002).

4.3 Thermal cameras

Thermal cameras are similar in design to conventional digital RGB cameras whereby the main components include a lens and detector or sensor. While the sensor of a conventional digital camera comprises a charge-coupled device, the sensor in a thermal camera is commonly an uncooled microbolometer (FLIR 2012). There is another type of sensor technology, known as quantum detectors, which are considerably more advanced (and expensive) than microbolometers and typically require cooling. Figure 5 provides a comparison of three currently available thermal cameras that are suitable for crop phenotyping. The vast majority of phenotyping studies will likely use thermal cameras with uncooled microbolometer sensors, as they are considerably more affordable (Fig. 5). Unlike conventional digital RGB cameras, the pixel resolution of thermal cameras is relatively low and typically limited to 640 × 480. Although a higher

Camera	Sensor type	Spectral range (μm)	Weight (kg)	Accuracy (°C)	NETD (°C)	Resolution	Response time	Pixel size (cm)[#]	Image swath (m)[#]	Papers	Approx. Cost (AUD)
(a) A65 	Uncooled microbolometer	7.5 – 13.0	0.2	±5 or ±5% of reading	< 0.05	640 by 512 (~0.3MP)	12 ms	7.8^	50 by 40^	Perich et al 2020	$3k
(b) A655sc 	Uncooled microbolometer	7.5 – 14.0	0.92	±2 or ±2% of reading	< 0.03	640 by 480 (~0.3MP)	8 ms	7.7^	50 by 37^	Deery et al 2016, 2019, Rutkoski et al 2016, Li et al 2019	$30k
(c) X8500sc 	Cooled quantum detector	7.5 (lower), 11.5 – 12.5 (upper)	7	±2 or ±2% of reading	< 0.04	1280 by 1024 (~1.3MP)	270 ns	2.8*	37 by 30*	?	$300k

(a) https://www.flir.com.au/products/a65/
(b) https://www.flir.com.au/products/a655sc/
(c) https://www.flir.com.au/products/x8500sc-sls-lwir/

[#] Distance: 60 m
^ Focal length: 13 mm (45 deg field of view)
* Focal length: 25 mm (34 deg field of view)

Figure 5 Comparison of currently available thermal cameras suitable for crop phenotyping. Information was sourced from FLIR®, as indicated for each camera. The estimates of pixel size and image swath are based on a 60 m distance to the scene and focal length as indicated.

resolution thermal camera is currently available (FLIR® X8500sc SLS shown in Fig. 5), with a resolution of 1280 × 1024, it is about ten times more expensive than the FLIR® A655sc uncooled thermal camera.

Thermal cameras are a powerful tool for phenotyping because of their ability to map the temperature over a given area in a single image, thereby enabling comparison between experimental treatments. Although thermal cameras are considerably more complicated than HHIRTs and GBSSs, the thermal resolution or the smallest temperature difference that can be quantified between pixels is typically quite low at approximately 0.03°C (often called noise equivalent temperature difference). This high pixel-to-pixel sensitivity is advantageous for making relative assessments between plant genotypes or treatments and therefore suitable for many phenotyping situations (Jones et al. 2009; Prashar and Jones 2014; Prashar et al. 2013; Takai et al. 2010).

4.4 Airborne thermography

Owing to the advent of relatively affordable thermal cameras (as shown in Fig. 5a and b), the use of thermal imaging approaches from airborne platforms is becoming increasingly common, either using manned (Deery et al. 2016, 2019; Li et al. 2019; Rutkoski et al. 2016) or unmanned aircraft (Anderegg et al. 2021; Perich et al. 2020). In comparison to the HHIRT and GBSS methods, airborne thermography enables acquisition of CT from a large area in a short time period, in the order of seconds, thereby reducing possible error caused by weather fluctuations. This feature has greatly enhanced the capacity to reliably phenotype large experiments and achieve far greater heritabilities (Deery et al. 2016, 2019; Rutkoski et al. 2016) in comparison to the HHIRT and GBSS methods. Similar to HHIRT, airborne thermography measurements should be taken when weather conditions are sunny and clear and the winds are light (≤20 km/h).

Unlike conventional RGB images, the processing methods for thermal images are still rather manual and therefore challenging (Malbéteau et al. 2021). Challenges can include vignetting and image distortion effects (Kelly et al. 2019). In the case of vignetting, a spurious border forms around the edges of an image, for example, where values at the edges of the image are lower than those in the centre. These challenges are often exacerbated when using uncalibrated thermal cameras which are often used on drones owing to their smaller size and lighter weight compared to those shown in Fig. 5 (which are radiometrically calibrated as part of the manufacturing process). Several studies have demonstrated methods for calibrating thermal cameras (e.g. Aragon et al. 2020; Budzier and Gerlach 2015; Nugent et al. 2013; Ribeiro-Gomes et al. 2017; Stutsel et al. 2021) to improve their accuracy, as well as incorporating ground control points (Kelly et al. 2019; Perich et al. 2020) (i.e.

highly visible infrared targets) to improve the georeferencing and processing of orthomosaics.

The manned-helicopter thermal image acquisition and processing method developed by the CSIRO group based in Canberra is an effective but rather manual process (Deery et al. 2016). Experiments are typically imaged at a height above ground level (AGL) sufficient to capture the entire experiment in a single flight pass (typically 60–90 m AGL). At 60 m AGL, an image swath of 50 × 37 m is obtained with a pixel size of 7.7 cm (Fig. 5), while at 90 m AGL, an image swath of 75 × 56 m is obtained with a pixel size of 11.7 cm. The trade-off between pixel size (also known as ground resolution or ground sampling distance) and image acquisition height AGL is particularly important for thermal cameras mounted on airborne platforms, owing to their limited resolution compared to conventional RGB cameras. Thermal images are acquired by the CSIRO system at a flight velocity of 25–35 knots (45–65 km/h) to avoid motion blur, noting that there is also a trade-off between speed of the aircraft and camera integration time (Jones and Sirault 2014). Image acquisition with this system typically takes <10 s for an experiment comprising *ca.* 1000 plots of 2 × 6 m, thereby minimising the chances of weather variation confounding the measured CT. Experimental plots are segmented from each thermal image using custom software which works on a frame-by-frame basis extracting data from the raw imagery, whereby the user navigates through the image stack to ensure that each plot in the experiment has been sampled. In this case, the plot segmentation is semi-automated whereby the user controls the area sampled within plots by the placement of bounding corners. More advanced methods are available, comprising automated segmentation of plots based on georeferencing (Anderegg et al. 2021; Perich et al. 2020) or image segmentation (Guo et al. 2018), as has been developed for RGB orthomosaics.

5 Traits measured

Thermography has been used as a field phenotyping tool for many years, principally to quantify the plant CT. The motivation for using CT is based on the fact that plant surfaces are cooled by evaporation, so that their temperatures fall in proportion to the evaporation rate. In this way, stomatal closure and a reduction in transpiration rate manifest as a warmer CT, while cooler CT is related to stomatal opening and higher transpiration rates.

Cooler CT has been associated with genetic gains in wheat yield, higher stomatal conductance and maximum photosynthetic rate under non-water-limited conditions (Fischer et al. 1998). Similarly, cooler CT has been associated with grain yield under warm irrigated conditions in Mexico (Amani et al. 1996; Ayeneh et al. 2002; Reynolds et al. 1994) and in a study comparing a selection of spring wheat cultivars from Australia and the International Maize and Wheat

Improvement Center (CIMMYT) (Rattey et al. 2011). Similar findings have been reported in water-limited environments, with cooler CT in wheat associated with increased yield (Blum et al. 1989; Olivares-Villegas et al. 2007; Rashid et al. 1999). When measured during grain filling during water limitation, cooler CT has been associated with increased rooting depth (Reynolds et al. 2007a), water use and yield (Lopes and Reynolds 2010). Conversely, warmer CT has been associated with conservative water use in different crops (e.g., Hatfield et al. 1987; Pinter et al. 1990).

There is extensive evidence in C3 crops of a concomitant increase in stomatal conductance with increased yields, attained through conventional plant breeding over the past several decades (Roche 2015). This provides a strong case for the development of CT phenotyping tools as a surrogate measure of stomatal conductance, particularly when water supply is plentiful (Fischer and Rebetzke 2018).

5.1 Heritability of the canopy temperature phenotype

The airborne thermography method has greatly increased the capacity for increased heritability of CT in comparison to the HHIRT method. To the best of our knowledge, when CT is measured using the HHIRT method, heritability of CT is typically small on a single-plot basis and often not reported. More commonly reported is the CT heritability estimated on a line mean basis from multiple environments. Studies reporting heritability of CT calculated on a line mean basis are often small to moderate in size for both diverse germplasm and related families such as recombinant inbred lines (RILs) and doubled-haploid (DH) lines. For example, narrow-sense heritabilities were small (0.12–0.32) on a single-plot basis and line mean heritability was moderate (0.38–0.91) for a study comprising three wheat populations of 144–178 DH lines grown in four environments (Rebetzke et al. 2013b). Additionally, broad-sense heritability of 0.49 was reported for CT measured during grain filling on a RIL wheat population comprising 167 lines grown in 6 field trials under drought and heat environments (Pinto et al. 2010). Under similar environmental conditions, in a separate study, moderate broad-sense heritabilities were reported for both a wheat population comprising 169 RILs (0.34) and 294 elite wheat lines from CIMMYT (0.38) (Lopes and Reynolds 2012). Further, other moderate line mean heritabilities have been reported for diverse wheat germplasm estimated from studies comprising multiple environments (e.g. Lopes et al. 2012; Rattey et al. 2011; Reynolds et al. 2007b).

In contrast to the above-mentioned studies where CT was measured using the HHIRT method, heritability estimates obtained using the airborne thermography method are consistently greater than 0.5 (Deery et al. 2016, 2019; Perich et al. 2020). Of note is a recent study at CIMMYT where heritabilities were particularly high, ranging from 0.65 to 0.95 (Rutkoski et al. 2016). The higher heritability estimates obtained using airborne thermography are attributable to the acquisition of thermal images for the whole experiment at essentially a single point in time, thereby avoiding any confounding changes in local weather conditions during sampling.

The prevailing weather conditions, especially the VPD and solar radiation, can influence the CT heritability estimates, with higher heritability more likely with higher VPD and solar radiation (Deery et al. 2019). The higher heritability observed under these conditions, which are more likely to occur in the afternoon and during the grain-filling period, is possibly (but not necessarily) because differences in stomatal conductance between genotypes were greater. The latter is plausible when considering energy balance theory (Jones and Vaughan 2010; Maes and Steppe 2012), whereby for a given stomatal conductance, CT is linearly related to VPD and solar radiation. Thus, any differences in stomatal conductance are likely to result in larger differences in CT with greater VPD and solar radiation.

6 Case studies

6.1 Thermography phenotyping in plant breeding

To the best of our knowledge, there are only a few studies in the literature explicitly investigating the use of CT phenotyping within a plant breeding programme. More common are studies which typically investigate the physiological and/or genotypic associations between CT and grain yield. There are however encouraging results from several studies highlighting the capacity of thermography in crop phenotyping to increase the accuracy of pedigree and/or genomic selection models (Rutkoski et al. 2016; Sun et al. 2017, 2019).

The utility of CT as an indirect selection tool was investigated within the CIMMYT wheat breeding programme using 504 advanced spring wheat breeding lines, grown in water-limited and heat stress environments (Mason and Singh 2014). HHIRT measurements were taken when all entries in the trial were at the full boot growth stage. The following six theoretical selection strategies were applied:

1 10% highest yielding;
2 10% coolest CT;
3 20% coolest CT and keeping 10% highest yielding;
4 10% warmest CT;
5 20% warmest CT and keeping 10% highest yielding; and
6 the mean of all entries (no selection).

In the heat stress treatment, they found that the yield benefit of the third strategy, selecting the 20% coolest CT entries and keeping the 10% highest yielding entries, was not significantly different from selecting the 10% highest yielding entries. They also found, in the heat treatment, that selecting the 10% warmest CT entries resulted in significantly lower grain yield than all other strategies, thereby indicating the potential for using high CT to discard low performing entries. In the drought treatment, selecting the 10% highest yielding entries resulted in a significantly higher yield than all other strategies. However, the strategy of selecting the 20% warmest CT and keeping the 10% highest yielding was not significantly different from the strategy of selecting the 20% lowest CT and keeping the 10% highest yielding. The yields from both strategies were significantly higher than the mean of all entries (no selection strategy). The authors concluded that the most useful application of CT within a breeding programme would occur in the early generations, where yield testing is not performed, and therefore an indirect selection method for yield would be beneficial (Mason and Singh 2014). The indirect selection for yield in early generations of cereal breeding programmes was recently reviewed in detail (Fischer and Rebetzke 2018), with airborne thermography, for quantifying detection of differences in stomatal conductance, identified as the most promising trait for testing in commercial breeding programs.

Studies at CIMMYT in the early 2000s designated 'stomatal aperture-related traits' (SATs) as shorthand for the surrogate measures of stomatal conductance, including leaf porosity and CT (Condon et al. 2004, 2007). Their study comprised five populations of F3-derived RILs, with parents selected from the CIMMYT wheat breeding programme, grown over three seasons. All populations were grown in 6 m^2 yield plots and in small plots (1.5 m × 2 rows), with the latter designed to simulate breeder's early generation plots. The genetic correlation between the small plot CT and the full plot yield was strong when averaged across all populations and years (−0.68). Yield gains from retrospective selection based on SATs were in the order of 5 10%. They concluded that appreciable yield gains could be made by combining visual selection with SATs in the breeding programme.

In a short conference paper, the same team reported an investigation augmenting breeder's visual selection with early generation measurements of CT in five crosses with parents selected from the CIMMYT wheat breeding programme (van Ginkel et al. 2004). They found that augmenting breeder's visual selection with CT measurements identified almost three times as many equally well-yielding lines as breeder's visual selection alone. The results varied by cross. In one cross, the mean yield of all lines identified with breeder selection augmented with CT measurements was higher than the mean yield for lines derived from breeder's visual selection alone. In two of the crosses, the highest yielding 24 and 3 lines, respectively, were derived from breeder selection augmented with CT measurements. In the remaining three crosses, a nearly equal number of lines ranked in the top ten for yield were derived from each of the two methods, breeder's visual selection alone or augmenting breeder's visual selection with CT.

Following the above-mentioned studies (Condon et al. 2004, 2007; van Ginkel et al. 2004), an economic analysis was undertaken (Brennan et al. 2007) to quantify the benefit of incorporating SATs as a selection tool within the CIMMYT breeding programme. A hypothetical case was presented based on the then CIMMYT breeding programme strategy of visually selecting, based on agronomic characteristics and disease, the best 30% of F4 lines (*ca.* 7500 lines) for advancement to F5 yield testing. From the *ca.* 7500 visually selected F4 lines, Brennan et al. (2007) proposed discarding the least suitable 20%, based on SATs and their physiological association with yield, so that only *ca.* 6000 F5 lines would be yield tested. After accounting for the cost of CT measurements on 25 000 F4 lines, Brennan et al. (2007) showed that this strategy would result in a cost-saving to the breeding programme of US$13 220 through sowing and harvesting *ca.* 1500 fewer F5 yield plots (based on US$ cost estimates *ca.* 2006). The economic analysis indicated that augmenting visual selection with CT measurements within a wheat breeding programme is likely to provide an economic benefit.

The cost structure of the newly developed airborne thermography method (see Section 4.4) is different from the HHIRT method evaluated by Brennan et al. (2007). The airborne thermography method requires fewer people than the HHIRT method, while the capital and operating expenses of the helicopter and thermal camera are far greater than the HHIRT method. The greater expense of the airborne thermography method may be offset by the appreciably greater heritability obtained from the airborne thermography method compared to the HHIRT method (Deery et al. 2016). Also, the airborne thermography method is scalable, potentially enabling screening of thousands of breeder's plots in a single day. Previous work in the literature indicates that the use of CT to augment early generation selection may benefit a plant breeding programme. Further work is required to quantify such a benefit, how it would be realised and whether it would be cost-effective.

6.2 Application of thermography phenotyping in Australian environments

Many of the above-mentioned studies relating cooler CT to grain yield in wheat typically comprised full irrigation, at CIMMYT's favourable Obregon environment (Yaqui Valley, Mexico), where clear skies and windless weather conditions often prevail. These are ideal conditions for obtaining quality CT measurements. The crop growing environments in the Yaqui Valley are in contrast to the mostly water-limited Australian wheat-growing environments (Fischer et al. 2014), where average annual rainfall usually ranges from 275 mm to 500 mm. It is reasonable to question whether the findings relating cooler CT to grain yield in wheat, undertaken in the Yaqui Valley, will apply to Australian environments.

To the best of our knowledge, few studies have evaluated CT and grain yield in wheat in Australian environments. Rattey et al. (2011) evaluated grain yield and CT, among a range of physiological traits, on 273 CIMMYT lines and 15 Australian cultivars in 27 environments in north-eastern Australia. The region of north-eastern Australia is summer rainfall dominant, whereby crops predominantly rely on stored soil moisture, with little in-season rainfall, and crops often experience water limitation during the grain-filling period. Rattey et al. (2011) found that the CIMMYT lines had cooler CT than the Australian lines across all environments, while grain yield for the CIMMYT lines was 5% and 4–10% greater than the Australian lines in higher and lower yielding environments, respectively. The performance of the CIMMYT lines and their association between cooler CT and grain yield accords with Mathews et al. (2007), where strong genetic correlations were reported between the region of north-eastern Australia and the Obregon environment, Yaqui Valley, Mexico.

In a separate study, Rebetzke et al. (2013b) assessed genotypic variation in CT and leaf porosity and their association with grain yield in four irrigated environments at Ginninderra Experiment Station, ACT, in south-eastern Australia. Three wheat populations were used, each containing 144–178 DH lines. Significant genetic correlations between CT and grain yield were reported for CT measurements pre- and post-flowering. For pre-flowering CT measurements, the genetic correlations with grain yield were positive. Conversely, for post-flowering CT measurements, the genetic correlations with grain yield were negative.

Sadras and Lawson (2011) evaluated a historic set of 13 South Australian wheat varieties, released between 1958 and 2007, for genetic progress in grain yield. The experiments were grown in three South Australian environments under water-limited conditions, and genetic progress in grain yield was positively associated with stomatal conductance and greater radiation-use efficiency. However, despite stomatal conductance increasing with year-of-release, Sadras and Lawson (2011) found that CT increased with year-of-release, possibly because height also decreased and aerodynamic resistance increased.

7 Main challenges

There are several considerations in order to successfully use thermography in crop phenotyping. The main challenges associated with airborne thermography in field conditions arise from the confounding effects of incomplete canopy cover and variation in plant height. A third challenge, arguably common to many phenotyping technologies, is the capacity to scale measurements, seamlessly, to the many thousands of individual plants within a plant breeding programme (discussed in Section 8).

In cases where the canopy ground cover is incomplete, it may be necessary to correct for the influence of background soil in order to obtain a more accurate estimate of the canopy foliage temperature (Jones and Sirault 2014). This issue is likely to become more important when evaluating genotype performance under water limitation, where greater incomplete ground cover is more probable, than in environments with high water supply. Further, this issue is also likely to depend on the variation in canopy ground cover. At least one study has shown that filtering the frequency distribution of the CT pixels to remove the influence of background soil did not improve broad-sense heritability (Deery et al. 2016). Nevertheless, the possibility of concurrently obtaining thermography and RGB images, from an airborne platform, holds promise for an automated method using the RGB to detect the soil and thereby using only plant-based information in such images. Such a method would require automated co-registration, pixel-to-pixel correspondence, between the thermal and RGB data (Jerbi et al. 2015).

Negative correlations between canopy height and CT have been reported in the literature (Giunta et al. 2008; Olivares-Villegas et al. 2007; Rattey et al. 2011; Rebetzke et al. 2013b; Sadras and Lawson 2011). This is because the CT also depends on height as well as the stomatal conductance (Maes and Steppe 2012). Additionally, plant morphology such as the presence and form of ears may also influence the resulting CT. The negative associations between CT and height, where taller canopies are cooler, are attributable to the greater coupling between the canopy and atmosphere (Jarvis and McNaughton 1986) for taller genotypes. Notably, even relatively small differences in height can influence CT. For example, despite only a 0.2 m range in height, significant associations between CT and height were reported (Deery and Jones 2021). That said, the inclusion of major height genes (as factors) together with quantitative variation in plant height as co-variates substantially improved the genetic correlation between CT and grain yield in populations varying in plant height (Rebetzke et al. 2013b). The need to account for the effect of canopy height on CT remains a critical consideration for the most effective use of CT as a phenotyping tool.

8 Summary and future trends

This chapter connects the fundamental principles of thermography with its application in crop phenotyping, including discussion of the key technologies, methods and their limitations. Thermography has been used extensively in crop phenotyping over the past decades, with most studies employing hand-held thermometers. The recent developments in thermal cameras and drones now make it possible to quantify the crop CT on many thousands of experimental plots effectively simultaneously. However, the processing and analysis of thermal data to obtain actionable information from an experiment is a current limitation. In thermography, and indeed other phenomics technology, there is a growing need for the development of software systems that enable the conversion of raw sensor data into biologically relevant information, georeferencing the data to individual experimental units within a given experimental design. The need for enabling software to facilitate the analysis and reusability of data has been highlighted previously (Roitsch et al. 2019). The creation of federated data services would enable crop scientists and plant breeders to leverage other enabling technologies, including the entire gamut of modelling and analysis techniques now available (van Eeuwijk et al. 2019).

9 Where to look for further information

9.1 Further reading

9.1.1 Infrared sensing fundamentals

- FLIR (2012). *The ultimate infrared handbook for research and development professionals*. Report. URL: www.FLIR.com.
- H. G. Jones and R. A. Vaughan (2010). *Remote sensing of vegetation: principles, techniques, and applications*. Oxford University Press.
- M. J. Ballico (2004). *Monograph 12 Radiation Thermometry - National Measurement Institute (NMI).*

9.1.2 Influential papers

- I. Amani, R. A. Fischer, and M. F. P. Reynolds (1996). "Canopy temperature depression association with yield of irrigated spring wheat cultivars in a hot climate". In: *Journal of Agronomy and Crop Science* 176.2, pp. 119-129. doi: 10.1046/j.1439-037x.1999.00335.x.
- H. G. Jones (2004). "Application of thermal imaging and infrared sensing in plant physiology and ecophysiology". In: *Advances in Botanical Research* 41, pp. 107–163. doi: 10.1016/S0065-2296(04)41003-9.
- W. H. Maes and K. Steppe (2012). "Estimating evapotranspiration and drought stress with ground-based thermal remote sensing in agriculture: a review." In: *Journal of experimental botany* 63 (13), pp. 4671–4712. doi: 10.1093/jxb/ers165.
- D. M. Deery, G. J. Rebetzke, J. A. Jimenez-Berni, R. A. James, et al. (2016). "Methodology for High-Throughput Field Phenotyping of Canopy

Temperature Using Airborne Thermography". In: *Frontiers in Plant Science* 7, p. 1808. doi: 10.3389/fpls.2016.01808.

- G. Perich et al. (2020). "Assessment of Multi-Image Unmanned Aerial Vehicle Based High-Throughput Field Phenotyping of Canopy Temperature". In: 11 (February), pp. 1–17. doi: 10.3389/fpls.2020.00150.

10 References

Amani, I., Fischer, R. A. and Reynolds, M. P. (1996). Canopy temperature depression association with yield of irrigated spring wheat cultivars in a hot climate. *Journal of Agronomy and Crop Science* 176(2), 119-129. doi: 10.1111/j.1439-037X.1996.tb00454.x.

Anderegg, J., Aasen, H., Perich, G., Roth, L., Walter, A. and Hund, A. (2021). Temporal trends in canopy temperature and greenness are potential indicators of late-season drought avoidance and functional stay-green in wheat. *Field Crops Research* 274, 108311. doi: 10.1016/J.FCR.2021.108311.

Aragon, B., Johansen, K., Parkes, S., Malbeteau, Y., Al-Mashharawi, S., Al-Amoudi, T., Andrade, C. F., Turner, D., Lucieer, A. and McCabe, M. F. (2020). A calibration procedure for field and UAV-based uncooled thermal infrared instruments. *Sensors* 20(11), 3316. doi: 10.3390/S20113316.

Ayeneh, A., van Ginkel, M., Reynolds, M. P. and Ammar, K. (2002). Comparison of leaf, spike, peduncle and canopy temperature depression in wheat under heat stress. *Field Crops Research* 79(2–3), 173–184.

Ballico, M. J. (2004). Monograph. In: *Radiation Thermometry*. Lindfield, NSW: National Measurement Institute (NMI), p. 12.

Blum, A., Shpiler, L., Golan, G. and Mayer, J. (1989). Yield stability and canopy temperature of wheat genotypes under drought-stress. *Field Crops Research* 22(4), 289–296.

Boltzmann, L. (1884). Derivation of Stefan's law concerning the dependence of the thermal radiation on the temperature of the electro-magnetic theory of light. *Annalen der Physik* 258(6), 291–294. doi: 10.1002/ANDP.18842580616.

Brennan, J. P., Condon, A. G., van Ginkel, M. and Reynolds, M. P. (2007). Paper presented at international workshop on increasing wheat yield potential, CIMMYT, Obregon, Mexico, 20–24 March 2006. An economic assessment of the use of physiological selection for stomatal aperture-related traits in the CIMMYT wheat breeding programme. *The Journal of Agricultural Science* 145(3), 187–194.

Budzier, H. and Gerlach, G. (2015). Calibration of uncooled thermal infrared cameras. *Journal of Sensors and Sensor Systems* 4(1), 187–197. doi: 10.5194/JSSS-4-187-2015.

Coast, O., Harden, S., Conaty, W. C., Brodrick, R. and Edwards, E. J. (2020). Canopy temperature of high-nitrogen water-stressed cotton. *Crop Science* 60(3), 1513–1529. doi: 10.1002/CSC2.20127.

Conaty, W. C., Burke, J. J., Mahan, J. R., Neilsen, J. E. and Sutton, B. G. (2012). Determining the optimum plant temperature of cotton physiology and yield to improve plant-based irrigation scheduling. *Crop Science* 52(4), 1828–1836. doi: 10.2135/CROPSCI2011.11.0581.

Conaty, W. C., Mahan, J. R., Neilsen, J. E. and Constable, G. A. (2014). Vapour pressure deficit aids the interpretation of cotton canopy temperature response to water deficit. *Functional Plant Biology* 41(5), 535–546. doi: 10.1071/FP13223.

Conaty, W. C., Mahan, J. R., Neilsen, J. E., Tan, D. K. Y., Yeates, S. J. and Sutton, B. G. (2015). The relationship between cotton canopy temperature and yield, fibre quality and water-use efficiency. *Field Crops Research* 183, 329–341. doi: 10.1016/J.FCR.2015.08.010.

Condon, A. G., Reynolds, M. P., Rebetzke, G. J., van Ginkel, M., Trethowan, R., Bonnett, D. G., Richards, R. A. and Farquhar, G. D. (2004). Physiological traits as indirect selection criteria for yield potential in bread wheat. In: Black, C., Panozzo, J. and Rebetzke, G. (Eds). *Cereals. Proceedings of the 54th Australian Cereal Chemistry Conference and 11th Wheat Breeders Assembly*, Canberra, 21–24 September 2004 (vol. 2004). Canberra, pp. 112–115.

Condon, A. G., Reynolds, M. P., Rebetzke, G. J., van Ginkel, M., Richards, R. A. and Farquhar, G. D. (2007). Chapter 12: Using stomatal aperture-related traits to select for high yield potential in bread wheat. In: Buck, H., Nisi, J. and Salom´on, N. (Eds). *Wheat Production in Stressed Environments. Proceedings of the 7th International Wheat Conference*, 27 November–2 December 2005, Mar del Plata, Argentina (vol. 12). Dordrecht: Springer Netherlands, pp. 617–624.

Deery, D. M. and Jones, H. G. (2021). Field phenomics: Will it enable crop improvement? *Plant Phenomics* 2021, 9871989. doi: 10.34133/2021/9871989.

Deery, D. M., Rebetzke, G. J., Jimenez-Berni, J. A., Bovill, W. D., James, R. A., Condon, A. G., Furbank, R. T., Chapman, S. C. and Fischer, R. A. (2019). Evaluation of the phenotypic repeatability of canopy temperature in wheat using continuous-terrestrial and airborne measurements. *Frontiers in Plant Science* 10, 875. doi: 10.3389/fpls.2019.00875.

Deery, D. M., Rebetzke, G. J., Jimenez-Berni, J. A., James, R. A., Condon, A. G., Bovill, W. D., Hutchinson, P., Scarrow, J., Davy, R. and Furbank, R. T. (2016). Methodology for high-throughput field phenotyping of canopy temperature using airborne thermography. *Frontiers in Plant Science* 7, 1808. doi: 10.3389/fpls.2016.01808.

Fischer, R. A., Byerlee, D. and Edmeades, G. O. (2014). *Crop Yields and Global Food Security: Will Yield Increase Continue to Feed the World?* Canberra: Australian Centre for International Agricultural Research.

Fischer, R. A. and Rebetzke, G. J. (2018). Indirect selection for potential yield in early-generation, spaced plantings of wheat and other small-grain cereals: A review. *Crop and Pasture Science* 69(5), 439–459. doi: 10.1071/CP17409.

Fischer, R. A., Rees, D., Sayre, K. D., Lu, Z.-M., Condon, A. G. and Saavedra, A. L. (1998). Wheat yield progress associated with higher stomatal conductance and photosynthetic rate, and cooler canopies. *Crop Science* 38(6), 1467–1475.

FLIR. (2012). *The Ultimate Infrared Handbook for Research and Development Professionals*. Report. www.FLIR.com.

Fuchs, M., Kanemasu, E. T., Kerr, J. P. and Tanner, C. B. (1967). Effect of viewing angle on canopy temperature measurements With infrared thermometers 1. *Agronomy Journal* 59(5), 494–496. doi: 10.2134/AGRONJ1967.00021962005900050040X.

Fuchs, M. and Tanner, C. B. (1966). Infrared thermometry of vegetation 1. *Agronomy Journal* 58(6), 597–601. doi: 10.2134/agronj1966.00021962005800060014x.

George, T., Gulati, S., Martin, S. and Nozaki, S. (2019). Comparison of mid wave infrared (MWIR) and long wave infrared (LWIR) imagery for precision agriculture applications. In: *IEEE Aerospace Conference* (vol. 2019), pp. 1–15. doi: 10.1109/aero.2019.8742183.

Giunta, F., Motzo, R. and Pruneddu, G. (2008). Has long-term selection for yield in durum wheat also induced changes in leaf and canopy traits? *Field Crops Research* 106(1), 68–76.

Guo, W., Zheng, B., Potgieter, A. B., Diot, J., Watanabe, K., Noshita, K., Jordan, D. R., Wang, X., Watson, J., Ninomiya, S. and Chapman, S. C. (2018). Aerial imagery analysis – Quantifying appearance and number of sorghum heads for applications in breeding and agronomy. *Frontiers in Plant Science* 871, 1544. doi: 10.3389/FPLS.2018.01544/BIBTEX.

Hatfield, J. L., Quisenberry, J. E. and Dilbeck, R. E. (1987). Use of canopy temperatures to identify water conservation in cotton germplasm. *Crop Science* 27(2), 269–273.

Huband, N. D. and Monteith, J. L. (1986). Radiative surface temperature and energy balance of a wheat canopy. *Boundary-Layer Meteorology* 36(1), 1–17. doi: 10.1007/BF00117455.

Idso, S. B., Jackson, R. D., Ehrler, W. L. and Mitchell, S. T. (1969). A method for determination of infrared emittance of leaves. *Ecology* 50(5), 899–902. doi: 10.2307/1933705.

Jarvis, P. G. and McNaughton, K. G. (1986). Stomatal control of transpiration: Scaling up from leaf to region. *Advances in Ecological Research* 15, 1–49. doi: 10.1016/S0065-2504(08)60119-1.

Jerbi, T., Wuyts, N., Cane, M. A., Faux, P. F. and Draye, X. (2015). High resolution imaging of maize (*Zea mays*) leaf temperature in the field: The key role of the regions of interest. *Functional Plant Biology* 42(9), 858–864. doi: 10.1071/FP15024.

Jones, H. and Sirault, X. (2014). Scaling of thermal images at different spatial resolution: The mixed pixel problem. *Agronomy* 4(3), 380–396. doi: 10.3390/agronomy4030380.

Jones, H. G. (2004). Application of thermal imaging and infrared sensing in plant physiology and ecophysiology. *Advances in Botanical Research* 41, 107–163. doi: 10.1016/S0065-2296(04)41003-9.

Jones, H. G., Hutchinson, P. A., May, T., Jamali, H. and Deery, D. M. (2018). A practical method using a network of fixed infrared sensors for estimating crop canopy conductance and evaporation rate. *Biosystems Engineering* 165, 59–69. doi: 10.1016/j.biosystemseng.2017.09.012.

Jones, H. G., Serraj, R., Loveys, B. R., Xiong, L., Wheaton, A. and Price, A. H. (2009). Thermal infrared imaging of crop canopies for the remote diagnosis and quantification of plant responses to water stress in the field. *Functional Plant Biology* 36(11), 978–989.

Jones, H. G., Stoll, M., Santos, T., de Sousa, C., Chaves, M. M. and Grant, O. M. (2002). Use of infrared thermography for monitoring stomatal closure in the field: Application to grapevine. *Journal of Experimental Botany* 53(378), 2249–2260. doi: 10.1093/jxb/erf083.

Jones, H. G. and Vaughan, R. A. (2010). *Remote Sensing of Vegetation: Principles, Techniques, and Applications*. New York: Oxford University Press.

Kelly, J., Kljun, N., Olsson, P., Mihai, L., Liljeblad, B., Weslien, P., Klemedtsson, L. and Eklundh, L. (2019). Challenges and best practices for deriving temperature data from an uncalibrated UAV thermal infrared camera. *Remote Sensing* 11(5), 567. doi: 10.3390/RS11050567.

Kirchhoff, G. (1860). Ueber das Verh¨altniss zwischen dem Emissionsverm¨ogen und dem Absorptionsverm¨ogen der K¨orper fu¨r W¨arme und Licht. *Annalen der Physik und Chemie* 185(2), 275–301. doi: 10.1002/ANDP.18601850205.

Li, X., Ingvordsen, C. H., Weiss, M., Rebetzke, G. J., Condon, A. G., James, R. A. and Richards, R. A. (2019). Deeper roots associated with cooler canopies, higher normalized difference vegetation index, and greater yield in three wheat populations grown on stored soil water. *Journal of Experimental Botany (May)* 70(18), 4963–4974. doi: 10.1093/jxb/erz232.

Lopes, M. S. and Reynolds, M. P. (2010). Partitioning of assimilates to deeper roots is associated with cooler canopies and increased yield under drought in wheat. *Functional Plant Biology* 37(2), 147–156.

Lopes, M. S. and Reynolds, M. P. (2012). Stay-green in spring wheat can be determined by spectral reflectance measurements (normalized difference vegetation index) independently from phenology. *Journal of Experimental Botany* 63(10), 3789–3798.

Lopes, M. S., Reynolds, M. P., Jalal-Kamali, M. R., Moussa, M., Feltaous, Y., Tahir, I. S. A., Barma, N., Vargas, M., Mannes, Y. and Baum, M. (2012). The yield correlations of selectable physiological traits in a population of advanced spring wheat lines grown in warm and drought environments. *Field Crops Research* 128, 129–136.

Lord, S. D. (1992). A new software tool for computing earth's atmospheric transmission of Near-and far-infrared radiation. *NASA Technical Memorandum* 103957.

Maes, W. H. and Steppe, K. (2012). Estimating evapotranspiration and drought stress with ground-based thermal remote sensing in agriculture: A review. *Journal of Experimental Botany* 63(13), 4671–4712. doi: 10.1093/jxb/ers165.

Mahan, J. R., Conaty, W., Neilsen, J., Payton, P. and Cox, S. B. (2010). Field performance in agricultural settings of a wireless temperature monitoring system based on a low-cost infrared sensor. *Computers and Electronics in Agriculture* 71(2), 176–181. doi: 10.1016/J.COMPAG.2010.01.005.

Malbéteau, Y., Johansen, K., Aragon, B., Al-Mashhawari, S. K. and McCabe, M. F. (2021). Overcoming the challenges of thermal infrared orthomosaics using a swath-based approach to correct for dynamic temperature and wind effects. *Remote Sensing* 13(16), 3255. doi: 10.3390/RS13163255.

Mason, R. E. and Singh, R. P. (2014). Considerations when deploying canopy temperature to select high yielding wheat breeding lines under drought and heat stress. *Agronomy* 4(2), 191–201.

Mathews, K. L., Chapman, S. C., Trethowan, R., Pfeiffer, W., van Ginkel, M., Crossa, J., Payne, T., Delacy, I., Fox, P. N. and Cooper, M. (2007). Global adaptation patterns of Australian and CIMMYT spring bread wheat. *TAG: Theoretical and Applied Genetics. Theoretische und Angewandte Genetik* 115(6), 819–835.

Nugent, P. W., Shaw, J. A. and Pust, N. J. (2013). Correcting for focal-plane-array temperature dependence in microbolometer infrared cameras lacking thermal stabilization. *Optical Engineering* 52(6), 061304. doi: 10.1117/1.OE.52.6.061304.

Olivares-Villegas, J. J., Reynolds, M. P. and McDonald, G. K. (2007). Drought-adaptive attributes in the Seri/babax hexaploid wheat population. *Functional Plant Biology* 34(3), 189–203.

O'Shaughnessy, S. A., Evett, S. R., Colaizzi, P. D. and Howell, T. A. (2011a). Using radiation thermography and thermometry to evaluate crop water stress in soybean and cotton. *Agricultural Water Management* 98(10), 1523–1535. doi: 10.1016/j.agwat.2011.05.005.

O'Shaughnessy, S. A., Hebel, M. A., Evett, S. R. and Colaizzi, P. D. (2011b). Evaluation of a wireless infrared thermometer with a narrow field of view. *Computers and Electronics in Agriculture* 76(1), 59–68. doi: 10.1016/j.compag.2010.12.017.

Pask, A. J. D., Pietragalla, J., Mullan, D. M. and Reynolds, M. P. (2012). Physiological breeding *II*: A field guide to wheat phenotyping. *Technical Report*. Mexico, DF: CIMMYT.

Peckham, J., O'Young, S. and Jacobs, J. T. (2015). Comparison of medium and long wave infrared imaging for ocean based sensing. *Journal of Ocean Technology* 10(3), 112–128.

Perich, G., Hund, A., Anderegg, J., Roth, L., Boer, M. P., Walter, A., Liebisch, F. and Aasen, H. (2020). Assessment of multi-image unmanned aerial vehicle based high-throughput field phenotyping of canopy temperature. *Frontiers in Plant Science* 11(February), 150. doi: 10.3389/fpls.2020.00150.

Pinter, P. J., Zipoli, G., Reginato, R. J., Jackson, R. D., Idso, S. B. and Hohman, J. P. (1990). Canopy temperature as an indicator of differential water use and yield performance among wheat cultivars. *Agricultural Water Management* 18(1), 35–48.

Pinto, R. S., Reynolds, M. P., Mathews, K. L., McIntyre, C. L., Olivares-Villegas, J. J. and Chapman, S. C. (2010). Heat and drought adaptive QTL in a wheat population designed to minimize confounding agronomic effects. *TAG: Theoretical and Applied Genetics. Theoretische und Angewandte Genetik* 121(6), 1001–1021.

Planck, M. (1900). Zur Theorie des Gesetzes der Energieverteilung im Normalspectrum. In: *Verhandlungen der Deutschen Physikalischen Gesellschaft 2*. English translation by D. ter Haar 1967. The Old Quantum Theory (Pergamon Press), pp. 237–245.

Prashar, A. and Jones, H. (2014). Infra-red thermography as a high-throughput tool for field phenotyping. *Agronomy* 4(3), 397–417. doi: 10.3390/agronomy4030397.

Prashar, A., Yildiz, J., McNicol, J. W., Bryan, G. J. and Jones, H. G. (2013). Infra-red thermography for high throughput field phenotyping in Solanum tuberosum. *PLoS ONE* 8(6), e65816. doi: 10.1371/journal.pone.0065816.

Rashid, A., Stark, J. C., Tanveer, A. and Mustafa, T. (1999). Use of canopy temperature measurements as a screening tool for drought tolerance in spring wheat. *Journal of Agronomy and Crop Science* 182(4), 231–238.

Rattey, A. R., Shorter, R. and Chapman, S. C. (2011). Evaluation of CIMMYT conventional and synthetic spring wheat germplasm in rainfed sub-tropical environments. II. Grain yield components and physiological traits. *Field Crops Research* 124(2), 195–204.

Rebetzke, G. J., van Herwaarden, A., Biddulph, B., Moeller, C., Richards, R., Rattey, A. and Chenu, K. (2013a). Field experiments in crop physiology. http://prometheuswiki; publish.csiro.au/tiki-index.php?page=Field+Experiments+in+Crop+Physiology (accessed on 22 January 2014).

Rebetzke, G. J., Rattey, A. R., Farquhar, G. D., Richards, R. A. and Condon, A. T. G. (2013b). Genomic regions for canopy temperature and their genetic association with stomatal conductance and grain yield in wheat. *Functional Plant Biology* 40(1), 14–33.

Reynolds, M., Balota, M., Delgado, M., Amani, I. and Fischer, R. (1994). Physiological and morphological traits associated with spring wheat yield under hot, irrigated conditions. *Functional Plant Biology* 21(6), 717–730.

Reynolds, M., Dreccer, F. and Trethowan, R. (2007a). Drought-adaptive traits derived from wheat wild relatives and landraces. *Journal of Experimental Botany* 58(2), 177–186.

Reynolds, M. P., Saint-Pierre, C., Saad, A. S. I., Vargas, M. and Condon, A. G. (2007b). Evaluating potential genetic gains in wheat associated with stress-adaptive trait expression in elite genetic resources under drought and heat stress. *Crop Science* 47 (Suppl.3), S172–S189.

Ribeiro-Gomes, K., Hernández-López, D., Ortega, J. F., Ballesteros, R., Poblete, T. and Moreno, M. A. (2017). Uncooled thermal camera calibration and optimization of the photogrammetry process for UAV applications in agriculture. *Sensors* 17(10), 2173. doi: 10.3390/S17102173.

Roche, D. (2015). Stomatal conductance is essential for higher yield potential of C3 crops. *Critical Reviews in Plant Sciences* 34(4), 429–453. doi: 10.1080/07352689.2015.1023677.

Roitsch, T., Cabrera-Bosquet, L., Fournier, A., Ghamkhar, K., Jiménez-Berni, J., Pinto, F. and Ober, E. S. (2019). Review: New sensors and data-driven approaches–A path to next generation phenomics. *Plant Science* 282, 2–10. doi: 10.1016/j.plantsci.2019.01.011.

Rutkoski, J., Poland, J., Mondal, S., Autrique, E., Pérez, L. G., Crossa, J., Reynolds, M. and Singh, R. (2016). Canopy temperature and vegetation indices from high-throughput phenotyping improve accuracy of pedigree and genomic selection for grain yield in wheat. *Genes–Genomes Genetics* 6(9), 2799–2808. doi: 10.1534/g3.116.032888.

Sadras, V. O. and Lawson, C. (2011). Genetic gain in yield and associated changes in phenotype, trait plasticity and competitive ability of South Australian wheat varieties released between 1958 and 2007. *Crop and Pasture Science* 62(7), 533–549.

Salehi, A., Jimenez-Berni, J., Deery, D. M., Palmer, D., Holland, E., Rozas-Larraondo, P., Chapman, S. C., Georgakopoulos, D. and Furbank, R. T. (2015). SensorDB: A virtual laboratory for the integration, visualization and analysis of varied biological sensor data. *Plant Methods* 11, 53. doi: 10.1186/s13007-015-0097-z.

Sobrino, J. A., Jiménez-Muñoz, J. C., Labed-Nachbrand, J. and Nerry, F. (2002). Surface emissivity retrieval from digital airborne imaging spectrometer data. *Journal of Geophysical Research: Atmospheres* 107(D23), ACL 24–ACL 21. doi: 10.1029/2002JD002197.

Sobrino, J. A., Raissouni, N. and Li, Z. L. (2001). A comparative study of land surface emissivity retrieval from NOAA data. *Remote Sensing of Environment* 75(2), 256–266. doi: 10.1016/S0034-4257(00)00171-1.

Stutsel, B., Johansen, K., Malbéteau, Y. M. and McCabe, M. F. (2021). Detecting plant stress using thermal and optical imagery from an unoccupied aerial vehicle. *Frontiers in Plant Science* 12, 734944. doi: 10.3389/FPLS.2021.734944.

Sun, J., Poland, J. A., Mondal, S., Crossa, J., Juliana, P., Singh, R. P., Rutkoski, J. E., Jannink, J. L., Crespo-Herrera, L., Velu, G., Huerta-Espino, J. and Sorrells, M. E. (2019). High-throughput phenotyping platforms enhance genomic selection for wheat grain yield across populations and cycles in early stage. *TAG: Theoretical and Applied Genetics. Theoretische und Angewandte Genetik* 132(6), 1705–1720. doi: 10.1007/s00122-019-03309-0.

Sun, J., Rutkoski, J. E., Poland, J. A., Crossa, J., Jannink, J. L. and Sorrells, M. E. (2017). Multitrait, random regression, or simple repeatability model in high-throughput phenotyping data improve genomic prediction for wheat grain yield. *The Plant Genome* 10(2). doi: 10.3835/plantgenome2016.11.0111.

Takai, T., Yano, M. and Yamamoto, T. (2010). Canopy temperature on clear and cloudy days can be used to estimate varietal differences in stomatal conductance in rice. *Field Crops Research* 115(2), 165–170. doi: 10.1016/j.fcr.2009.10.019.

van Eeuwijk, F. A., Bustos-Korts, D., Millet, E. J., Boer, M. P., Kruijer, W., Thompson, A., Malosetti, M., Iwata, H., Quiroz, R., Kuppe, C., Muller, O., Blazakis, K. N., Yu, K., Tardieu, F. and Chapman, S. C. (2019) Modelling strategies for assessing and increasing the effectiveness of new phenotyping techniques in plant breeding. *Plant Science* 282, 23–39. doi: 10.1016/j.plantsci.2018.06.018.

van Ginkel, M., Reynolds, M., Trethowan, R. and Hernandez, E. (2004). Can Canopy Temperature Depression measurements help breeders in selecting for yield in wheat under irrigated production conditions? In: Fischer, R. A., et al (Eds). *New Directions for a Diverse Planet: Proceedings for the 4th International Crop Science Congress*, Brisbane, Australia, 26 September. Brisbane: The Regional Institute Ltd, p. 3.4.6.

Wien, W. (1896). Concerning the energy distribution in the emission spectrum of a blackbody. *Annalen der Physik und Chemie* 294(8), 662–669. doi: 10.1002/ANDP.18962940803.

Zadoks, J. C., Chang, T. T. and Konzak, C. F. (1974). A decimal code for the growth stages of cereals. *Weed Research* 14(6), 415–421.

Chapter 5

Advances in the use of X-ray computed tomography in crop phenotyping

Stefan Gerth, Norman Uhlmann and Michael Salamon, Fraunhofer EZRT, Germany

1 Introduction

Over the last few decades, X-ray technology has been utilized for the non-destructive visualization of optically inaccessible plant structures [1-7]. This technology was previously mainly used for medical imaging [8, 9] but is currently used as a standard tool in industrial applications for material analysis or inspection [9-12]. In contrast to optical imaging, X-rays are able to penetrate solid objects, being attenuated within the material [13]. X-ray computed tomography (CT) enables the 3D volume information of objects to be reconstructed using X-ray projections of the object obtained from different angles [14]. Thus, with CT, it is possible to perform non-destructive analysis of optically inaccessible structures below- and aboveground. The focus of these technological developments is the visualization of the belowground growth processes of roots [3, 4, 15-20] or, e.g. potato tubers [21-23]. Due to the fact that X-ray imaging necessitates access from both sides of the object, belowground structures are at the moment only accessible in pots or special setups. Observing these structures introduces new possibilities for crop phenotyping, as they are essential for the plant to utilize water or nutrients or store starch [23, 24]. Additionally, the root system provides anchorage of the

http://dx.doi.org/10.19103/AS.2022.0102.06

plant to the soil, and the root system architecture (RSA) is an important factor in enhancing plant resilience to biotic and abiotic stresses [6, 16].

It is important to note that data from an X-ray CT sensor will require post-processing. To observe root growth or the development of belowground structures, they must first be segmented. In this step, the structure of interest is separated from the opaque soil and additional traits can be subsequently extracted from the segmented structure using distinct algorithms [25–34]. Several software packages have been developed for trait extraction within the RSA, which allow the quantitative evaluation of the captured sensor data. The huge variation of scales from the xylem level within the root, root hairs, to lateral and seminal roots, up to the storage organs, such as tubers or big storage roots, span approximately seven magnitudes. Thus, depending on the sensor used and the phenotyping purpose, different algorithms are suitable for distinct traits.

Depending on the resolution, the pot size and the required throughput, there are several different X-ray CT designs available. However, the majority of crop phenotyping systems are versatile micro-CT systems from the non-destructive testing domain. This type of system is not specifically adapted for crop phenotyping; however, there are CT systems adapted for plant phenotyping and directly integrated into a conveyor system for the automatic feeding of plants. These types of systems are not particularly flexible in resolution alteration and are typically optimized for a distinct purpose. Reducing the versatility of these CT systems increases the throughput and allows the automatic assessment of quantitative traits in larger quantities. However, all of these systems are complex and consist of distinct technologies needed to derive quantitative phenomic traits from the acquired data. Thus, we will address the main parts of a CT system influencing the quality of the acquired data, the CT system design and the data processing in different sections. This will help understand how each component of a CT system, from the X-ray source to the traits of interest, is interconnected. We will first address the X-ray source itself and explain how X-rays are generated and what type of X-ray tube is used within different applications. After the generation of the X-ray radiation itself, we will cover the interaction with the plant and discuss how a digital projection is acquired. Additionally, we will summarize different ways to acquire a complete set of projections for a CT reconstruction and discuss how the phenotypic trait is gathered from the reconstructed data. Thus, we will cover the whole technological range of crop phenotyping CT systems and present end-use cases for above- and belowground crop phenotyping.

2 X-ray sources

Within an X-ray tube, a high voltage is applied which accelerates electrons from the cathode to the anode. The interior of the X-ray tube is evacuated to a pressure of 1×10^{-6} bar to allow electrons to travel between the cathode and anode. In the majority of crop phenotyping systems, the anode is made from tungsten, which defines the composition of the resulting X-ray energy spectrum. This spectrum consists of two components, the Bremsstrahlung and the characteristic spectral lines. The Bremsstrahlung is generated from the deceleration and directional changes of the high-energy electrons in close proximity to the atoms (more specifically to the electrons surrounding the atom) radiating the energy difference in the shape of a broad spectrum. In contrast, the characteristic lines originate from the energy difference between different electron shells of the anode atoms. In this case, the high-energy electrons will remove an electron from the shell, this hole will be filled from another shell and the energy difference is released as a photon of characteristic wavelength or X-ray radiation. While in the majority of cases we describe the radiation as an electromagnetic wave with a wavelength, in some cases it is necessary to see the photon as a particle with a distinct energy. The fact that the radiation can be interpreted either as an electromagnetic wave or as a particle is called wave-particle duality [35, 36]. Due to the fact that the energy difference between the shells is characteristic for the anode material, the emitted radiation, with its characteristic wavelengths, is called characteristic spectral line(s). In contrast to medical applications, the acceleration voltage ranges up to 225 kV, especially for belowground phenotyping applications. Technologically, there are variations in the design of X-ray tubes with each having unique benefits; however, each tube uses the same physical principle [37].

The electrons required to generate the X-ray beam are emitted from a filament similar to that in a light bulb. The interior of the X-ray tube is evacuated in order to enable the acceleration of these free electrons in an electromagnetic field, as it is necessary to avoid collisions with other particles such as gas molecules. The vacuum can be sealed during manufacturing or permanently pumped out in a system known as 'open.' The open design has the advantage that the worn-out parts inside the X-ray tube can be replaced, resulting in the tube having an almost unlimited lifetime. However, the disadvantage is that the maintenance and operation require experienced personnel to handle the complex systems. Open type X-ray tubes offer additional features such as a focusing unit that minimizes the size of the focal spot. Smaller focal spot sizes allow higher optical magnifications, overcoming the limited resolution of the detector; however, the thermal stress on the anode material is much higher. These high-resolution X-ray tubes are more demanding in terms of maintenance. Sealed tubes benefit from low maintenance and lower system complexity, which enhance the robustness of the X-ray inspection, with the trade-off of relatively high replacement costs. Furthermore, the good lifetime predictability of sealed tubes allows a reliable cost estimate. The choice of a suitable X-ray tube is primarily determined by the application and the resulting requirements for penetration capability, scanning time and the focal spot size. Table 1 gives an overview of the different types of X-ray tubes, classified by the focal spot size.

In a laboratory environment, with varying samples and manual operation by experienced users, the flexibility of a microfocus X-ray tube is balanced against the higher complexity and lower reliability of the tube. In production environments such as an automated greenhouse, robustness and ease of maintenance are critical. Here, a further design aspect of X-ray tubes comprises the high voltage (HV) supply, which defines the maximum acceleration energy of the tube in kV. This also directly limits the maximum photonic energy (kVp) of the X-ray spectrum. Higher X-ray energy increases the variety of applications but places demands on the system size and complexity, which derive mainly from the insulation properties of the HV supply and its peripherals, e.g. cables preventing short circuits and flashovers. The high voltage can be utilized by two electrical designs. Unipolar X-ray tubes consist of one HV supply that delivers the full voltage range but requires high internal and peripheral insulation. This often results in a large oil vessel encapsulating the HV electronics and a large-diameter HV cable connecting the cathode to the supply. Bipolar tubes consist of two HV supplies, each delivering half of the total potential. While in unipolar tubes the anode is grounded, in bipolar tubes it is set to the negative value of the cathode voltage, leading in total to double the voltage difference obtained by one HV supply. This doubles the number of parts but the insulation required and electrical circuit complexity are dramatically reduced, leading to lower costs. The high-volume sectors in particular, such as dental imaging and food and cargo inspection, have forced the use of simple and cheap components to drive the costs down. The reduction of peripheral components such as HV cables has led to the so-called 'monoblock' type X-ray tube dominating the mass inspection market. This type of X-ray tube combines the insulation oil vessel with the X-ray tube in one housing, reducing maintenance and peripherals prone to failure. The X-ray tube has become a simple black box that is easily interchangeable with standard electrical interfaces instead of HV cabling. Although this type of X-ray tube has many advantages, there are limitations in terms of achievable

Table 1 Different categories of X-ray tubes determined by the focal spot size ranging from 150 nm up to 1 mm

	Nanofocus	Microfocus	Minifocus
Focal spot size	150 nm–1 μm	1–300 μm	300–1000 μm
Energy	<100 kVp	<300 kVp	<600 kVp
Power	~1 W	~50 W	~500–2000 W
Vacuum type	open	open/sealed	sealed

power and beam characteristics. The choice of an appropriate X-ray tube is mandatory to fulfil individual customer requirements.

3 Interaction of X-rays with material

The basis for X-ray imaging is the interaction of the X-rays with matter. If you have an X-ray flux with a primary intensity I_0 that penetrates a material of a certain thickness l, the intensity behind this material is given by the Lambert-Beer law:

$$I = I_0 e^{-\mu l}$$

where μ is the absorption coefficient and depends on the material's atomic number, its density, and also on the energy of the X-ray photon [14, 38, 39]. More X-rays are absorbed by the material with increasing path length (l) of the X-rays traveling through the material and with a higher atomic number and higher density of the material. Additionally, the penetration capability of the photons increases with higher X-ray energy. Therefore, a more detailed form of the Lambert-Beer Law is given by:

$$I = \int_0^{E_{max}} I_0(E) e^{-\int_z \mu(z,E)dz} dE.$$

Due to the harmful effect of X-rays interacting with biological materials, X-ray imaging devices need to be shielded to avoid radial damage to the surroundings. Materials with a high atomic number and high density are mostly used for shielding. Lead is a popular material for shielding due to its relatively low price, and its availability and formability, which is important in the construction of shielding.

The absorption of X-rays is due to their interaction with matter. X-rays interact with different interaction branches depending on their energy and properties of the material.

3.1 Photoelectric absorption

In this case, the X-ray photon interacts with the electrons within the atomic hull in matter and is completely absorbed by transferring all its energy to the bound hull electron. The kinetic energy of the bound electron leaving the shell after the interaction is the energy of the X-ray photon minus the binding energy [14, 38–40]. This interaction branch is the favored type for X-ray imaging because the photon is completely absorbed, resulting in ideal image contrast.

3.2 Compton scattering

With Compton scattering or incoherent scattering, the X-ray photon interacts with a hull electron, transfers a part of its energy to the electron and leaves the atom at a different angle and with less energy. The scattering angle θ and the energy (E_γ) of the incident photon contribute to the residual energy of the scattered photon [38–40]. The Compton formula is used to calculate this residual energy of the scattering process:

$$E_\gamma^{'} = \frac{E_\gamma}{1 + \frac{E_\gamma\left(1 - \cos^{\sim}\ \right)}{m_0 c^2}}$$

This interaction decreases the X-ray imaging quality. As the scattered X-rays do not have the trajectory of origin as they hit the detector, they cannot be interpreted in the same way as the absorbed X-ray photons.

3.3 Rayleigh scattering

The third interaction branch is Rayleigh scattering or coherent scattering. The X-ray photon interacts with a hull electron and is scattered in a different direction but no energy is transferred to the electron [14, 38–40]. The X-ray photon leaves the atom at a different angle from the original path vector.

3.4 Pair production

At X-ray energies higher than 1.022 MeV, the X-ray photon can interact directly with the field of the atomic core and is converted into an electron and a positron [38–40]. The positron travels through the surrounding material and annihilates with another electron into two X-ray photons, each with an energy of the electron rest mass (511 keV).

In Fig. 1, the different absorption coefficients μ for the different interaction branches of X-ray photons with matter are given as a function of the X-ray energy.

In the Lambert-Beer Law, the four absorption coefficients for each interaction branch sum up simply to the overall absorption coefficient μ:

$$\mu(E, \rho, Z) = \mu_{\mathrm{ph}} + \mu_{\mathrm{coh}} + \mu_{\mathrm{inc}} + \mu_{\mathrm{pp}}$$

where μ_{ph} denotes the photoelectric absorption, μ_{coh} the Rayleigh scattering, μ_{incoh} the Compton scattering and μ_{pp} the pair production. Using the Lambert-Beer Law, the intensity I_a of the X-ray beam behind a specific material *a* with a thickness l_a is given by $I_a = I_0 e^{-\mu_a l_a}$, where μ_a is the absorption coefficient of the specific material for a given X-ray spectrum. The contrast between the different intensities I_0 and I_a is defined by:

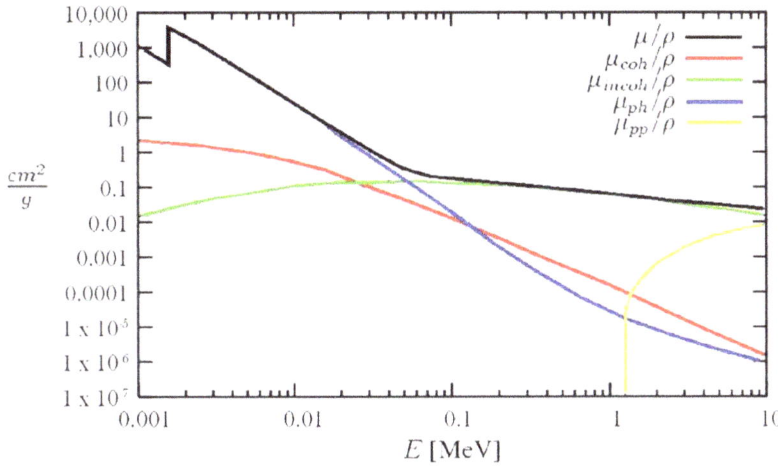

Figure 1 Energy dependence of the cross-sections for coherent and incoherent scattering, photoelectric absorption and pair production.

$$C = \frac{(I_0 - I_a)}{(I_0 + I_a)} \text{ with } I_0 > I_a .$$

If the X-ray beam penetrates multiple materials of differing thicknesses (l_a, l_b), the intensity of the beam behind the materials can be calculated by $I = I_0 e^{-(\mu_a l_a + \mu_b l_b + \ldots)}$. During X-ray imaging, the X-rays from the X-ray source penetrate all the different materials of the object with their subsequent, different path lengths. Thus, the intensity is attenuated by the given formula, resulting in reduced intensities at the detector for each of the different paths through the object. It must be highlighted that the intensities given by the Lambert-Beer Law cannot be easily calculated due to several reasons.

One reason is that the X-ray spectrum emitted from the source changes along the path through the object. Not only is the intensity attenuated due to absorption, the energy spectrum also changes; the mean energy increases and thus the attenuation behavior for subsequent materials varies as a function of the path length. This effect is called beam hardening, as lower energies are more attenuated in matter, compared with the higher energies of the spectrum, and have a higher penetration capability. Another reason the Lambert-Beer Law is not easy to calculate is that when the X-rays interact with the material of the object, some of the X-ray intensity is scattered (Compton or Rayleigh scattering) and hits the detector apart from its prior path, resulting in a false intensity at a given point on the detector. This is of particular importance as the water within the plant and soil is a very strong source of scattering. Furthermore, the detector's internal scattering and noise (electronics, statistics in the scintillator or sensor layer) further contribute to a false intensity measurement. All these factors need to be addressed when the images are used in later stages for a 3D reconstruction, or analyzed for density or path length calculations.

While knowledge regarding the interaction between the radiation and material is essential for the quality of the reconstruction, especially for crop phenotyping, the influence of the radiation on the plant itself is important. In contrast to medical CT, the influence of a distinct radiation dose on the resulting phenotype is not completely understood [41, 42]. However, at large doses of radiation - generated from high resolution or long measurements - negative effects on plants have been reported, depending on the variety and X-ray spectrum used [42–46]. Thus, to leverage CT as a non-destructive method in crop phenotyping, the measurement frequency and scanning duration need to be considered in the experimental design [47, 48]. This prevents the extracted traits being influenced by the additional stress originating from the dose. Due to the constant exposure of the plant to direct sunlight, all plants have mechanisms to regulate stress from reactive oxygen [49, 50]. This could explain the plant variety-dependent reaction to increased radiation and the improved handling of ionizing radiation compared with a human. However, X-ray radiation will increase the generation of reactive oxygen, and when the plant is unable to regulate it, this will result in tissue and DNA damage.

4 Detector

The basic principle for all imaging X-ray detectors is the same. X-ray photons emitted by the source penetrate the object and are attenuated according to their energy, the material and path length in the object. They continue their travel and hit the detector, i.e. they interact with the detection-sensitive layer either by photoabsorption, Compton scattering or, at higher energies, pair production [39]. All these interactions produce a high-energy electron emitted from the atomic shell which deposits its energy by producing more electrons along its path through the sensor layer. This deposited energy in the sensor layer is used to generate an electronic signal that can be measured and stored as a digital image. Within this image, the values are stored in a pixel matrix and the detector is often described with the value of the pixel size or the pixel pitch, respectively.

There are two main types of detectors; the direct conversion detector, in which the sensor layer is a semi-conductor [51, 52] or a high-pressure gas, and the indirect conversion detector, in which a scintillator is used as the sensor layer [53, 54].

The sensor layer of the indirect conversion detector consists of scintillator material where the deposited energy of the X-ray interaction is converted into optically visible photons, which travel through the scintillator until they are detected by a photodiode coupled to the scintillator. The photodiode converts the optical signal into an electrical signal. The advantage of the indirect conversion detector is that scintillators can be produced in relatively large plates and photo diode grids can be coupled with low effort. This results in detectors of large area and relatively low prices with good efficiency. The operational mode is one in which, over a certain period of time, all interactions are detected and the charge on the photodiodes is integrated. After the integration period, the collected charge is measured, which correlates with the intensity of X-rays and the corresponding energy hitting the detector pixel.

The sensor layer of the direct conversion detector consists of a semiconductor in which the deposited energy in the form of secondary electrons is collected by a high bias voltage and directly converted into an electrical signal. There are two possible modes of direct conversion detectors. One is the integration mode, which is comparable to the indirect conversion detector, in which the charge is collected over a certain period of time (integration time) and the strength of the generated electrical signal represents the intensity of the X-ray hitting the detector in this time slot. The other operation mode is the single-counting mode which requires more sophisticated electronics for each pixel. In the photon-counting mode, the charge of each X-ray interaction in the sensor layer pixel is collected and converted into a digital signal. Depending on the electronic capability, in each pixel the signal is compared via different thresholds resulting in the increase of different counters, discriminating the number of photons hitting the detector and the energy of the photons. Direct conversion detectors feature a higher efficiency at lower energies compared with indirect conversion detectors but are more expensive and only available with smaller sensitive areas.

5 Computed tomography systems for crop phenotyping

A CT system for crop phenotyping consists of the X-ray source, a manipulation system rotating the plant or tissue structure to phenotype and the detector. The distances from the source to the plant, and from the source to the detector, are called the focus object distance (FOD) and focus detector distance (FDD), respectively. Geometrically, the object is magnified to the detector (see Fig. 2). The magnification factor is determined by $M = \mathrm{FDD}/\mathrm{FOD}$ [14]. As the detector is built from individual pixels, several parts contribute to the final resolution of the CT, i.e. the pixel size of the detector, the magnification and the focal spot size. Additionally, the contrast in the sample is limiting whether or not a structure is visible.

There are two main possibilities to build a system dedicated for plant phenotyping. The first approach is via leveraging the geometrical magnification

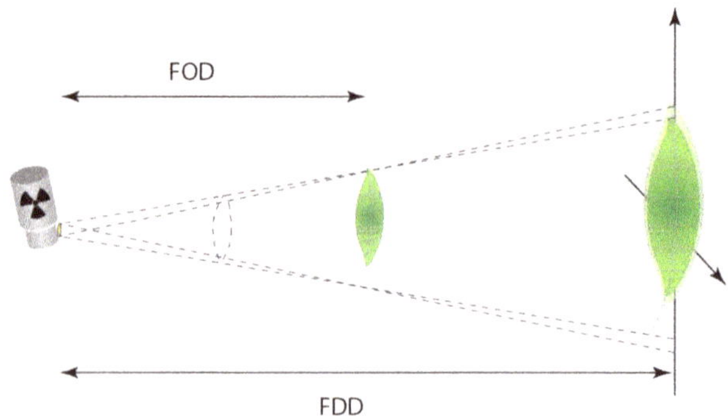

Figure 2 Schematic representation of the geometric magnification in a computed tomography system. The distances between the focal spot, object and detector determine the magnification on the detector.

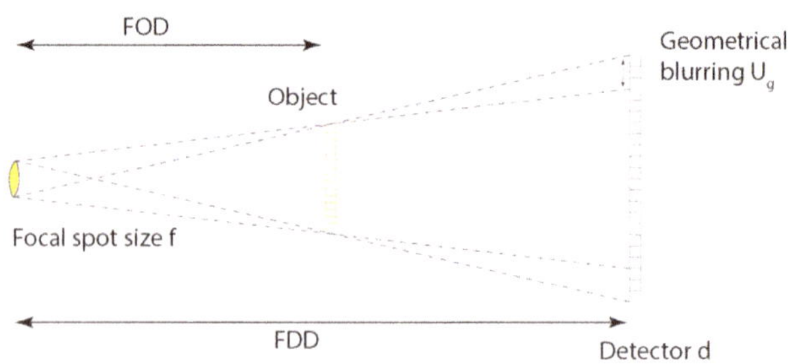

Figure 3 Visualization of the geometrical blurring dependent on the magnification *M*. The focal spot size contributes to the geometrical blurring and increases with the resolution.

M. Typically, this results in larger systems to allow a wide range of differing magnifications. In these systems, detectors feature higher pixel pitch with high quantum efficiency and X-ray sources with small focal spot sizes *f*. These small spot sizes are needed to reduce the blurring U_g resulting from the combination of focal spot size and high magnification (see Fig. 3). This blurring is calculated using $U_g = f * (M - 1)$. However, due to the small focal spot size, the maximum power - and thus the maximum intensity - is limited. These systems are highly versatile and have good sensor data quality. However, compared with other approaches, they need longer scanning times due to the large FDD and the small focal spot size.

The second approach positions the plant directly next to the detector, resulting in the magnification *M* of the CT system being close to 1 (see Fig. 4). Thus, the blurring $U_g = f * (M - 1)$ is not correlated with the focal spot size. However, in this arrangement of source, object and detector, the achievable

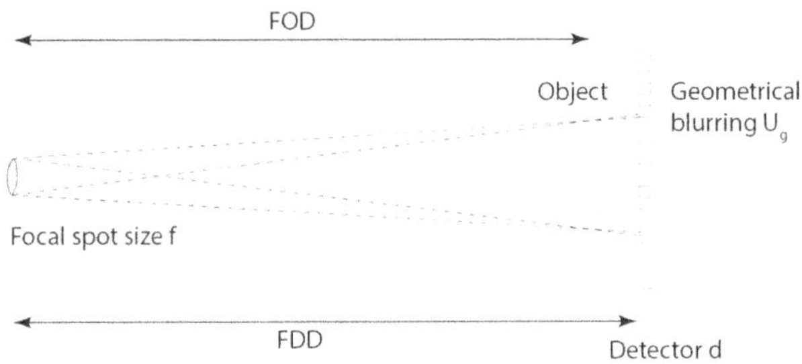

Figure 4 Visualization of the geometrical blurring for a magnification close to 1. The size of the focal spot hardly contributes to geometrical blurring.

resolution is solely dependent on the detector pixel pitch. The majority of these systems are built for high-throughput measurements utilizing a high-power X-ray source with a large focal spot. Typically, this results in higher scanning speeds but limits the resolution. This appears to be contrary to the reason mentioned above regarding the system resolution not being dependent on the focal spot size. However, reducing the pixel size of the detector reduces the sensitivity of each pixel. The majority of detectors operate with quadratic pixel sizes of 200 μm or even higher, resulting in high sensitivity. These types of detectors are typically used in CT systems relying on geometric magnification with small focal spot sizes. In contrast, quadratic pixel sizes as low as 45 μm are used on higher throughput systems relying on a geometric magnification close to 1.

A CT scan is acquired by projecting the attenuation image of the object from 360°, which can be achieved by rotating the object or rotating the imaging system on a so-called 'gantry' system. Importantly, the mathematical reconstruction relies on only the scanning angle changing during the scanning process. In medical radiology, the gantry type dominates the application due to the unfavorable rotation of the patient with approximately 3–4 Hz [55], whereas in industrial applications, the object is rotated while the imaging system remains in position. The latter arrangement offers much higher flexibility due to the simple adjustment of magnification and reduction of costs. The costly core of a medical gantry system - a massive rotor with its slip-ring power and data transmission allowing operation of the X-ray tube, detector and all peripherals while rotating at high speed - is not necessary. For crop phenotyping, either approach might be needed. Scanning aboveground plant parts, a gantry system would be advantageous to prevent fragile structures like stems, fruits and even leaves from moving due to the rotation. Contrary to the medical gantry systems, these devices rotate in the horizontal plane, allowing the plant to remain in an upright orientation. For stable objects like roots or tubers in substrate-filled pots, for which the rotation has negligible influence on the position of the imaged structures, a fixed imaging system is preferred.

The scanning itself can be done in different ways depending on the required image quality and throughput. Over the last few years, the so-called 'FlyBy' or continuous scan has established a reduced time-consuming positioning for each of the several hundreds or thousands of scan angles [14]. In this approach, the full rotation is carried out continuously at a very slow speed. The image acquisition is done simultaneously and is adapted to the speed in a way that the blurring caused by the motion remains smaller than the pixel size and thus is not visible. The number of angles (NOA) needed for a reconstruction depends on the number of detector pixels (NOP) on the detector covered by the object projection. With the following rule, derived from the Nyquist Theorem $NOA = NOP * \pi / 2$, the highest quality in terms of geometrical accuracy is achieved [56]. In practice, the rule is reduced to $NOA = NOP * \pi / 4$. Thus, in typical applications with 2000 detector pixels, the NOA of 1200–1600 is applied. Besides the FlyBy operation, a Stop & Go mode may also be used. In this mode, each of the angular steps is approached individually. Its application is for measurements with very long scan times, in which each image requires a long integration time and it might be necessary to further average each of the images. Thus, the total scan time is dominated by the acquisition itself and the positioning effort can be neglected. In these cases, the NOA value can be greatly underestimated, but good reconstruction can still be achieved due to the clear and sharp angular images. The Stop & Go measurement mode is typically used for high-resolution measurements of stationary objects like seeds or organs displayed in pots. In contrast, for the higher throughput measurements, a FlyBy mode is applied.

The scanning trajectories and beam geometries are further aspects to be noted in modern CT applications. The beam geometry is given by the focal spot point and detector edges. Using a so-called 'linear detector array' with just one pixel row, the beam geometry forms a fan shape. The predominant beam geometry today is the cone beam, which results from the use of a 2D detector array (DDA). Due to the simultaneous acquisition of a square area, the scanning of large-sized objects is achieved using one rotation, whereas using fan-beam geometry, each row of the object requires an individual scan and rotation.

In addition to the beam geometry, the trajectory describes the imaging path around the object used for acquisition and reconstruction of the data. The original trajectory is the 360° rotation with equidistant steps, which is also applied in the most popular reconstruction algorithm announced in 1984 by Feldkamp et al. [57]. Today, this type of trajectory and reconstruction is also known as 3D-CT. Further trajectories have evolved during decades of research and development, the most relevant being spiral or helical scanning [58, 59]. This trajectory adds a transversal movement to the rotation and thereby allows continuous scanning of longitudinal objects. In medical applications, this is a full scan of a human body. In crop phenotyping, this can be done when acquiring combined above- and belowground measurements of a whole plant with a height exceeding the vertical field of view of the CT system. Additionally, this trajectory enables overcoming the so-called 'Feldkamp' artifact [60], which derives from the underrepresented Tuy criterion. This criterion defines the degrees of freedom needed in a trajectory to acquire full information of the object. In a rotational cone-beam 3D-CT trajectory, this criterion is not matched, resulting in strong artifacts at planes perpendicular to the rotation axis and distant from the central beam path. Using helical scanning, this effect can be avoided and an overall accurate volume is reconstructed.

6 From sensor to data

For an individual scan, each CT system generates a set of 2D projections along a given trajectory; however, in most cases it is not possible to directly extract phenotypic features from these sets of 2D images. The acquired sensor data needs to be processed to extract the relevant information from the scan. For this, several steps are needed:

1. **Data normalization and correction**: As a first step, all acquired 2D projections are offset-, gain- and bad-pixel-corrected in order to equalize different detector properties in a way that the absorption behavior in the 2D projection is not affected by the type and inhomogeneity of the detector [14]. In most systems, this is done automatically, during a regular maintenance procedure or directly before scanning. Some systems allow further corrections in this pre-processing step, e.g. a scattering correction or a beam-hardening correction [61–66]. The scattering correction will counteract a decrease of image quality mainly due to water scattering from the plant or wet soil. The beam-hardening correction will decrease artifacts arising from a change of the spectral composition in compact soil.
2. **3D reconstruction**: Pre-processed projections are used for the reconstruction. Many different reconstruction algorithms are available. In the majority of systems, a filtered back projection is used to calculate a 3D volume from the set of 2D projections [14]. However, depending on the trajectory of the system, several other algorithms exist [67–70]. Each reconstruction algorithm requires well-defined system geometry, where for each projection the position of the X-ray tube and the detector with respect to the sample is known. In the end, the 3D volume consists of voxels, a 3D pixel containing the absorption of X-rays at this position in the volume.
3. **Segmentation**: Starting from the reconstructed volume, the next step is separating the structure of interest from the remainder within the 3D

data cube. In most systems available for CT crop phenotyping, this step is not part of the scope of delivery from the manufacturer. However, there are commercial software toolkits available from the domain of non-destructive testing that support the user in the segmentation. This can be conducted either with thresholding [27, 71–74] or with region-growing algorithms starting from seed points [75–77]. Additionally, there are dedicated crop-specific algorithms available that allow the automatic segmentation of the reconstructed data [25–34]. As a result of this step, there is often a binary volume containing the structure of interest only as a mask for the original data cube.

4 **Feature calculation**: After segmentation, feature extraction is the last step in the data processing pipeline. Having the structure of interest extracted first, morphological features can be subsequently extracted, such as, e.g. volumes, shapes, surface diameters or lengths. Additionally, the gray value information from the data cube can be used to evaluate the density of different, specific plant organs or tissues. In most cases, the feature calculation and the segmentation are iterative processes.

It is obvious that conducting a CT measurement for crop phenotyping does not result directly in an extracted trait. For most CT systems, the first two steps are included and, after a scan, it is just a matter of time to visualize the 3D volume of the reconstructed crop. However, this alone will not allow quantitative trait analysis. For crop phenotyping, there are several solutions available to segment the reconstructed volume and analyze the features within. It must be noted that not every algorithm is capable of processing every reconstructed volume due to data type incompatibilities or due to the scanner resolution and resulting file size.

Several solutions are available for belowground root phenotyping. Most of these algorithms are based on the work of Frangi et al. (1998) [78], and were originally designed to segment blood vessels within the human body leveraged from the Eigenvalues of the Hessian matrix. Due to the similarity of the RSA and blood vessels, these or further extensions to the work of Frangi et al. are often described as 'vesselness' approaches [27, 79, 80]. However, at each junction within the RSA, algorithms based purely on vesselness will disconnect the individual roots. Thus, for example, Schulz et al. [80] first detected vesselness structures and subsequently reconnected the separated roots using a shortest path estimation. Alternative approaches, mostly from the area of computer vision, rely on tracking the root based on the level set segmentation method [34]. Slice by slice, the root is tracked based on grayscale intensities and shape information on the boundaries or leveraged from a local feature detected in the tubular shapes [29]. With modern developments in the area of machine learning, the self-similarity of root systems is used for the segmentation. A so-called 'U-net' was recently suggested [31, 81] and encoder-decoder architectures [82] for root segmentation were introduced. Originally, the U-net convolutional network was developed for biomedical applications and was built for a fast and precise image segmentation [83]. In theory, via heavy downscaling and upscaling in this network, arbitrary image sizes can be processed. The application is very fast compared with sliding window network architectures. Encoder–decoder networks are typically used in natural language processing in order to understand connected structures. In the area of image segmentation, the idea is to learn the connectivity of structures, e.g., the RSA. Once a crop-specific network is trained, these machine learning approaches allow the rapid segmentation of huge data sets. However, in contrast to conventional image segmentation algorithms, machine learning approaches [31, 81, 82] demand a considerable amount of manually segmented training data. These are very laborious to generate and require a well-established and normalized data generation pipeline from the sensor data to the reconstructed data set to avoid sensor bias during segmentation.

7 Case studies: Phenotyping using computed tomography

CT systems for crop phenotyping are used in circumstances in which non-destructive trait extraction is challenging due to occlusion effects. For belowground structures, this is due to the surrounding soil or growth substrate. Thus, most CT-based phenotyping approaches are focused on the visualization of occluded belowground storage organs or the RSA. However, occlusion can also be a hindering factor in aboveground phenotyping studies of plant structures. In many cases, assessing the ripening of different fruits, seeds or cobs is optically not possible. We will present two different case studies, one aboveground for post-harvest wheat phenotyping, and the other for belowground potato tuber phenotyping. Both case studies present potato tuber phenotyping and wheat ear phenotyping, leveraged from the non-destructive screening of occluded structures. While CT-based wheat ear phenotyping drastically reduces manual labor, potato tuber phenotyping additionally leverages from the ability to analyze tuber growth over time. Thus, it is possible to connect extracted structural traits in tuber development to environmental conditions at a given time in the growth process.

7.1 Belowground phenotyping of potato tubers

In this study, the belowground growth of potato tubers was analyzed using a dedicated CT system for crop phenotyping [21]. This system was built to achieve fast scanning times of less than 5 minutes per pot maintaining a good resolution to track tuber growth, beginning at a size of a few millimeters.

Thus, tuber development from initiation until harvest could be monitored and analyzed. In total, more than 2500 measurements were conducted by monitoring different genotypes during a heat and drought scenario after tuber initiation. In this setting, we used a magnification close to 1 to use high-power X-ray sources with high intensities, operating them on the large focal spot with powers up to 1600 W. The resolution was achieved via an XEye detector with an adapted scintillator and small pixel sizes (see Fig. 5).

Tuber growth was monitored by scanning three times per week and analyzing the growth of each individual tuber. A unique possibility with this form of non-destructive tuber analysis was the timing of the stress treatment itself. We started the stress when all plants had detectable tubers, increasing the temperature for 2 weeks while decreasing the daily water supply by 50%. During this stress period, we determined that the tuber growth was inhibited within a week. However, the tubers were able to resume growth at the end of the stress period and reached approximately the same size as those in the control group at the end of the experimental period.

In order to scan a plant three times per week, we needed to generate a semi-automatic imaging pipeline for the analysis. As mentioned, the scan itself generated no phenotypic data. Sophisticated post-processing in a very standardized way was needed to extract phenotypic information such as number, density, shape and size of each tuber in each pot at each time point. Manual data segmentation and analysis were not feasible. The generated data analysis platform was realized in a special C++ application for reasons of performance and data size. Normalization of the raw sensor information is key for a semi-automatic data analysis pipeline. With this procedure, we were able to segment the tubers from the surrounding soil by using only one set of parameters for the data analysis pipeline for each genotype and watering condition (dry soil and wet soil). Analyzing the same pot several times during

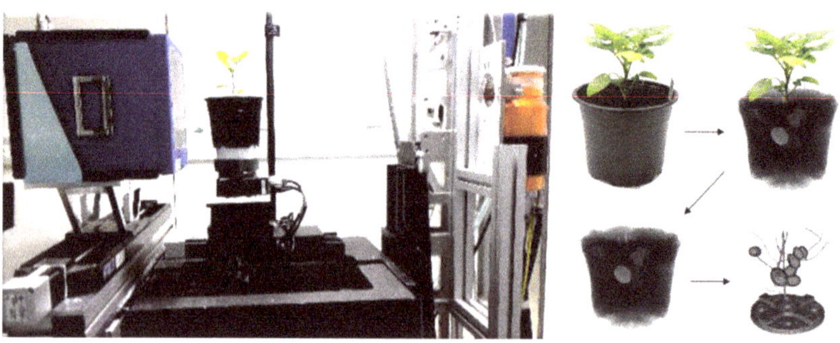

Figure 5 On the left-hand side, the first prototype of the tuber CT system is shown featuring a magnification close to 1 and a high-power X-ray tube for fast tuber phenotyping. On the right-hand side, a schematic representation of the data processing pipeline from the plant to the segmented tuber system is shown.

the experimental phase, we were able to further increase the segmentation of the individual tubers. Leveraging from the added information from the time series allowed differentiating between tubers and other segmented objects like stones and parts of unconnected stolons. For example, a tuber has to grow over time, or the aspect ratio additionally discriminates tubers and stolons. Identification of the tuber in the pot was realized by calculating the center of mass for each individual potato tuber and comparing it with the position at the previous time point; however, it was crucial to assure the same orientation of the reconstructed CT volume.

We showed that with this analysis pipeline, high-throughput tracking of tuber growth was possible. However, the combination of a dedicated CT system for crop phenotyping, the normalization of raw data and automated data processing was the key to generating quantitative phenotypic data.

7.2 Aboveground phenotyping of wheat ears

The phenotyping of yield data is typically obtained by machine threshing in the field [84] or via hand threshing in small-scale experiments. For wheat in particular, this phenotyping process results in lost seeds during machine threshing (about 10%) or in seed breaking (4–6% during machine threshing and about 1% in samples threshed by hand) [85, 86]. However, hand threshing is laborious and thus costly and time-consuming. For wild accessions and hulled landraces in particular, threshing is very challenging. Thus, the phenotypic evaluation of yield-related seed traits will require additional manual labor and further restricts the throughput for high-precision seed phenotyping [87–89]. This case study shows an accurate, non-destructive method, measuring the grains of wheat ears under different abiotic stresses via a post-harvest CT-based crop phenotyping system [2].

The system we used (CTportable160.90) features a high-resolution flat-panel detector with a quadratic pixel size of only 49.5 µm and 2304 pixels in the horizontal direction. Thus, scanning the ears close to the detector allows a large field of view while maintaining voxel sampling with a cubic size of 31.25 µm. In a helical scanning mode (see Section 5), we could simultaneously scan four intact ears with lengths up to 20 cm. With this approach, we could maintain a relatively high throughput of 7 minutes per ear. With the high resolution, we could reconstruct the morphological deformations, symptomatic of wheat grown under drought and frost stress.

After data generation, data processing generated four individual sub volumes containing one ear. The EarS segmentation algorithm was then used to process each individual 3D volume of an ear (see Fig. 6). Additional to the segmentation of each individual grain, oriented 2D cross-sections were generated and each virtually threshed grain was stored in an individual 3D volume for precise feature

Figure 6 From left to right: Photograph of the portable CT system for wheat ear phenotyping. 3D rendering of an individual ear with a length of up to 20 cm. Virtual 2D cross-section with an overlay of the segmented wheat grains. Overview image with each extracted grain automatically oriented alongside the largest diameter.

extraction. The features were calculated at the ear level (total seed weight per ear, seed number per ear and seed weight along the ear) and at the single grain level (single seed weight, seed size, seed shape, seed surface area and physical density of seeds). The extracted quantitative characteristics were used to estimate the performance of diverse wheat accessions under two stress regimes. In the comparison between the drought treatment and the combined drought and heat treatment, the single seed weight and the seed size were reduced under combined stress. Due to the high resolution, we could analyze the morphological shape of the individual seeds and observed an elongated shape for the combined stress treatment and, at the same time, a higher surface area. Moreover, the attenuation coefficient within the seed was similar for both treatments. This attenuation coefficient represents the physical density of the seed and combined with the detailed analysis of the seed volume, we could conclude that seeds exposed to combined stress were on average 0.02 g lighter compared with seeds exposed to drought stress only. Additionally, ears of plants exposed to drought stress contained on average larger and rounder seeds. These larger seeds were in most cases fully developed. In contrast, for the combined stress scenario, the seeds were not fully developed and showed morphological differences (large cavities and surface areas).

Using this combination of a CT system and a data processing pipeline, we could phenotype individual seeds in great detail without threshing the ear. Utilizing this approach to extract morphological features from CT data, it was possible to extract phenotypic traits at ear and seed level, and provide insights into the phenotypic differences of various genotypes in two abiotic stress regimes.

8 Summary and future trends

CT systems are currently frequently used for both below- and aboveground phenotyping. With the current available technology, it is possible to visualize optically opaque structures in high detail. Depending on the phenotypic trait of interest, different resolutions are required, and the resulting throughput of these systems varies. Reported measurement times range from a few minutes to an hour, illustrating the fact that there is no distinct CT crop phenotyping system available. Either the systems are general purpose CT systems from the domain of non-destructive testing and are built for a huge range of applications, or they are specifically designed to assess a set of phenotypic traits via high throughput. Additionally, the CT system never directly generates a quantitative trait. Thus, it is equally important to figure out the best combination of sensor data generation and post-processing. Only in combination with the appropriate data segmentation and analysis algorithms is it possible to extract quantitative phenotypic traits. Currently, a comparison between extracted phenotypic traits originating from different data sources is almost impossible due to the diversity of CT systems, post-processing algorithms and their combinations. However, the use of X-rays allows the non-destructive analysis of growth processes. For the direct evaluation of changing biotic or abiotic stresses in particular, this is a valuable tool in combination with genotyping approaches.

In future, standardization of data generation and post-processing algorithms will be needed to compare phenotypic experiments conducted at differing locations with different systems. This standardization has already been conducted in medical applications normalizing the CT systems to the proportions of a human body. Due to the diversity of pot sizes, plant architectures and watering regimes, any generalization is a challenging task in crop phenotyping and there is no standard solution available. With standardization of the data processing pipelines and access to the generated data from CT crop phenotyping systems following the FAIR principles (findable, accessible, interoperable and reusable), future artificial intelligence (AI) approaches will allow faster and easier data segmentation and analysis. Additionally, to further increase throughput of the whole phenotyping process, combined sensor acquisition may be relevant in the future. For example, during the CT measurements, it might be possible to capture the aboveground crop architecture with optical scanners and at the same time measure the hyperspectral reflectance of the plant. Thus, the analysis time per plant would be reduced and additional information by fusing the sensor data might become available for further phenotyping. Again, these fused data are a great field of application for the currently refined AI-based algorithms.

Technologically, the future of CT crop phenotyping involves pulsed high-power sources, as they are used for medical applications, or high-power sources with small focal spot sizes, as they can be used to further increase throughput. Combined with new single-photon detectors with small pixel pitches, fast scanning times can be realized. Currently, also from the medical CT imaging field, single-photon detectors discriminating different energy levels of the X-ray spectrum are used. With these approaches, beam-hardening artifacts can be reduced while increasing the contrast.

9 Where to look for further information

9.1 Further reading

- A comprehensive overview on computed tomography is given by P. R. T. Munro, 'Computed Tomography: From Photon Statistics to Modern Cone-Beam CT, by Thorsten M. Buzug,' *Contemp. Phys.*, vol. 51, no. 6, pp. 554–555, Nov. 2010, doi: 10.1080/00107514.2010.482334.
- A detailed explanation about the interaction of X-rays with material is given by G. Knoll, 'Radiation Detection and Measurement.' John Wiley & Sons, 1989.
- A description of root phenotyping approaches using computed tomography is given by J. Pfeifer, N. Kirchgessner, T. Colombi, and A. Walter, 'Rapid phenotyping of crop root systems in undisturbed field soils using X-ray computed tomography,' *Plant Methods*, vol. 11, no. 1, Aug. 2015, doi: 10.1186/s13007-015-0084-4.
- Several different approaches for algorithms in crop phenotyping are presented by Z. Xu, C. Valdes, and J. Clarke, 'Existing and Potential Statistical and Computational Approaches for the Analysis of 3D CT Images of Plant Roots,' *Agronomy*, vol. 8, no. 5, 2018, doi: 10.3390/agronomy8050071.
- Computed tomography for root phenotyping was discussed by J. A. Atkinson, M. P. Pound, M. J. Bennett, and D. M. Wells, 'Uncovering the hidden half of plants using new advances in root phenotyping,' *Curr. Opin. Biotechnol.*, vol. 55, pp. 1–8, Feb. 2019, doi: 10.1016/j.copbio.2018.06.002.

9.2 Key journals/conferences

- IPPN is the international plant phenotyping network organizing the IPPS (International Plant Phenotyping Symposium). The conference is well attended by researchers and technology developers at the same time. URL: https://www.plant-phenotyping.org/.
- NAPPN is the North American plant phenotyping network hosting conferences presenting modern technological ways of phenotyping. URL: https://www.nappn2021.org/.

- PAG (Plant and Animal Genome) holds regular meetings and hosts a phenotyping session. URL: https://www.intlpag.org/.
- *Plant Methods* is a journal publishing many approaches for crop phenotyping. In particular, the combination of sensor technology and algorithms is highlighted. URL: https://plantmethods.biomedcentral.com/.
- *IEEE Transactions on Image Processing* is a journal mainly focused on algorithm development. Several of the crop phenotyping algorithms are published here. URL: https://ieeexplore.ieee.org/.
- *Frontiers in Plant Science* is a journal publishing technological approaches on crop phenotyping. URL: https://www.frontiersin.org/journals/plant-science.

10 References

1. Liu, W., Liu, C., Jin, J., Li, D., Fu, Y. and Yuan, X. High-throughput phenotyping of morphological seed and fruit characteristics using X-ray computed tomography. *Front. Plant Sci.* 11, 601475, 2020, doi: 10.3389/fpls.2020.601475.
2. Schmidt, J., Claussen, J., Wörlein, N., Eggert, A., Fleury, D., Garnett, T. and Gerth, S. Drought and heat stress tolerance screening in wheat using computed tomography. *Plant Methods* 16(1), 15, 2020, doi: 10.1186/s13007-020-00565-w.
3. Metzner, R., Eggert, A., van Dusschoten, D., Pflugfelder, D., Gerth, S., Schurr, U., Uhlmann, N. and Jahnke, S. Direct comparison of MRI and X-ray CT technologies for 3D imaging of root systems in soil: Potential and challenges for root trait quantification. *Plant Methods* 11(1), 17, 2015, doi: 10.1186/s13007-015-0060-z.
4. Gregory, P. J., Hutchison, D. J., Read, D. B., Jenneson, P. M., Gilboy, W. B. and Morton, E. J. Non-invasive imaging of roots with high resolution X-ray micro-tomography. *Plant Soil* 255(1), 351–359, 2003, doi: 10.1023/A:1026179919689.
5. Paya, A. M., Silverberg, J. L., Padgett, J. and Bauerle, T. L. X-ray computed tomography uncovers root-root interactions: Quantifying spatial relationships between interacting root systems in three dimensions. *Front. Plant Sci.* 6, 1–12, 2015, doi: 10.3389/fpls.2015.00274.
6. Pfeifer, J., Kirchgessner, N., Colombi, T. and Walter, A. Rapid phenotyping of crop root systems in undisturbed field soils using X-ray computed tomography. *Plant Methods* 11(1), 41, 2015, doi: 10.1186/s13007-015-0084-4.
7. Heeraman, D. A., Hopmans, J. W. and Clausnitzer, V. Three dimensional imaging of plant roots in situ with X-ray computed tomography. *Plant Soil* 189(2), 167–179, 1997, doi: 10.1023/B:PLSO.0000009694.64377.6f.
8. Kalender, W. A. X-ray computed tomography. *Phys. Med. Biol.* 51(13), R29–R43, 2006, doi: 10.1088/0031-9155/51/13/R03.
9. Villarraga-Gómez, H., Herazo, E. L. and Smith, S. T. X-ray computed tomography: From medical imaging to dimensional metrology. *Precis. Eng.* 60, 544–569, 2019, doi: 10.1016/j.precisioneng.2019.06.007.
10. Hanke, R., Fuchs, T. and Uhlmann, N. X-ray based methods for non-destructive testing and material characterization. *Nucl. Instrum. Methods Phys. Res. A* 591(1), 14–18, 2008, doi: 10.1016/j.nima.2008.03.016.

11. du Plessis, A., le Roux, S. G. and Guelpa, A. Comparison of medical and industrial X-ray computed tomography for non-destructive testing. *Case Stud. Nondestruct. Test. Eval.* 6, 17–25, 2016, doi: 10.1016/j.csndt.2016.07.001.
12. Jolly, M., Prabhakar, A., Sturzu, B., Hollstein, K., Singh, R., Thomas, S., Foote, P. and Shaw, A. Review of non-destructive testing (NDT) techniques and their applicability to thick walled composites. *Procedia CIRP* 38, 129–136, 2015, doi: 10.1016/j.procir.2015.07.043.
13. Carlsson, C. A. and Carlsson, G. A. *Basic Physics of X-Ray Imaging*, 1996. Retrieved from Linköping University Electronic Press website: http://urn.kb.se/resolve?urn=urn:nbn:se:liu:diva-51634
14. Munro, P. R. T. Computed tomography: From photon statistics to modern cone-beam CT, by Thorsten M. Buzug. *Contemp. Phys.* 51(6), 554–555, 2010, doi: 10.1080/00107514.2010.482334.
15. Rogers, E. D., Monaenkova, D., Mijar, M., Nori, A., Goldman, D. I. and Benfey, P. N. X-ray computed tomography reveals the response of root system architecture to soil texture. *Plant Physiol.* 171(3), 2028–2040, 2016, doi: 10.1104/pp.16.00397.
16. Mawodza, T., Burca, G., Casson, S. and Menon, M. Wheat root system architecture and soil moisture distribution in an aggregated soil using neutron computed tomography. *Geoderma* 359, 113988, 2020, doi: 10.1016/j.geoderma.2019.113988.
17. Mairhofer, S., Pridmore, T., Johnson, J., Wells, D. M., Bennett, M. J., Mooney, S. J. and Sturrock, C. J. X-ray computed tomography of crop plant root systems grown in soil. *Curr. Protoc. Plant Biol.* 2(4), 270–286, 2017, doi: 10.1002/cppb.20049.
18. Xu, Z., Valdes, C. and Clarke, J. Existing and potential statistical and computational approaches for the analysis of 3D CT images of plant roots. *Agronomy* 8(5), 2018, doi: 10.3390/agronomy8050071.
19. Mooney, S. J., Pridmore, T. P., Helliwell, J. and Bennett, M. J. Developing X-ray computed tomography to non-invasively image 3-D root systems architecture in soil. *Plant Soil* 352(1–2), 1–22, 2012, doi: 10.1007/s11104-011-1039-9.
20. Tracy, S. R., Roberts, J. A., Black, C. R., McNeill, A., Davidson, R. and Mooney, S. J. The X-factor: Visualizing undisturbed root architecture in soils using X-ray computed tomography. *J. Exp. Bot.* 61(2), 311–313, 2010, doi: 10.1093/jxb/erp386.
21. Van Harsselaar, J. K., Claußen, J., Lübeck, J., Wörlein, N., Uhlmann, N., Sonnewald, U. and Gerth, S. X-ray CT phenotyping reveals bi-phasic growth phases of potato tubers exposed to combined abiotic stress. *Front. Plant Sci.* 12, 613108, 2021, doi: 10.3389/fpls.2021.613108.
22. Pérez-Torres, E., Kirchgessner, N., Pfeifer, J. and Walter, A. Assessing potato tuber diel growth by means of X-ray computed tomography. *Plant Cell Environ.* 38(11), 2318–2326, 2015, doi: 10.1111/pce.12548.
23. Ferreira, S. J., Senning, M., Sonnewald, S., Keßling, P. M., Goldstein, R. and Sonnewald, U. Comparative transcriptome analysis coupled to X-ray CT reveals sucrose supply and growth velocity as major determinants of potato tuber starch biosynthesis. *BMC Genomics* 11(1), 93, 2010, doi: 10.1186/1471-2164-11-93.
24. Ahmed, S., Klassen, T. N., Keyes, S., Daly, M., Jones, D. L., Mavrogordato, M., Sinclair, I. and Roose, T. Imaging the interaction of roots and phosphate fertiliser granules using 4D X-ray tomography. *Plant Soil* 401(1–2), 125–134, 2016, doi: 10.1007/s11104-015-2425-5.

25. Strange, H., Zwiggelaar, R., Sturrock, C., Mooney, S. J. and Doonan, J. H. Automatic estimation of wheat grain morphometry from computed tomography data. *Funct. Plant Biol.* 42(5), 452–459, 2015, doi: 10.1071/FP14068.
26. Mairhofer, S., Zappala, S., Tracy, S. R., Sturrock, C., Bennett, M., Mooney, S. J. and Pridmore, T. RooTrak: Automated recovery of three-dimensional plant root architecture in soil from X-ray microcomputed tomography images using visual tracking. *Plant Physiol.* 158(2), 561–569, 2012, doi: 10.1104/pp.111.186221.
27. Gerth, S., Claußen, J., Eggert, A., Wörlein, N., Waininger, M., Wittenberg, T. and Uhlmann, N. Semiautomated 3D root segmentation and evaluation based on X-ray CT imagery. *Plant Phenom.* 2021, 8747930, 2021, doi: 10.34133/2021/8747930.
28. Flavel, R. J., Guppy, C. N., Rabbi, S. M. R. and Young, I. M. An image processing and analysis tool for identifying and analysing complex plant root systems in 3D soil using non-destructive analysis: Root1. *PLoS ONE* 12(5), e0176433, 2017, doi: 10.1371/journal.pone.0176433.
29. Gao, W., Schlüter, S., Blaser, S. R. G. A., Shen, J. and Vetterlein, D. A shape-based method for automatic and rapid segmentation of roots in soil from X-ray computed tomography images: Rootine. *Plant Soil* 441(1–2), 643–655, 2019, doi: 10.1007/s11104-019-04053-6.
30. Tabb, A., Duncan, K. E. and Topp, C. N. Segmenting root systems in X-ray computed tomography images using level sets. In: *IEEE Winter Conference on Applications of Computer Vision (WACV)*, March 2018, pp. 586–595, doi: 10.1109/WACV.2018.00070.
31. Smith, A. G., Petersen, J., Selvan, R. and Rasmussen, C. R. Segmentation of roots in soil with U-Net. *Plant Methods* 16(1), 13, 2020, doi: 10.1186/s13007-020-0563-0.
32. Douarre, C., Schielein, R., Frindel, C., Gerth, S. and Rousseau, D. *Deep Learning Based Root-Soil Segmentation From X-Ray Tomography. bioRxiv*, p. 071662, 2016, doi: 10.1101/071662.
33. Phalempin, M., Lippold, E., Vetterlein, D. and Schlüter, S. An improved method for the segmentation of roots from X-ray computed tomography 3D images: Rootine v.2. *Plant Methods* 17(1), 39, 2021, doi: 10.1186/s13007-021-00735-4.
34. Mairhofer, S., Johnson, J., Sturrock, C. J., Bennett, M. J., Mooney, S. J. and Pridmore, T. P. Visual tracking for the recovery of multiple interacting plant root systems from X-ray μ CT images, *Mach. Vis. Appl.* 27(5), 721–734, 2016, doi: 10.1007/s00138-015-0733-7.
35. Brown, G. B. The evolution of physics: The growth of ideas from the early concepts to relativity and quanta. By Albert Einstein and Leopold Infeld. (Cambridge: At the University Press, 1938. Pp. x + 319. Price 8s. 6d.), *Philosophy* 14(54), 242–242, 1939, doi: 10.1017/S0031819100011670.
36. De Broglie, L. Waves and quanta. *Nature* 112(2815), 540, 1923, doi: 10.1038/112540a0.
37. Hayashi, H., Kimoto, N., Asahara, T., Asakawa, T., Lee, C. and Katsumata, A. (eds) Generation of X-rays. In: *Photon Counting Detectors for X-Ray Imaging*. Springer International Publishing 2021, pp. 1–29, https://doi.org/10.1007/978-3-030-62680-8.
38. Authier, A. Dynamical theory of X-ray diffraction. In: Shmueli, U. (eds) *International Tables for Crystallography Volume B: Reciprocal Space*. International Tables for Crystallography, vol B. Dordrecht: Springer 2006, pp. 534–551.
39. Knoll, G. *Radiation Detection and Measurement*. Ann Arbor, MI: John Wiley & Sons 1989.

40. Sivia, D. S. *Elementary Scattering Theory*. New York: Oxford University Press 2013.
41. McCormick, J. F. and Platt, R. B. Effects of ionizing radiation on a natural plant community. *Radiat. Bot.* 2(3–4), 161–186, 1962, doi: 10.1016/S0033-7560(62)80100-8.
42. Zappala, S., Helliwell, J. R., Tracy, S. R., Mairhofer, S., Sturrock, C. J., Pridmore, T., Bennett, M. and Mooney, S. J. Effects of X-ray dose on rhizosphere studies using X-ray computed tomography. *PLoS ONE* 8(6), e67250, 2013, doi: 10.1371/journal.pone.0067250.
43. Blaser, S. R. G. A., Schlüter, S. and Vetterlein, D. How much is too much?–Influence of X-ray dose on root growth of faba bean (*Vicia faba*) and barley (*Hordeum vulgare*). *PLoS ONE* 13(3), e0193669, 2018, doi: 10.1371/journal.pone.0193669.
44. Gianoncelli, A., Vaccari, L., Kourousias, G., Cassese, D., Bedolla, D. E., Kenig, S., Storici, P., Lazzarino, M. and Kiskinova, M. Soft X-ray microscopy radiation damage on fixed cells investigated with synchrotron radiation FTIR microscopy. *Sci. Rep.* 5, 10250, 2015, doi: 10.1038/srep10250.
45. Beetz, T. and Jacobsen, C. Soft X-ray radiation-damage studies in PMMA using a cryo-STXM. *J. Synchrotron Radiat.* 10(3), 280–283, 2003, doi: 10.1107/S0909049503003261.
46. Hornsey, S. The effect of X-irradiation on the length of the mitotic cycle in *Vicia faba* roots. *Exp. Cell Res.* 11(2), 340–345, 1956, doi: 10.1016/0014-4827(56)90110-0.
47. Jones, M. W. M., Hare, D. J., James, S. A., De Jonge, M. D. and McColl, G. Radiation dose limits for bioanalytical X-ray fluorescence microscopy. *Anal. Chem.* 89(22), 12168–12175, 2017, doi: 10.1021/acs.analchem.7b02817.
48. Jones, M. W. M., Kopittke, P. M., Casey, L., Reinhardt, J., Blamey, F. P. C. and Van Der Ent, A. Assessing radiation dose limits for X-ray fluorescence microscopy analysis of plant specimens. *Ann. Bot.* 125(4), 599–610, 2020, doi: 10.1093/aob/mcz195.
49. Huang, H., Ullah, F., Zhou, D. X., Yi, M. and Zhao, Y. Mechanisms of ROS regulation of plant development and stress responses. *Front. Plant Sci.* 10, 800, 2019, doi: 10.3389/fpls.2019.00800.
50. Apel, K. and Hirt, H. Reactive oxygen species: Metabolism, oxidative stress, and signal transduction. *Annu. Rev. Plant Biol.* 55, 373-399, 2004, doi: 10.1146/annurev.arplant.55.031903.141701.
51. Knoll, G. F. and McGregor, D. S. Fundamentals of semiconductor detectors for ionizing radiation. *MRS Proc.* 302(1), 3–17, 1993, doi: 10.1557/PROC-302-3.
52. Hall, T., Bowman, H. R. and Hyde, E. K. Semiconductor detectors. *Science* 153(3733), 320–321, 1966, doi: 10.1126/science.153.3733.320.
53. Chotas, H. G., Dobbins, J. T. and Ravin, C. E. Principles of digital radiography with large-area, electronically readable detectors: A review of the basics. *Radiology* 210(3), 595-599, 1999, doi: 10.1148/radiology.210.3.r99mr15595.
54. Siewerdsen, J. H., Antonuk, L. E., El-Mohri, Y., Yorkston, J., Huang, W. and Cunningham, I. A. Signal, noise power spectrum, and detective quantum efficiency of indirect-detection flat-panel imagers for diagnostic radiology. *Med. Phys.* 25(5), 614–628, 1998, doi: 10.1118/1.598243.
55. Bley, T. A., Ghanem, N. A., Foell, D., Uhl, M., Geibel, A., Bode, C. and Langer, M. Computed tomography coronary angiography with 370-millisecond gantry rotation time: evaluation of the best image reconstruction interval. *J. Comput. Assist. Tomogr.* 29(1), 1–5, 2005, doi: 10.1097/01.rct.0000149234.09647.8c.

56. Yester, M. V. and Barnes, G. T. Geometrical limitations of computed tomography (CT) scanner resolution. In: *Appl. Opt. Instrum. Med. VI*, 0127(27), 296–303, 1977, doi: 10.1117/12.955953.
57. Feldkamp, L. A., Davis, L. C. and Kress, J. W. Practical cone-beam algorithm. *J. Opt. Soc. Am. A* 1(6), 612, 1984, doi: 10.1364/JOSAA.1.000612.
58. Gakkai, D. J. T. (Japan) and 電子情報通信学会 (Japan), Denshi jōhō Tsūshin Gakkai *ronbunshi. D*-II, Jōhō, shisutemu II--jōhō shori = The transactions of the Institute of Electronics, Information and Communication Engineers. *D*-II. *Denshi Jōhō Tsūshin Gakkai*, 1989.
59. Bresler, Y. and Skrabacz, C. J. Optimal interpolation in helical scan 3D computerized tomography. In: *ICASSP, IEEE International Conference on Acoustics, Speech and Signal Processing, Proceedings* 3, 1472–1475, 1989, doi: 10.1109/icassp.1989.266718.
60. Tang, X., Krupinski, E. A., Xie, H. and Stillman, A. E. On the data acquisition, image reconstruction, cone beam artifacts, and their suppression in axial MDCT and CBCT - A review. *Med. Phys.* 45(9), e761–e782, 2018, doi: 10.1002/mp.13095.
61. Joseph, P. M. and Spital, R. D. The effects of scatter in X-ray computed tomography. *Med. Phys.* 9(4), 464–472, 1982, doi: 10.1118/1.595111.
62. Dremel, K. and Fuchs, T. Scatter simulation and correction in computed tomography: A reconstruction-integrated approach modelling the forward projection, *NDT E Int.* 86, 132–139, 2017, doi: 10.1016/j.ndteint.2016.12.002.
63. Jin, J. Y., Ren, L., Liu, Q., Kim, J., Wen, N., Guan, H., Movsas, B. and Chetty, I. J. Combining scatter reduction and correction to improve image quality in cone-beam computed tomography (CBCT). *Med. Phys.* 37(11), 5634–5644, 2010, doi: 10.1118/1.3497272.
64. Wang, J., Mao, W. and Solberg, T. Scatter correction for cone-beam computed tomography using moving blocker strips: A preliminary study. *Med. Phys.* 37(11), 5792–5800, 2010, doi: 10.1118/1.3495819.
65. Wiegert, J., Bertram, M., Rose, G. and Aach, T. Model based scatter correction for cone-beam computed tomography. *J. Med. Imaging* 5745, 271, 2005, doi: 10.1117/12.594520.
66. Reims, N., Schoen, T., Boehnel, M., Sukowski, F. and Firsching, M. Strategies for efficient scanning and reconstruction methods on very large objects with high-energy X-ray computed tomography. *Dev. X-Ray Tomogr. IX* 9212(11), 921209, 2014, doi: 10.1117/12.2062002.
67. Three-dimensional helical-scan computed tomography using cone-beam projections. https://inis.iaea.org/search/search.aspx?orig_q=RN:23014546.
68. Hiller, J., Kasperl, S., Schön, T., Schröpfer, S. and Weiss, D. *Comparison of Probing Error in Dimensional Measurement by Means of 3D Computed Tomography With Circular and Helical Sampling*, 2010.
69. Kong, H. and Yu, H. Analytic reconstruction approach for parallel translational computed tomography. *J. Xray Sci. Technol.* 23(2), 213–228, 2015, doi: 10.3233/XST-150482.
70. Schön, T., Fuchs, T., Hanke, R. and Dremel, K. A translation-based data acquisition method for computed tomography: Theoretical analysis and simulation study, *Med. Phys.* 40(8), 081922, 2013, doi: 10.1118/1.4813896.

71. Houston, A. N., Otten, W., Baveye, P. C. and Hapca, S. Adaptive-window indicator kriging: A thresholding method for computed tomography images of porous media. *Comput. Geosci.* 54, 239–248, 2013, doi: 10.1016/j.cageo.2012.11.016.
72. Chisena, R. S., Engstrom, S. M. and Shih, A. J. Automated thresholding method for the computed tomography inspection of the internal composition of parts fabricated using additive manufacturing. *Addit. Manuf.* 33, 101185, 2020, doi: 10.1016/j.addma.2020.101185.
73. Zhang, P., Lu, S., Li, J., Zhang, P., Xie, L., Xue, H. and Zhang, J. Multi-component segmentation of X-ray computed tomography (CT) image using multi-Otsu thresholding algorithm and scanning electron microscopy. *Energy Explor. Exploit.* 35(3), 281–294, 2017, doi: 10.1177/0144598717690090.
74. Inohara, K., Sumita, Y. I., Ohbayashi, N., Ino, S., Kurabayashi, T., Ifukube, T. and Taniguchi, H. Standardization of thresholding for binary conversion of vocal tract modeling in computed tomography. *J. Voice* 24(4), 503–509, 2010, doi: 10.1016/j.jvoice.2008.10.013.
75. Metz, C., Schaap, M., Van Der Giessen, A., Van Walsum, T. and Niessen, W. Semi-automatic coronary artery centerline extraction in computed tomography angiography data. In: *Proceedings 4th IEEE International Symposium on Biomedical Imaging: From Nano to Macro*, Vol. 2007, pp. 856–859, doi: 10.1109/ISBI.2007.356987.
76. Pan, Z. and Lu, J. A Bayes-based region-growing algorithm for medical image segmentation. *Comput. Sci. Eng.* 9(4), 32–38, 2007, doi: 10.1109/MCSE.2007.67.
77. Revol-Muller, C., Peyrin, F., Carrillon, Y. and Odet, C. Automated 3D region growing algorithm based on an assessment function. *Pattern Recognit. Lett.* 23(1–3), 137–150, 2002, doi: 10.1016/S0167-8655(01)00116-7.
78. Frangi, A. F., Niessen, W. J., Vincken, K. L. and Viergever, M. A. Multiscale vessel enhancement filtering. In: *Lect. Notes Comput. Sci. (Including Subseries Lecture Notes in Artificial Intelligence and Lecture Notes in Bioinformatics) 1496*, 130–137, 1998, doi: 10.1007/BFb0056195.
79. Lo, P., Van Ginneken, B. and De Bruijne, M. Vessel tree extraction using locally optimal paths. In: *Proceedings 7th IEEE International Symposium on Biomedical Imaging: From Nano to Macro, ISBI 2010*, 2010, pp. 680–683, doi: 10.1109/ISBI.2010.5490083.
80. Schulz, H., Postma, J. A., Van Dusschoten, D., Scharr, H. and Behnke, S. 3D reconstruction of plant roots from MRI images. In: *VISAPP 2012 Proceedings of the International Conference on Computer Vision Theory and Applications* (vol. 2), 2012, pp. 24–33, doi: 10.5220/0003869800240033.
81. Zhao, Y., Wandel, N., Landl, M., Schnepf, A. and Behnke, S. 3D U-net for segmentation of plant root MRI images in super-resolution. In: *ESANN Proceedings, 28th Europa Symposium Artifact Neural Networks, Comput. Intell. Mach. Learn*, 2020, pp. 463–468. http://arxiv.org/abs/2002.09317.
82. Soltaninejad, M., Sturrock, C. J., Griffiths, M., Pridmore, T. P. and Pound, M. P. Three dimensional root CT segmentation using multi-resolution encoder-decoder networks. *IEEE Trans. Image Process.* 29, 6667–6679, 2020, doi: 10.1109/TIP.2020.2992893.
83. Ronneberger, O., Fischer, P. and Brox, T. U-net: Convolutional networks for biomedical image segmentation. In: *Lect. Notes Comput. Sci. (Including Subseries Lecture Notes*

in Artificial Intelligence and Lecture Notes in Bioinformatics) 9351, 234–241, 2015, doi: 10.1007/978-3-319-24574-4_28.
84. Macdonald, S. The progress of the early threshing machine. *Agric. Hist. Rev.* 23(1), 63–77. http://www.jstor.org/stable/40273647.
85. Tuff, D. W. and Telford, H. S. Wheat fracturing as affecting infestation by *Cryptolestes ferrugineus*. *J. Econ. Entomol.* 57(4), 513–516, 1964, doi: 10.1093/jee/57.4.513.
86. Basavaraja, H., Mahajanashetti, S. B. and Udagatti, N. C. *Economic Analysis of Post-Harvest Losses in Food Grains in India: A Case Study of Karnataka*. doi: 10.22004/AG.ECON.47429, 2007.
87. Tzarfati, R., Saranga, Y., Barak, V., Gopher, A., Korol, A. B. and Abbo, S. Threshing efficiency as an incentive for rapid domestication of emmer wheat. *Ann. Bot.* 112(5), 829–837, 2013, doi: 10.1093/aob/mct148.
88. Denčić, S., Kastori, R., Kobiljski, B. and Duggan, B. Evaluation of grain yield and its components in wheat cultivars and landraces under near optimal and drought conditions. *Euphytica* 113(1), 43–52, 2000, doi: 10.1023/A:1003997700865.
89. Nicol, J. M. and Ortiz-Monasterio, I. Effects of the root-lesion nematode, *Pratylenchus thornei*, on wheat yields in Mexico. *Nematology* 6(4), 485–493, 2004, doi: 10.1163/1568541042665223.

Part 3

Carrier/delivery systems

Chapter 6

Field robots for plant phenotyping

Rick van de Zedde, Wageningen University and Research, The Netherlands; and Lili Yao, Visiting Researcher – Wageningen University and Research, The Netherlands

1 Introduction

Plant phenotypes are collections of information concerning the composition, responses, performance and growth of plants based on the monitoring of their physical appearance or characteristics (e.g. height, leaf inclination, colour, yield) (Furbank and Tester, 2011; Gjuvsland et al., 2013; Freimer and Sabatti, 2003; Houle et al., 2010). Phenotyping is an essential manifestation of the synergy between genotyping and the environment. Whereas genotyping concerns the final appearance of plant growth and development, the results of phenotyping can differ according to several environmental factors. The main factors that interfere with the process and results of plant growth are transcriptome, proteome and metabolome, as well as the induction of mutations in specific genes within the plant genome (Pieruschka and Poorter, 2012). As a result, plants of the same genotype can exhibit varying phenotypic features in different environments. The analysis of phenotypes can provide valuable information for the functional analysis of genomes and molecular breeding, thus helping to enhance yields with plants that are resistant to drought, salinization, disease and other consequences of climate change (Tester and Langridge, 2010; Yang et al., 2013; Pan, 2015; Araus et al., 2018).

The traditional estimation method for phenotyping relies primarily on human observation, manual measurement or destructive sampling tests (Pask et al., 2012; Maiti et al., 2004; Qiu et al., 2018). In addition to being time-consuming and labour-intensive, such processes produce test results with low

http://dx.doi.org/10.19103/AS.2022.0102.08

accuracy. As phenotyping in plant breeding has advanced, a large number of scientific research institutions and organizations have been investing in research on the acquisition of phenotypic information. Sensor technologies have been advancing at the same time. The field now has access to a variety of sensors, including ground spectrometers and colour, thermal, multispectral and hyperspectral imaging devices, in addition to light detection and ranging (LiDAR), amongst other possibilities (Zhang et al., 2019; Chéné et al., 2012; Andújar et al., 2015; Stovall et al., 2017; Sakamoto et al., 2012; Falco et al., 2015; Stoll et al., 2008). Research investments in these devices have brought about a qualitative leap in the development of phenotyping monitoring while also driving the construction of sensor-carrying platforms - within the context of this chapter, 'field robots'. Phenotyping field robots are autonomous, self-driving devices that are equipped with a range of sensors that are able to reach each spot on the field and carry higher payloads than are possible with unmanned aerial vehicles (UAVs). Testing methods that rely on manual, handheld sensors obstruct the ability to meet the demand for high-throughput information in phenomics research. New directions within this field include multi-sensor deployment, continuity, synchronicity and high throughput. Developments have also included the introduction of field robots, the Internet of Things (IoT), large vehicle platforms and remote-sensing platforms (e.g. UAVs and satellites) (Shakoor et al., 2017; Hunt et al., 2005; Thomason et al., 2011; Henry et al., 2002; Mirik et al., 2011). A conceptual depiction of the integrated system created by the Netherlands Plant Eco-phenotyping Centre (NPEC) is presented in Fig. 1.

Methods of phenotype monitoring have been enriched by the application of the IoT. This technology uses a variety of sensors with different functions to process phenotypic data and to achieve mutual communication between sensors based on the technology of wireless sensor networks (WSN). Finally, plant-phenotyping parameters are obtained by applying the inversion model between sensor data and plant phenotypes (Fan et al., 2021). The common application of open-field IoT consists largely of deploying fixed monitoring nodes. For example, Liu et al. propose intelligent sensing nodes for farmlands and the associated field-deployment method, which is capable of efficiently sensing various parameters, including the leaf area index (LAI) of rice and wheat crops (Liu et al., 2015). In addition, Bauer et al. (2019) designed low-cost sensors based on the technology of WSN and interconnected them with *in situ* agricultural sensor networks in order to estimate crop LAI (Bauer et al., 2019). Furthermore, Avotins et al. (2020) used IoT technology to design a camera for recording the single-image normalized difference vegetation index (SI-NDVI). The device can be used to continuously obtain NDVI data on greenhouse tomatoes and diagnose tomato quality according to diurnal variations in NDVI (Kviesis et al., 2020). Another new hotspot of agricultural research involving IoT technology has to do with the combination of fixed nodes with UAVs and

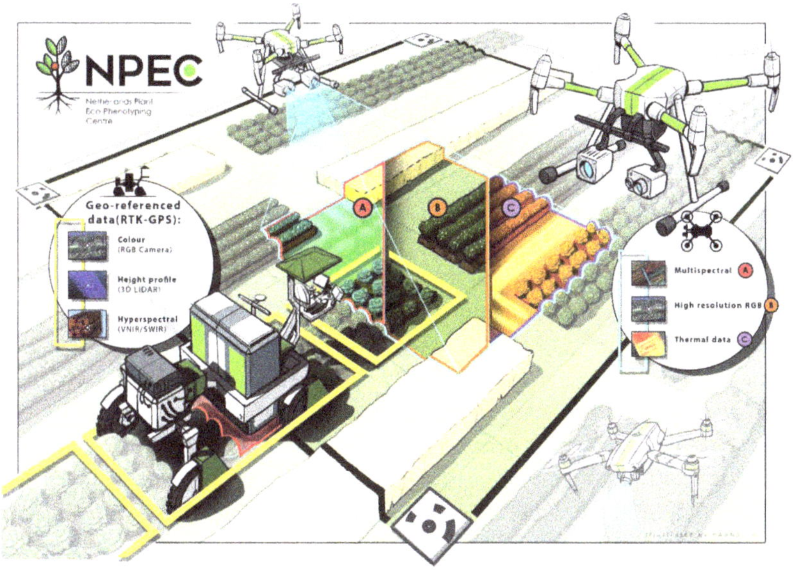

Figure 1 Artist's impression of a field robot and several drones collecting phenotypic information in a field. Courtesy: Netherlands Plant Eco- phenotyping Centre (NPEC) (2021).

unmanned ground vehicles (UGVs). In this regard, Gao et al. (2020) designed a cooperation scheme between UAV and fixed IoT sensor nodes in the field, in which the IoT nodes are used to obtain meteorological parameters. The UAV serves as an image-acquisition node in order to obtain hyperspectral crop images for the diagnosis of pests (Gao et al., 2020). With regard to the methods of data collection, Uddin et al. (2018) propose a UAV dynamic method that can efficiently and intelligently select fixed IoT field nodes for collecting data through mobile UAV nodes. The method could potentially reduce the energy consumption of fixed IoT nodes and extend their service life (Uddin et al., 2018). As demonstrated by these studies, the application of agricultural IoT technology can facilitate the acquisition of phenotypic information on plants. In the open-field environment, however, sensors and mechanical devices are exposed to harsh weather conditions, thereby decreasing the lifetime of any IoT instrument and requiring frequent maintenance. Although fixed nodes can be used to cover a large area at the same time, their use in full-range testing is severely limited, as fixed nodes cannot move by themselves.

A research team at the American University of Rome in Italy started investigating variable-rate fertilizer applicators and their supporting technologies based on optical-sensing information in 1993. This research led to the 2002 development of the real-time variable fertilizer applicator known as 'GreenSeeker', which uses spectral reflectance to monitor crop-growth information and calculate nutritional status in crops (Lukina et al., 2001). This system was followed by the vehicle-mounted crop growth sensor Yara N-Sensor, which was first tested in 1998 and mounted on agricultural tractors in 2002 (Link et al., 2003). Since then, digital sensors have increasingly been used for open-field vehicle platforms. This study traces the early development of the tractor as a vehicle platform in the process of acquiring high-throughput data. In the actual testing process, however, it is easy to cause irreversible damage to fields and field crops due to the large weight, low chassis and lack of a field-control robot model. It is thus not conducive to sustainable development.

Aerial remote-sensing platforms allow the rapid, real-time acquisition of large-area crop-phenotypic information. One common type, the UAV, offers the advantages of high efficiency, flexibility, convenient operation and strong terrain applicability. Combined with its lower cost and greater accessibility, the recent increase in the popularity of UAVs has led to the widespread use of this technology for plant phenotyping (Sagan et al., 2019). A large body of research results has been accumulated with regard to the use of UAVs to acquire information on the phenotypic characteristics of plants (Xie and Yang, 2020). The characteristic phenotypic parameters that can be obtained using UAVs include LAI, nutritional parameter status, biotic and abiotic threats (Su et al., 2019; Buchaillot et al., 2019; Chivasa et al., 2020). As the demand for agricultural monitoring has increased, a variety of satellite sensors have been introduced to the market. New monitoring methods have emerged (e.g. fluorescence, multispectral and hyperspectral measurements). The classification of plants and the measurement of their phenotypic information can be achieved effectively through the analysis of medium-resolution or high-resolution satellite images. Commonly used satellite series include the Sentinel-1/2 satellites, the WorldView satellite and the Landsat satellite. More specific applications of satellite images are referenced in an overview published by Zhang et al. (2020). Although high-altitude satellite images cover the most comprehensive range, they are subject to three important disadvantages: some of these images are expensive; some have only low resolution and, in many cases, the image quality is diminished by clouds, thereby requiring extensive optical correction.

Lightweight field robots offer a number of advantages over large tractors or UAVs: flexible operation, minimal impact on soil and crops, durability and the ability to accommodate larger payloads (Michaels et al., 2015a,b). It is more difficult for field robots to obtain phenotypic data on large-scale plant groups simultaneously than it is for drones and satellites. They are nevertheless capable of providing all-around, multi-angle and automatic monitoring of the mounted sensors, which makes them highly suitable for integrating state-of-the-art AI technologies (Grigorescu et al., 2020; Zhou et al., 2018). Based on the advantages of field robots in phenotyping monitoring, more mature platforms have been introduced on the market, and several have performed well in field experiments. It is important to note, however, that different countries and regions have different plant species and varieties. Most existing field robots focus only on specific plant species with similar growth characteristics, and none of the existing phenotypic robots is capable of the comprehensive general monitoring of multiple plant species.

2 Specific challenges associated with field robots

Based on their operating environments, phenotyping field robots can be roughly classified suitable for use indoors (controlled environment) or in the field (uncontrolled environment) (Kirchgessner et al., 2016; Pieruschka and Schurr, 2019). Research on indoor robots pre-dates that on field robots. In 2003, Belgian CropDesign pioneered the development of a high-throughput phenotyping monitoring platform known as TraitMill (Reuzeau et al., 2006), which uses a plant-to-sensor working mode and which can automatically obtain phenotypic information on potted plants in a greenhouse. Since that time, scientific research institutions in Germany, Australia, the Czech Republic and other countries have also successfully launched numerous platforms for acquiring phenotypic information indoors (Deery et al., 2014; Yang et al., 2013). Several of these platforms are displayed in Fig. 2.

Most of the indoor high-throughput phenotyping platforms and automated robot facilities mentioned above focus on monitoring changes in above-ground plant characteristics. To observe below-ground characteristics, researchers have explored innovative monitoring solutions for soil and plant roots involving rhizotrons and Ecotrons. Rhizotrons are amongst the earliest non-destructive below-ground sensing platforms. Due to limitations in sensor technology, early rhizotron facilities relied on human inspection through cellars or underground corridors with transparent glass windows. By entering such facilities, researchers

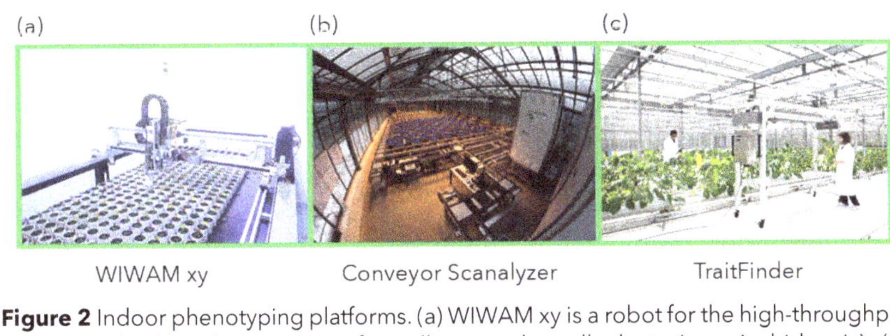

Figure 2 Indoor phenotyping platforms. (a) WIWAM xy is a robot for the high-throughput and reproducible phenotyping of seedlings and small plants (e.g. Arabidopsis). (b) Conveyor Scanalyzer is an automated high-throughput platform for indoor environments (e.g. glasshouses, growth rooms, climate chambers). (c) TraitFinder is a powerful tool that automates plant phenotyping and screening applications.

could directly observe root phenotyping and soil condition (Klepper and Kaspar, 1994). Despite their advantages, massive underground observation facilities are expensive, are troublesome to maintain and have major destructive effects on the soil structure. The rapid advancement of sensor technology and the continuous improvement of automated robotics technology have allowed the continuing development of small, highly accurate observation devices. The research and development of Ecotrons have at least partially facilitated the study of the relationship between above-ground phenotyping, below-ground phenotyping and environmental characteristics. The device's more intelligent control is able to monitor soil status, plant-root phenotyping and the above-ground phenotypic characteristics of plants, while simulating different natural environmental conditions in order to control changes in ecological environmental factors to induce different stress conditions in plants (Granjou and Walker, 2016; Roy et al., 2021). Indoor platforms for plant phenotyping monitoring offer high accuracy and reliable repeatability, and they are not easily disturbed by the external environment. These approaches to monitoring below-ground aspects have nevertheless been criticized as deviating from the actual growth environment, such that the phenotypic information extracted with these systems is negatively affected by the artificial growth conditions to which plants are exposed (Jiang et al., 2018).

Phenotyping open-field robots are subject to even more challenges than is the case for other robots. First, farmland environments are challenging and complex, often with relatively harsh terrain characterized by stones and uneven surfaces. In addition, the spacing of plants is often too narrow for robots to move between rows. Because of these conditions, robots must be able to respond quickly to differences in terrain and ensure highly accurate auto-navigation. Second, the growth of plants is directly affected by the environment and growth period, and plant morphology changes accordingly. Moreover, due to the diversity of sensing principles applied in phenotyping, the sensors used for phenotyping require a variety of monitoring heights and angles. The acquisition of phenotypic information by field robots should be able to adapt to sensor-monitoring requirements and changes in plant morphology. Finally, field-planting areas are larger than those of greenhouses. Mobile robots are often expected to work in teams or to work in conjunction with other tools (e.g. IoT devices and drones), in order to achieve high-efficiency monitoring. Fieldwork also places higher demands with regard to safety and the communication between various devices.

3 Currently available field robots for phenotyping

The development of field-phenotyping robots begins with the transformation of agricultural machinery. Phenotypic information is obtained by mounting sensors on tractors or sprinklers, as is the case with applications of the GreenSeeker airborne sensor developed by the NT Company (Wójtowicz et al., 2016) and the Yara N-sensor from the Agri Con GmbH Company (Link et al., 2003). Although the load capacity of these agricultural machinery platforms is sufficient to meet the needs of high-throughput testing, they are highly likely to cause damage to plants and soil during operation, and it is difficult to perform all-around and multi-view testing for different sensors. With the development of automation technology and robot technology, agricultural machinery platforms have been gradually replaced by more field-specific robots.

3.1 The design of the current field robots

After a period of development, the design of field robots has become more diverse. Depending on chassis and operating modes, they can be roughly divided into the categories of track, crawler, foot and wheeled robots, as shown in Fig. 3.

Track robots often use the gantry structure constructed with two supports and a connecting beam in the middle. The sensors are fixed on the beam with the vision field perpendicular to the crop canopy. Rails or tracks on both sides of the field plots enable the gantry to move repeatedly along a fixed trajectory. Sensors can move along the beam to obtain phenotypic data. Another type of track robot is the rope camera, which often uses hanging ropes and electric hinges to drive sensors along the plant canopy (Kirchgessner et al., 2016; Bai et al., 2019). The PhenoField gantry robot (co-designed by Robepec and Meca3d and located in France) can test the abiotic stress of wheat (Beauchêne

Robot types	Robots	Advantages	Disadvantages
Track robots		Strong load capacity High repeatability Extremely convenient control	Requires long-term maintenance Low moving flexibility High investment Relatively small monitoring area Complex structure
Crawler robots		Strong ability to cross obstacles Powerful terrain adaptability High moving stability	Slow speed and low efficiency Loud/noisy Majordamage to crops
Legged robots		Good manoeuvrability Small ground-touch area Low damage to crops	Complex control method Poor moving stability Slow speed and low efficiency
Wheeled robots		Easy to control High mobility High working efficiency Conducive to sensor deployment	Poor ability to cross obstacles Easy to slip on muddy ground

Figure 3 Field robots with different structural designs (Beauchêne et al., 2019; Virlet et al., 2016; Susko et al., 2018; Basu et al., 2020; Young et al., 2019; Zhang et al., 2013; Dorhout, 2021; Shafiekhani et al., 2017; De Solan et al., 2015).

et al., 2019). In 2016, the University of Arizona's Maricopa Agricultural Center built the world's largest gantry robot, the Field Scanalyzer, which has 200-metre steel rails and can acquire phenotypic information on energy sorghum across an areas of more than 1.5 acres (Virlet et al., 2016). The structure of the FieldScan robot (manufactured by Phenospex) is similar to that of other track robots (Vadez et al., 2015). In general, the width and height of track robots depends on the size of the plots on which they are used. One important advantage of these devices is that they are able to carry large numbers of sensors for continuously monitoring a specific area, with reliable repeatability. Given that the platform is exposed to the natural environment for a long time, however, it requires frequent maintenance. Moreover, the track structure does not provide sufficient flexibility, and the monitoring area is relatively small.

Crawler robots offer the advantages of strong obstacle-crossing ability and stable operation. They are often used in agriculture for purposes of automatic land preparation prior to sowing and mechanical harvesting (as reported by Zhang et al., 2013; Noguchi and Barawid Jr., 2011; and Takai et al., 2013). Given the shortcomings of the crawler structure's large contact area, low ground clearance and inability to perform cross-row operations, it causes considerable damage to the plants and soil during movement. Several applications have been developed for the acquisition of phenotypic information. For example, Young et al. (Young et al., 2019; Baharav et al., 2017) developed a small-tracked field robot–the Transportation Energy Resource from Renewable Agriculture Mobile Energy-crop Phenotyping Platform (TERRA-MEPP)–for the purpose of monitoring energy sorghum. This platform is able to run between planting rows, with the sensor perspective extended through sensor support in order to obtain phenotypic information on sorghum. Another small crawler robot was developed by Ibex Automation Ltd. for the purpose of weed identification and precision weeding (Basu et al., 2020). This robot nevertheless causes damage to plants, due to the low ground clearance of its chassis. This problem is difficult to resolve.

Research on legged robots has only recently started, and the related technologies are still under active investigation with regard to sophisticated control, unstable movement and other matters. Dorhout (2021) designed a legged robot–the 'metal crab'–that can explore field conditions using a camera and other sensors mounted and assisted by agricultural equipment in such operations as sowing, fertilizing, planting, dispersing herbicides or picking fruit. Although a report has been published on agricultural auxiliary equipment (Toyama and Yamamoto, 2009), we have not found any successful application for obtaining phenotypic information. In addition to the field of agriculture, research on legged robots has been conducted in industrial settings. The Legged Squad Support System (LS3) introduced by Boston Dynamics in recent years has been used to perform various functions, including moving motion control, obstacle crossing and navigation. For example, the Spot robot was used in a sheep-herding demonstration in 2020.[1] Due to the combination of more intelligent AI systems, the robot can also cope with various sudden situations, and it performs outstanding teamwork (Raibert et al., 2008). This technology will soon have excellent applications within the context of agriculture and plant phenotyping.

Wheeled robots are the most common amongst the field robots that are currently used for plant phenotyping, due to their simple control and stable operation. Research on wheeled robots is also attracting considerable attention from scientific research institutions and enterprises around the world. Related research has been conducted by Shafiekhani et al. (Shafiekhani et al., 2017), Weiss and Biber (Weiss and Biber, 2011) and Reina (Reina et al., 2018) on wheeled robots with low ground clearance, including the Husky A200 and the Volksbot RT3. These robots have exhibited good performance in the phenotyping of plants with large row spacing (e.g. sorghum, apple, peach). Wheeled robots with ground clearance higher than 0.5 m–such as the Ladybird developed by the University of Sydney (Underwood et al., 2017), the Bonirob co-developed by Bosch (Pretto et al., 2020), and the Robotti by Agrointelli–can operate effectively with dwarf-row plants (e.g. soybeans, lettuce, potatoes). The Proximal sensing cart, a wheeled robot developed by White et al. (Bai et al., 2016; Yuan et al., 2018; White and Conley, 2013), has a high ground clearance, but movement must be assisted by human thrust. This simple robot is actually a modified bicycle, which requires human power to move over fields. It has been used successfully in wheat fields. The structure of the PhenoMobile® Lite designed by the Australian Plant Phenomics Facility is similar to that of the proximal sensing cart. It is also equipped with sensors–including a global positioning system (GPS) and an inertial measurement unit (IMU)–which allow it to perform automatic phenotyping monitoring (De Solan et al., 2015). Wheeled robots with higher ground clearance (>1.2m), including the Phenomobile developed by the Australian Plant Phenomics Facility (Deery et al., 2014) and the Gecko developed by the University of Queensland, can be used for the phenotyping of taller plants. These robots are equipped with height-adjustable supports, with a broad adjustment range that is able to accommodate the different test heights of sensors.

4 Sensors and technologies for phenotyping field robots

Physiological and biochemical changes in plants cause differences in their phenotypic traits, which subsequently result in changes in their absorption, reflection and refraction of certain spectral features (Feng et al., 2008a;

1 https://blog.rocos.io/rocos-partners-with-boston-dynamics

Thenkabail et al., 2000). Such changes in data on spectral features offer an efficient, non-destructive and accurate means of retrieving phenotypic information. Many scientific research institutions have carried out related research on spectrum-sensing technology and published results with reference significance. For example, information on red-light reflection provides good inversion results for the LAI and chlorophyll (Thenkabail et al., 2000); the near-infrared band (740–1100 nm) provides an effective means of monitoring biomass (Shibayama and Akiyama, 1989, 1991); and the combination of the near-infrared band and the red-edge band can be used to detect nitrogen change (Yao et al., 2019; Zhu et al., 2008).

Based on the aforementioned principles, various non-imaging and imaging sensors have been applied to research on the use of field robots in phenotyping monitoring, as shown in Table 1. These instruments can acquire information that cannot be seen by the naked eye. The most prevalent non-imaging sensors are portable ground spectrometers. Examples include ASD Fieldspec (Feng et al., 2008b; Wang et al., 2012), GreenSeeker (Osborne, 2007), RapidScan (Miller et al., 2018) and SPAD (Uddling et al., 2007). Applications of imaging sensors on plant-phenotyping platforms include visible-light cameras, multispectral cameras, hyperspectral imagers, time-of-flight cameras, fluorescence cameras, thermal-infrared cameras and ground-based LiDAR.

Portable ground spectrometers adopt photoelectric sensing technology and use photosensitive components to convert photoelectric information. For example, electrical signals can be processed in order to calculate the characteristic band spectral vegetation index and then to invert specific phenotyping information. Commonly used vegetation indices include the normalized difference vegetation index (NDVI) (Tucker, 1979), the ratio vegetation index (RVI) (Ni et al., 2017), the soil-adjusted vegetation index (SAVI) (Huete, 1988) and the enhanced vegetation index (EVI) (Liu and Huete, 1995). The retrievable phenotypic parameters include nitrogen nutrition, biomass, chlorophyll content and water content.

Compared to non-imaging instruments, imaging sensors can obtain information on plant image, spectrum or three-dimensional structure with a high volume of visualization information. With regard to the acquisition of phenotypic information, in addition to relying on the vegetation index for inversion, the image sensors can intuitively extract leaf colour, flowering period, plant height, crown coverage, crop water status and other parameters according to the differences between the imaging-sensor perception method and the data-processing methods. The extracted phenotypic information can be combined with AI technologies to perform the rapid identification of plant diseases, biological stress and abiotic stress (Valente et al., 2019). For example, Sakurai et al. (2018) propose a fully convolutional network (FCN) (Long et al., 2015) for semantic plant-image segmentation. Shi et al. (2019) combine deep

Table 1 Phenotyping sensors for field robots

Phenotyping sensors (sensing principle)	Sensor type	Data type	Extractable parameters	Phenotyping information
Image sensors (photoelectric imaging sensing)	RGB cameras	RGB images	Vegetation index based on RGB band; classification and target extraction based on RGB features	Plant height, flowering number, spike number, nitrogen status, biomass, diseases and stress with RGB colour characteristics
	Multi-spectral cameras	Spectral images of several specific bands	Vegetation index based on specific bands; targeted diseases, biological and abiotic stress features	Emergence rate, flowering rate, nitrogen status, diseases and stress with specific spectral-band features
	Hyperspectral cameras	Spectral images of continuous bands	Spectral data on each pixel of the image; multiple vegetation indices; multiple diseases, biological and abiotic stress features	Chlorophyll content, water content, protein content, nitrogen status, biomass, multiple diseases and stress
	Time of flight cameras	Depth images and RGB images	The distance from each pixel to the camera	Plant height, stem thickness, leaf inclination
	Thermal cameras	Thermal images	The temperature value of each pixel in the image	Crop water status, stomatal conductance, canopy and leaf temperature
	Fluorescence cameras	Fluorescence images	Visible fluorescence images in different spectral bands	Photosynthetic productivity, nitrogen status, biomass, diseases and stress with fluorescence features
	Lidar	3D point clouds images	High-resolution distance information from every point to the device	Plant height, stem thickness, flowering number, spike number, leaf inclination
Non-image sensors (photoelectric conversion sensing)	Ground hyperspectral spectrometers	Spectral data of continuous specific bands	Multiple vegetation indices based on continuous spectral bands	Chlorophyll content, water content, protein content, nitrogen status, biomass
	Ground multi-spectral spectrometers	Spectral data of several specific bands	Vegetation indices based on specific bands	Nitrogen status, biomass

Source: Li et al. (2014); Perez-Sanz et al. (2017); Fiorani and Schurr (2013).

learning with multi-view images to generate a 3D point-cloud representation of a plant. Mohanty et al. (2016) have trained a deep convolutional neural network (CNN) to identify 14 crop species and 26 diseases. Fuentes et al. (2017) present a deep-learning-based approach for detecting diseases and pests in tomato plants using images captured in place by a camera with various resolutions. Li et al. (2014) report on the acquisition of current images with AI technologies, equipment and applications, and Perez-Sanz et al. (2017) summarize various image-processing methods. New imaging sensors–including X-ray computed tomography (CT), positron emission tomography (PET) and magnetic resonance imaging (MRI) (Jahnke et al., 2009)–that have been applied in other fields are gradually being used in plant phenotyping monitoring. Due to their investment costs and physical limitations, however, these instruments have thus far been used in only a limited number of institutes and only in indoor environments.

In addition to phenotypic information sensors, phenotyping robots need a variety of auxiliary functional sensors to support automatic field operations. They include GPS/real-time kinematic (RTK), IMU, LiDAR, cameras, Hall-effect sensors, ultrasonic sensors and ranging infrared sensors. For row crops, phenotyping robots require high-precision automatic navigation and path planning in order to improve their operational efficiency and avoid damage to the plants during movement. Unlike navigation in an open environment, automatic navigation in a plant-field environment requires the coordination of multiple sensors, including those listed above. While GPS/RTK can provide robots with error position information at the centimetre level, IMU can obtain a robot's three-axis attitude angle and acceleration. In addition, LiDAR and cameras can sense environmental details in real time and then cooperate with GPS/RTK and IMU to achieve simultaneous localization and mapping (SLAM). Derived from autonomous driving technology, SLAM has recently been used in the automatic navigation of agricultural robots. When a field robot starts operating from any location in the field, SLAM can use information from the sensors listed above to observe and locate its position, posture and trajectory during its movement. A SLAM system thus builds an incremental map based on its position to achieve simultaneous positioning, field map construction and automatic navigation (Durrant-Whyte and Bailey, 2006; Montemerlo et al., 2002). Two components that are essential to SLAM are LiDAR and vision technology. Despite its high accuracy, LiDAR (the core method applied during early research on SLAM) is also subject to a number of shortcomings, including high price, large volume and lack of intuitive vision information. As the popularity of consumer-level RGB-D cameras has increased, vision-based SLAM has also been applied successfully in several research fields (e.g. autonomous driving, AR and intelligent robots), with the camera being used as the primary sensor and the video stream serving as the input for achieving simultaneous positioning and construction (Jiang et al., 2017; Miro et al., 2006).

The advantages of this system include the fact that it is inexpensive and able to capture images with high resolution in order to classify objects. Its data depth is nevertheless not as good as that of LiDAR.

It is also possible to use GPS/RTK to add geo-referencing to phenotypic information on plants in order to generate more conducive results in analysis, data classification and mapping (Xiu et al., 2010; Tamaki et al., 2013). The pitch angle and roll angle provided by the IMU can also be used to calibrate the attitude of phenotyping monitoring sensors. Hall-effect sensors can perform the robot's speed test and improve control over the robot, thereby providing precise speed change and steering.

5 Robotic arms for fruit phenotyping and harvesting

Fruit harvesting is the final link between plant cultivation and the most relevant process for yielding economic benefits. Traditional fruit harvesting depends on the manual assessment of the maturity and quality of fruit. Differences in experience with such assessment lead to wide variations in picking results across operators, with a direct impact on fruit sales. The recognition of fruit phenotypic features is an important part of plant phenotyping research. Fruit picking based on the recognition of such features can be combined with eye-hand coordination to achieve the full automation of fruit harvesting, thereby improving the accuracy and work efficiency of fruit picking (Hashimoto, 2003). Results of studies on the method of acquiring fruit phenotypic information on plants are consistent with those on the method of recognition for other parts of the plant. Spectral reflectance monitoring, thermal infrared imaging monitoring and the extraction of colour features and textures have also demonstrated good recognition effects for fruit phenotypic features (Bulanon et al., 2002; Arivazhagan et al., 2010, Jiménez et al., 1999). In addition to the basic phenotyping monitoring method, AI techniques can be used to detect the position and shape of the fruit under occlusion. For example, Blok et al. (2021) use Mask R-CNN to detect the position and shape of broccoli heads for autonomous harvesting. Zhao et al. (2016) provide a systematic review of control technology based on visual sensors, as used in harvesting robots, including recognition technology for fruits and control technology for manipulators.

With regard to fruit picking, the earliest mechanized picking method relies primarily on mechanical or pneumatic methods to shake plants. However, because this non-selective picking method does not involve the identification of fruit phenotyping characteristics, it is more harmful to plants and has low work efficiency. As precision and automated fruit-picking methods have become more popular, most fruit-picking robotic arm mechanisms have come to be designed as two parts: a moving mechanism and an end effector. The mobile mechanism is used to adjust the position of the end effector, and the end effector completes the final picking task. In current industrial production, the design of manipulators has become more mature, with structures including the cylindrical-coordinate type, the rectangular-coordinate type, the spherical-coordinate type and the joint type (Almurib et al., 2011). When used for open-field robots, manipulators are subject to a number of complex problems, including the random growth of fruit position and occlusion of fruit. For this reason, most of the manipulators that are used for fruit picking are based on the joint type of design. Motion angle is a key factor in the design of a joint manipulator. Although joints with a low degree of freedom (DOF) lack flexibility, an increase in degrees greatly increases the difficulty of control. Kondo et al. (1996a,b) designed a 7-DOF articulated manipulator that can be used for tomato harvesting. The end effector of the manipulator is designed with Type B suction cups. After locating the fruit by binocular vision, the suction cup is used to grasp and harvest the tomato. Van Henten et al. (2003) designed a 7-DOF cucumber-harvesting manipulator. The end effector is designed with scissors that can cut the stem of cucumber while clamping it with a suction cup/gripper. De-An et al. (2011) designed a 5-DOF apple-harvesting manipulator with a spoon-shaped end effector, which can firmly fix apples and cut stems with an electric cutting device. Many studies have been conducted on various fruits to be harvested, including strawberries (Xiong et al., 2020), grapes (Badeka et al., 2020) and pumpkins (Roshanianfard and Noguchi, 2018). The design of the mobile mechanisms of these manipulators is relatively similar. Most differences have to do with size and degrees of freedom. The design of the end effector is based primarily on the comprehensive consideration of the morphological structure and texture of the target fruit.

6 Conclusion and future trends

Current studies have indicated that the use of field robots for phenotyping can improve the efficiency, quality, stability and accuracy of acquiring phenotypic information (Tsukor et al., 2012; Zhou et al., 2017). Existing phenotyping robots nevertheless continue to be designed for monitoring specific plant-growth periods or plant species. The development of different platforms for different plants is costly and time-consuming. The construction of a universal phenotyping robot platform is therefore needed. The Thorvald II robot provides several ideas for developments in this direction (Grimstad and From, 2017). The platform for this robot is designed with various drive chassis and sensor-support modules. It can adjust the height, width and drive mode of the chassis to adapt to differing field-cultivation environments and plant objects. When developing the design of phenotyping robots, however, greater attention should be paid to intelligent operations. With the support of image sensors and AI technologies, robots should be capable of recognizing the row distance and height of plants. They should then have the ability to automatically calculate and adjust the mechanical structure in real time, thereby achieving adaptive object monitoring and autonomous operations.

With regard to the acquisition of phenotypic information, field robots and UAVs offer different advantages. For example, UAVs are more suitable for obtaining canopy-level phenotypic information, and they can efficiently monitor dynamic changes in plant populations (Holman et al., 2016). In contrast, given their limited test height, field robots are more suitable for obtaining phenotypic information on individual plants. Given that both group and individual characteristics are clearly of great significance to breeding, studies that combine the use of UAVs and field robots are highly valuable. Future developments in this direction could focus on drones that are capable of obtaining macro-scale information on the canopy traits of plant populations (e.g. plant emergence rate, ground coverage, LAI and nitrogen content (Burud et al., 2017; Schirrmann et al., 2016; Liu et al., 2017; Pretto et al., 2020)). The macro-scale data are used to command the phenotyping robots to go to specific areas to obtain individual phenotypic information on individual representative plants (e.g. plant height, protein content, stomatal conductance and nitrogen assimilation rate). Given the large monitoring coverage and high efficiency of drones, phenotyping robots can be operated in teams, although this obviously increases the requirements for collaboration between different software systems. The combination of information at different scales can be used to establish the relationship between phenotypic information on individual plants and group characteristics. This could improve the prediction of dynamic changes in the growth processes of individual plants and groups of plants, in addition to providing effective guidance for breeding and cultivation, thus ultimately generating maximum-yield production benefits.

When implementing phenotyping sensors, it is important to note that some sensors are still difficult to use in open-field environments, due to volume, costs, scalability and measurement requirements (e.g. PET, CT and MRI). These problems are likely to be resolved as the costs and size of sensor chips decrease and further improvements in sensor integration are achieved. In addition, the phenotyping monitoring sensors that are currently carried by field robots are used primarily to monitor the phenotypic characteristics of the above-ground parts of plants. The current technology for below-ground plant phenotyping and the evaluation of the soil environment relies on static measurement. There are no specialized methods for combining below-ground phenotyping monitoring sensors and soil-environment information sensors with moving robots. The related sensing technology and intelligent equipment have yet to be developed. Data interaction between different sensors is also important with regard to the integration of sensor data. Suppose that it is possible to assemble results from 2D data tests, 3D data tests, spectral tests and temperature tests quickly and then systematically display them through the human-computer interaction interface. This could usher in a new era for research on plant phenotyping monitoring. At present, however, major differences exist in the control software, test methods and data-aggregation methods of different sensor platforms. There are also differences in the analytic methods, programming languages and software platforms that are used for the analysis of sensing results. To a certain extent, these differences pose obstacles to data interaction between sensor devices.

With the advent of 5G (or even 6G), the combination of AI technology and phenotyping robots based on 5G communication offers excellent potential. Limited by the speed of data transmission, most current phenotyping robots rely on a self-mounted host-computer platform to realize the calculation and storage of high-throughput data. This method increases the load of the robot, and it does not help researchers to track plant phenotyping information in real time. Researchers often need to extract data stored by the host computer for post-processing after the robots have obtained all of the results. As a result, the results lag behind the real-time growth state of the plant. Wageningen University & Research has conducted preliminary research on the use of 5G technology and achieved real-time control of weeds in sugar-beet fields (Visser, 2020). The high bandwidth and low latency characteristics of 5G technology make it possible to receive newly obtained phenotyping information on cloud platforms and remote computers. After the obtained data has been calculated through the aforementioned platforms, the results can be fed back to the field robots in real time, cooperating with the actuator to achieve totally unmanned farm management.

7 Where to look for further information

The following article provides a good overview of the subject:

- Atefi, A., Ge, Y., Pitla, S. and Schnable, J. 2021. Robotic technologies for high-throughput plant phenotyping: Contemporary reviews and future perspectives. *Frontiers in Plant Science* 12:611940.

Key research in this area can be found in the following organizations:

- Netherlands Plant Eco-phenotyping Centre (NPEC) at Wageningen University & Research (www.npec.nl).
- Australian Centre for Field Robotics – University of Sydney (https://www.sydney.edu.au/engineering/our-research/robotics-and-intelligent-systems/australian-centre-for-field-robotics.html).
- UMT CAPTE, a scientific research unit based in Avignon (France), a collaboration between INRAE, ARVALIS and HIPHEN (https://umt-capte.fr/).

- Jülich Plant Phenotyping Centre (JPPC) – https://www.fz-juelich.de/ibg/ibg-2/EN/_organisation/JPPC/JPPC_node.html.
- EMPHASIS, an organization that enables researchers to use facilities, resources and services for plant phenotyping across Europe (https://emphasis.plant-phenotyping.eu/).

8 References

A science-fiction-like creature trundling through sorghum crops in Queensland is a crucial new tool for unlocking the genetic secrets behind the development of higher-performing, climate-resilient crop varieties [Online]. Available at: https://grdc.com.au/resources-and-publications/groundcover/ground-cover-issue-122-may-jun-2016/all-seeing-tractor-takes-crop-science-to-the-future.

Almurib, H. A., Al-Qrimli, H. F. and Kumar, N. 2011. A review of application industrial robotic design. 2011 Ninth International Conference on ICT and Knowledge Engineering, 2012. IEEE, 105–112.

Andújar, D., Fernández-Quintanilla, C. and Dorado, J. 2015. Matching the best viewing angle in depth cameras for biomass estimation based on poplar seedling geometry. *Sensors* 15(6), 12999–13011.

Araus, J. L., Kefauver, S. C., Zaman-Allah, M., Olsen, M. S. and Cairns, J. E. 2018. Translating high-throughput phenotyping into genetic gain. *Trends in Plant Science* 23(5), 451–466.

Arivazhagan, S., Shebiah, R. N., Nidhyanandhan, S. S. and Ganesan, L. 2010. Fruit recognition using color and texture features. *Journal of Emerging Trends in Computing Information Sciences* 1, 90–94.

Avotins, A., Kviesis, K., Bicans, J., Alsina, I. and Dubova, J. 2020. Experimental analysis of IoT based camera SI-NDVI values for tomato plant health monitoring application. *Agronomy Research* 18(S2), 1138–1146. DOI: https://doi.org/10.15159/AR.20.087.

Badeka, E., Vrochidou, E., Papakostas, G. A., Pachidis, T. and Kaburlasos, V. G. 2020. Harvest crate detection for grapes harvesting robot based on YOLOv3 model. 2020 Fourth International Conference on Intelligent Computing in Data Sciences (ICDS). IEEE, 1-5.

Baharav, T., Bariya, M. and Zakhor, A. 2017. Computing height and width of in situ sorghum plants using 2.5 d infrared images. Proceedings of the IS&T International Symposium on Electronic Imaging.

Bai, G., Ge, Y., Hussain, W., Baenziger, P. S. and Graef, G. 2016. A multi-sensor system for high throughput field phenotyping in soybean and wheat breeding. *Computers and Electronics in Agriculture* 128, 181–192.

Bai, G., Ge, Y., Scoby, D., Leavitt, B., Stoerger, V., Kirchgessner, N., Irmak, S., Graef, G., Schnable, J. and Awada, T. 2019. NU-Spidercam: a large-scale, cable-driven, integrated sensing and robotic system for advanced phenotyping, remote sensing, and agronomic research. *Computers and Electronics in Agriculture* 160, 71–81.

Basu, S., Omotubora, A., Beeson, M. and Fox, C. 2020. Legal framework for small autonomous agricultural robots. *AI and Society* 35(1), 113–134.

Bauer, J., Jarmer, T., Schittenhelm, S., Siegmann, B. and Aschenbruck, N. 2019. Processing and filtering of leaf area index time series assessed by in-situ wireless sensor networks. *Computers and Electronics in Agriculture* 165, 104867.

Beauchêne, K., Leroy, F., Fournier, A., Huet, C., Bonnefoy, M., Lorgeou, J., De Solan, B., Piquemal, B., Thomas, S. and Cohan, J. P. 2019. Management and characterization of abiotic stress via PhénoField®, a high-throughput field phenotyping platform. *Frontiers in Plant Science* 10, 904.

Blok, P. M., Van Evert, F. K., Tielen, A. P., Van Henten, E. J. and Kootstra, G. 2021. The effect of data augmentation and network simplification on the image-based detection of broccoli heads with Mask R-C N N. Journal of Field Robotics 38, 85-104.

Buchaillot, M. L., Gracia-Romero, A., Vergara-Diaz, O., Zaman-Allah, M. A., Tarekegne, A., Cairns, J. E., Prasanna, B. M., Araus, J. L. and Kefauver, S. C. 2019. Evaluating maize genotype performance under low nitrogen conditions using RGB UAV phenotyping techniques. *Sensors* 19(8), 1815.

Bulanon, D. M., Kataoka, T., Ota, Y. and Hiroma, T. 2002. A color model for recognition of apples by a robotic harvesting system. *Journal of the Japanese Society of Agricultural Machinery* 64, 123-133.

Burud, I., Lange, G., Lillemo, M., Bleken, E., Grimstad, L. and Johan From, P. 2017. Exploring Robots and UAVs as Phenotyping Tools in Plant Breeding. *IFAC-PapersOnLine* 50(1), 11479-11484.

Chéné, Y., Rousseau, D., Lucidarme, P., Bertheloot, J., Caffier, V., Morel, P., Belin, É. and Chapeau-Blondeau, F. 2012. On the use of depth camera for 3D phenotyping of entire plants. *Computers and Electronics in Agriculture* 82, 122-127.

Chivasa, W., Mutanga, O. and Biradar, C. 2020. UAV-based multispectral phenotyping for disease resistance to accelerate crop improvement under changing climate conditions. *Remote Sensing* 12(15), 2445.

Conveyor-based Scanalyzer Solutions for glasshouses, growth rooms, climate chambers, cultivation halls, or indoor farms [Online]. Available at: https://www.lemnatec.com/products/conveyor-scanalyzer/.

De-An, Z., Jidong, L., Wei, J., Ying, Z. and Yu, C. 2011. Design and control of an apple harvesting robot. *Biosystems Engineering* 110(2), 112-122.

De Solan, B., Baret, F., Thomas, S., Beauchêne, K., Comar, A., Fournier, A., Weiss, M. and Gouache, D. 2015. PHENOMOBILE-V1: a fully automated high throughput phenotyping system. EPPN Plant Phenotyping Symposium. Barcelona, Spain.

Deery, D., Jimenez-Berni, J., Jones, H., Sirault, X. and Furbank, R. 2014. Proximal remote sensing buggies and potential applications for field-based phenotyping. *Agronomy* 4(3), 349-379.

Dorhout, D. 2021. Automated agricultural helpers ripe for robots [Online]. Available at: https://www.cropscience.bayer.com/innovations/data-science/a/ripe-robots.

Durrant-Whyte, H. and Bailey, T. 2006. Simultaneous localization and mapping: part I. *IEEE Robotics and Automation Magazine* 13(2), 99-110.

Falco, N., Benediktsson, J. A. and Bruzzone, L. 2015. Spectral and spatial classification of hyperspectral images based on ICA and reduced morphological attribute profiles. *IEEE Transactions on Geoscience and Remote Sensing* 53(11), 6223-6240.

Fan, J., Zhang, Y., Wen, W., Gu, S., Lu, X. and Guo, X. 2021. The future of internet of things in agriculture: plant high-throughput phenotypic platform. *Journal of Cleaner Production* 280, 123651.

Feng, W., Yao, X., Tian, Y., Cao, W. and Zhu, Y. 2008a. Monitoring leaf pigment status with hyperspectral remote sensing in wheat. *Australian Journal of Agricultural Research* 59(8), 748-760.

Feng, W., Yao, X., Zhu, Y., Tian, Y. C. and Cao, W. X. 2008b. Monitoring leaf nitrogen status with hyperspectral reflectance in wheat. *European Journal of Agronomy* 28(3), 394–404.

Fiorani, F. and Schurr, U. 2013. Future scenarios for plant phenotyping. *Annual Review of Plant Biology* 64, 267-291.

Freimer, N. and Sabatti, C. 2003. The human phenome project. *Nature Genetics* 34(1), 15–21.

Fuentes, A., Yoon, S., Kim, S. C. and Park, D. S. 2017. A robust deep-learning-based detector for real-time tomato plant diseases and pests recognition. *Sensors* 17(9).

Furbank, R. T. and Tester, M. 2011. Phenomics-technologies to relieve the phenotyping bottleneck. *Trends in Plant Science* 16(12), 635–644.

Gao, D., Sun, Q., Hu, B. and Zhang, S. 2020. A framework for agricultural pest and disease monitoring based on internet-of-things and unmanned aerial vehicles. *Sensors* 20(5), 1487.

Gjuvsland, A. B., Vik, J. O., Beard, D. A., Hunter, P. J. and Omholt, S. W. 2013. Bridging the genotype-phenotype gap: what does it take? *The Journal of Physiology* 591(8), 2055–2066.

Granjou, C. and Walker, J. 2016. Promises that matter: reconfiguring ecology in the ecotrons. *Science and Technology Studies* 29(3), 49–67.

Grigorescu, S., Trasnea, B., Cocias, T. and Macesanu, G. 2020. A survey of deep learning techniques for autonomous driving. *Journal of Field Robotics* 37(3), 362–386.

Grimstad, L. and From, P. 2017. The Thorvald II agricultural robotic system. *Robotics* 6(4), 24.

Hashimoto, K. 2003. A review on vision-based control of robot manipulators. *Advanced Robotics* 17(10), 969–991.

Henry, J.-B., Malet, J.-P., Maquaire, O. and Grussenmeyer, P. 2002. The use of small-format and low-altitude aerial photos for the realization of high-resolution DEMs in mountainous areas: application to the Super-Sauze earthflow (Alpes-de-Haute-Provence, France). *Earth Surface Processes Landforms: The Journal of the British Geomorphological Research Group* 27(12), 1339–1350.

Holman, F. H., Riche, A. B., Michalski, A., Castle, M., Wooster, M. J. and Hawkesford, M. J. 2016. High throughput field phenotyping of wheat plant height and growth rate in field plot trials using UAV based remote sensing. *Remote Sensing* 8(12), 1031.

Houle, D., Govindaraju, D. R. and Omholt, S. 2010. Phenomics: the next challenge. *Nature Reviews. Genetics* 11(12), 855–866.

Huete, A. R. 1988. A soil-adjusted vegetation index (SAVI). *Remote Sensing of Environment* 25(3), 295–309.

Hunt, E. R., Cavigelli, M., Daughtry, C. S. T., McMurtrey, J. E. and Walthall, C. L. 2005. Evaluation of digital photography from model aircraft for remote sensing of crop biomass and nitrogen status. *Precision Agriculture* 6(4), 359–378.

Jahnke, S., Menzel, M. I., Van Dusschoten, D., Roeb, G. W., Bühler, J., Minwuyelet, S., Blümler, P., Temperton, V. M., Hombach, T., Streun, M., Beer, S., Khodaverdi, M., Ziemons, K., Coenen, H. H. and Schurr, U. 2009. Combined MRI-PET dissects dynamic changes in plant structures and functions. *Plant Journal: For Cell and Molecular Biology* 59(4), 634–644.

Jiang, Y., Li, C., Paterson, A. H., Sun, S., Xu, R. and Robertson, J. 2017. Quantitative analysis of cotton canopy size in field conditions using a consumer-grade RGB-D camera. *Frontiers in Plant Science* 8, 2233.

Jiang, Y., Li, C., Robertson, J. S., Sun, S., Xu, R. and Paterson, A. H. 2018. GPhenoVision: a ground mobile system with multi-modal imaging for field-based high throughput phenotyping of cotton. *Scientific Reports* 8(1), 1213.

Jiménez, A. R., Jain, A. K., Ceres, R. and Pons, J. L. 1999. Automatic fruit recognition: a survey and new results using range/attenuation images. *Pattern Recognition* 32(10), 1719-1736.

Kirchgessner, N., Liebisch, F., Yu, K., Pfeifer, J., Friedli, M., Hund, A. and Walter, A. 2016. The ETH field phenotyping platform FIP: a cable-suspended multi-sensor system. *Functional Plant Biology* 44(1), 154-168.

Klepper, B. and Kaspar, T. C. 1994. Rhizotrons: their development and use in agricultural research. *Agronomy Journal* 86(5), 745-753.

Kondo, N., Monta, M. and Fujiura, T. 1996a. Fruit harvesting robots in Japan. *Advances in Space Research* 18(1-2), 181-184.

Kondo, N., Nishitsuji, Y., Ling, P. P. and Ting, K. C. 1996b. Visual feedback guided robotic cherry tomato harvesting. *Transactions of the ASAE* 39(6), 2331-2338.

Kviesis, K., Dubova, L., Bicans, J., Avotins, A. and Alsina, I. 2020. Experimental analysis of IoT based camera SI-NDVI values for tomato plant health monitoring application. *Agronomy Research* 18, 1138-1146.

Li, L., Zhang, Q. and Huang, D. 2014. A review of imaging techniques for plant phenotyping. *Sensors* 14(11), 20078-20111.

Link, A., Panitzki, M. and Reusch, S. 2003. Hydro N-Sensor: tractor-mounted remote sensing for variable nitrogen fertilization. Proceedings of the 6th International Conference on Precision Agriculture and Other Precision Resources Management, Minneapolis, MN, USA, 14-17 July, 2002. American Society of Agronomy, 1012-1017.

Liu, H., Zhu, H. and Wang, P. 2017. Quantitative modelling for leaf nitrogen content of winter wheat using UAV-based hyperspectral data. *International Journal of Remote Sensing* 38(8-10), 2117-2134.

Liu, H. Q. and Huete, A. 1995. A feedback based modification of the NDVI to minimize canopy background and atmospheric noise. *IEEE Transactions on Geoscience and Remote Sensing* 33(2), 457-465.

Liu, N., Cao, W., Zhu, Y., Zhang, J., Pang, F. and Ni, J. 2015. The node deployment of intelligent sensor networks based on the spatial difference of farmland soil. *Sensors* 15(11), 28314-28339.

Long, J., Shelhamer, E. and Darrell, T. 2015. Fully convolutional networks for semantic segmentation. Proceedings of the IEEE Conference on Computer Vision and Pattern Recognition (CVPR), 3431-3440.

Lukina, E. V., Freeman, K. W., Wynn, K. J., Thomason, W. E., Mullen, R. W., Stone, M. L., Solie, J. B., Klatt, A. R., Johnson, G. V., Elliott, R. L. and Raun, W. R. 2001. Nitrogen fertilization optimization algorithm based on in-season estimates of yield and plant nitrogen uptake. *Journal of Plant Nutrition* 24(6), 885-898.

Maiti, D., Das, D. K., Karak, T. and Banerjee, M. 2004. Management of nitrogen through the use of leaf color chart (LCC) and soil plant analysis development (SPAD) or chlorophyll meter in rice under irrigated ecosystem. *TheScientificWorldJournal* 4, 838-846.

Michaels, A., Haug, S. and Albert, A. 2015a. Vision-based high-speed manipulation for robotic ultra-precise weed control. 2015 IEEE/RSJ International Conference on Intelligent Robots and Systems (IROS). IEEE, 5498-5505.

Michaels, A., Haug, S., Albert, A. and Grzonka, S. 2015b. Vision-based manipulation for weed control with an autonomous field robot. *VDI-Berichte*, 289–294.

Miller, J. J., Schepers, J. S., Shapiro, C. A., Arneson, N. J., Eskridge, K. M., Oliveira, M. C. and Giesler, L. J. 2018. Characterizing soybean vigor and productivity using multiple crop canopy sensor readings. *Field Crops Research* 216, 22–31.

Mirik, M., Jones, D. C., Price, J. A., Workneh, F., Ansley, R. J. and Rush, C. M. 2011. Satellite remote sensing of wheat infected by wheat streak mosaic virus. *Plant Disease* 95(1), 4–12.

Miro, J. V., Zhou, W. and Dissanayake, G. 2006. Towards vision based navigation in large indoor environments. IEEE/RSJ International Conference on Intelligent Robots and Systems.

Mohanty, S. P., Hughes, D. P. and Salathé, M. 2016. Using deep learning for image-based plant disease detection. *Frontiers in Plant Science* 7, 1419.

Montemerlo, M., Thrun, S., Koller, D. and Wegbreit, B. 2002. FastSLAM: a factored solution to the simultaneous localization and mapping problem. Innovative Applications of Artificial Intelligence Conference.

Ni, J., Yao, L., Zhang, J., Cao, W., Zhu, Y., Tai, X. 2017. Development of an Unmanned Aerial Vehicle-Borne Crop-Growth Monitoring System. *Sensors* 17, 502. https://doi.org/10.3390/s17030502.

Noguchi, N. and Barawid Jr., O. C. 2011. Robot farming system using multiple robot tractors in Japan agriculture. *IFAC Proceedings Volumes* 44(1), 633–637.

Osborne, S. L. 2007. Utilization of existing technology to evaluate spring wheat growth and nitrogen nutrition in South Dakota. *Communications in Soil Science and Plant Analysis* 38(7–8), 949–958.

Pan, Y. 2015. Analysis of concepts and categories of plant phenome and phenomics. *Acta Agronomica Sinica* 41(2), 175–186.

Pask, A., Pietragalla, J., Mullan, D. and Reynolds, M. 2012. Physiological breeding *II*: a field guide to wheat phenotyping. *Cimmyt*.

Perez-Sanz, F., Navarro, P. J. and Egea-Cortines, M. 2017. Plant phenomics: an overview of image acquisition technologies and image data analysis algorithms. *GigaScience* 6(11), 1–18.

Pieruschka, R. and Poorter, H. 2012. Phenotyping plants: genes, phenes and machines. *Functional Plant Biology* 39(11), 813–820.

Pieruschka, R. and Schurr, U. 2019. Plant phenotyping: past, present, and future. *Plant Phenomics* 2019, 7507131.

Pretto, A., Aravecchia, S., Burgard, W., Chebrolu, N., Dornhege, C., Falck, T., Fleckenstein, F. V., Fontenla, A., Imperoli, M., Khanna, R., Liebisch, F., Lottes, P., Milioto, A., Nardi, D., Nardi, S., Pfeifer, J., Popovic, M., Potena, C., Pradalier, C., Rothacker-Feder, E., Sa, I., Schaefer, A., Siegwart, R., Stachniss, C., Walter, A., Winterhalter, W., Wu, X. and Nieto, J. 2020. Building an aerial-ground robotics system for precision farming: an adaptable solution. *IEEE Robotics and Automation Magazine* 28(3), 29–49. DOI: 10.1109/MRA.2020.3012492.

Qiu, R., Wei, S., Zhang, M., Li, H., Sun, H., Liu, G. and Li, M. 2018. Sensors for measuring plant phenotyping: a review. *International Journal of Agricultural and Biological Engineering* 11(2), 1–17.

Raibert, M., Blankespoor, K., Nelson, G. and Playter, R. 2008. Bigdog, the rough-terrain quadruped robot. *IFAC Proceedings Volumes* 41(2), 10822–10825.

Reina, G., Galati, R. and Milella, A. 2018. All-terrain estimation for mobile robots in precision agriculture. 2018 IEEE International Conference on Industrial Technology (ICIT). IEEE, 63–68.

Reuzeau, C., Frankard, V., Hatzfeld, Y., Sanz, A., Van Camp, W., Lejeune, P., De Wilde, C., Lievens, K., De Wolf, J., Vranken, E., Peerbolte, R. and Broekaert, W. 2006. Traitmill™: a functional genomics platform for the phenotypic analysis of cereals. *Plant Genetic Resources* 4(1), 20–24.

Roshanianfard, A. and Noguchi, N. 2018. Characterization of pumpkin for a harvesting robot. *IFAC-PapersOnLine* 51(17), 23–30.

Roy, J., Rineau, F., De Boeck, H. J., Nijs, I., Pütz, T., Abiven, S., Arnone, J. A., Barton, C. V. M., Beenaerts, N., Brüggemann, N., Dainese, M., Domisch, T., Eisenhauer, N., Garré, S., Gebler, A., Ghirardo, A., Jasoni, R. L., Kowalchuk, G., Landais, D., Larsen, S. H., Leemans, V., Le Galliard, J. F., Longdoz, B., Massol, F., Mikkelsen, T. N., Niedrist, G., Piel, C., Ravel, O., Sauze, J., Schmidt, A., Schnitzler, J. P., Teixeira, L. H., Tjoelker, M. G., Weisser, W. W., Winkler, B. and Milcu, A. 2021. Ecotrons: powerful and versatile ecosystem analysers for ecology, agronomy and environmental science. *Global Change Biology* 27(7), 1387–1407.

Sagan, V., Maimaitijiang, M., Sidike, P., Eblimit, K., Peterson, K. T., Hartling, S., Esposito, F., Khanal, K., Newcomb, M., Pauli, D., Ward, R., Fritschi, F., Shakoor, N. and Mockler, T. 2019. UAV-based high resolution thermal imaging for vegetation monitoring, and plant phenotyping using ICI 8640 P, FLIR Vue Pro R 640, and thermoMap Cameras. *Remote Sensing* 11(3), 330.

Sakamoto, T., Gitelson, A. A., Nguy-Robertson, A. L., Arkebauer, T. J., Wardlow, B. D., Suyker, A. E., Verma, S. B. and Shibayama, M. 2012. An alternative method using digital cameras for continuous monitoring of crop status. *Agricultural and Forest Meteorology* 154–155, 113–126.

Sakurai, S., Uchiyama, H., Shimada, A., Arita, D. and Taniguchi, R.-I. 2018. Two-step transfer learning for semantic plant segmentation. *ICPRAM* 2018, 332–339.

Schirrmann, M., Giebel, A., Gleiniger, F., Pflanz, M., Lentschke, J. and Dammer, K.-H. 2016. Monitoring agronomic parameters of winter wheat crops with low-cost UAV imagery. *Remote Sensing* 8(9), 706.

Shafiekhani, A., Kadam, S., Fritschi, F. B. and Desouza, G. N. 2017. Vinobot and vinoculer: two robotic platforms for high-throughput field phenotyping. *Sensors* 17(1), 214.

Shakoor, N., Lee, S. and Mockler, T. C. 2017. High throughput phenotyping to accelerate crop breeding and monitoring of diseases in the field. *Current Opinion in Plant Biology* 38, 184–192.

Shelhamer, E., Long, J. and Darrell, T. 2017. Fully Convolutional Networks for Semantic Segmentation. *IEEE Transactions on Pattern Analysis and Machine Intelligence* 39(4), 640–651.

Shi, W., van de Zedde, R., Jiang, H. and Koostra, G. 2019. Plant-part segmentation using deep learning and multi-view vision. *Biosystems Engineering* 187, 81–95. DOI: https://doi.org/10.1016/j.biosystemseng.2019.08.014.

Shibayama, M. and Akiyama, T. 1989. Seasonal visible, near-infrared and mid-infrared spectra of rice canopies in relation to LAI and above-ground dry phytomass. *Remote Sensing of Environment* 27(2), 119–127.

Shibayama, M. and Akiyama, T. 1991. Estimating grain yield of maturing rice canopies using high spectral resolution reflectance measurements. *Remote Sensing of Environment* 36(1), 45–53.

Stoll, M., Schultz, H. R., Baecker, G. and Berkelmann-Loehnertz, B. 2008. Early pathogen detection under different water status and the assessment of spray application in vineyards through the use of thermal imagery. *Precision Agriculture* 9(6), 407–417.

Stovall, A. E. L., Vorster, A. G., Anderson, R. S., Evangelista, P. H. and Shugart, H. H. 2017. Non-destructive aboveground biomass estimation of coniferous trees using terrestrial LiDAR. *Remote Sensing of Environment* 200, 31–42.

Su, J., Liu, C., Hu, X., Guo, L. and Chen, W.-H. 2019. Spatio-temporal monitoring of wheat yellow rust using UAV multispectral imagery. *Computers and Electronics in Agriculture* 167. DOI: https://doi.org/10.1016/j.compag.2019.105035.

Susko, A. Q., Gilbertson, F., Heuschele, D. J., Smith, K. and Marchetto, P. 2018. An automatable, field camera track system for phenotyping crop lodging and crop movement. *HardwareX* 4, e00029.

Takai, R., Yang, L. and Noguchi, N. 2013. Development of crawler-type robot tractor based on GNSS and IMU. *IFAC Proceedings Volumes* 46(4), 95–98.

Tamaki, K., Nagasaka, Y., Nishiwaki, K., Saito, M., Kikuchi, Y. and Motobayashi, K. 2013. A robot system for paddy field farming in Japan. *IFAC Proceedings Volumes* 46(18), 143–147.

Tester, M. and Langridge, P. 2010. Breeding technologies to increase crop production in a changing world. *Science* 327(5967), 818–822.

Thenkabail, P. S., Smith, R. B. and De Pauw, E. 2000. Hyperspectral vegetation indices and their relationships with agricultural crop characteristics. *Remote Sensing of Environment* 71(2), 158–182.

Thomason, W. E., Phillips, S. B., Davis, P. H., Warren, J. G., Alley, M. M. and Reiter, M. S. 2011. Variable nitrogen rate determination from plant spectral reflectance in soft red winter wheat. *Precision Agriculture* 12(5), 666–681.

Toyama, S. and Yamamoto, G. 2009. Development of Wearable-Agri-Robot~ mechanism for agricultural work~. 2009 IEEE/RSJ International Conference on Intelligent Robots and Systems. IEEE, 5801–5806.

TraitFinder easy plant phenotyping [Online]. Available at: https://phenospex.com/products/plant-phenotyping/traitfinder-for-lab-and-greenhouse-phenotyping-automation/.

Tsukor, V., Klose, R. and Ruckelshausen, A. 2012. Multi-sensor simulation method for outdoor plant phenotyping based on autonomous field robots. Proceedings of the CIGR-AgEng conference, Valencia, 9–12.

Tucker, C. J. 1979. Red and photographic infrared linear combinations for monitoring vegetation. United States patent application. *Remote Sensing of Environment* 8(2), 127–150.

Uddin, M. A., Mansour, A., Jeune, D. L., Ayaz, M. and Aggoune, E.-H. M. 2018. UAV-assisted dynamic clustering of wireless sensor networks for crop health monitoring. *Sensors* 18(2), 555.

Uddling, J., Gelang-Alfredsson, J., Piikki, K. and Pleijel, H. 2007. Evaluating the relationship between leaf chlorophyll concentration and SPAD-502 chlorophyll meter readings. *Photosynthesis Research* 91(1), 37–46.

Underwood, J., Wendel, A., Schofield, B., Mcmurray, L. and Kimber, R. 2017. Efficient in-field plant phenomics for row-crops with an autonomous ground vehicle. *Journal of Field Robotics* 34(6), 1061–1083.

Vadez, V., Kholová, J., Hummel, G., Zhokhavets, U., Gupta, S. K. and Hash, C. T. 2015. LeasyScan: a novel concept combining 3D imaging and lysimetry for high-throughput phenotyping of traits controlling plant water budget. *Journal of Experimental Botany* 66(18), 5581-5593.

Valente, J., Doldersum, M., Roers, C. and Kooistra, L. 2019. Detecting Rumex obtusifolius weed plants in grasslands from uav rgb imagery using deep learning. ISPRS Annals of Photogrammetry, Remote Sensing and Spatial Information Sciences, 4.

Van Henten, E. J., Van Tuijl, B. A. J., Hemming, J., Kornet, J. G., Bontsema, J. and Van Os, E. A. 2003. Field test of an autonomous cucumber picking robot. *Biosystems Engineering* 86(3), 305-313.

Virlet, N., Sabermanesh, K., Sadeghi-Tehran, P. and Hawkesford, M. J. 2016. Field Scanalyzer: an automated robotic field phenotyping platform for detailed crop monitoring. *Functional Plant Biology* 44(1), 143-153.

Visser, C. C. D. 2020. 5G network successfully tested in agricultural practice [Online]. Available at: https://www.wur.nl/en/Research-Results/Research-Institutes/plant-research/show-wpr/5G-network-successfully-tested-in-agricultural-practice.htm [Accessed 05/09/2019].

Wang, W., Yao, X., Yao, X., Tian, Y., Liu, X., Ni, J., Cao, W. and Zhu, Y. 2012. Estimating leaf nitrogen concentration with three-band vegetation indices in rice and wheat. *Field Crops Research* 129, 90-98.

Weiss, U. and Biber, P. 2011. Plant detection and mapping for agricultural robots using a 3D LIDAR sensor. *Robotics and Autonomous Systems* 59(5), 265-273.

White, J. W. and Conley, M. M. 2013. A flexible, low-cost cart for proximal sensing. *Crop Science* 53(4), 1646-1649.

WIWAM xy [Online]. Available at: https://www.wiwam.be/phenotyping-systems/wiwam-xy/.

Wójtowicz, M., Wójtowicz, A. and Piekarczyk, J. 2016. Application of remote sensing methods in agriculture. *Communications in Biometry and Crop Science* 11, 31-50.

World's largest robotic field scanner now in Place [Online]. Available at: https://uanews.arizona.edu/story/world-s-largest-robotic-field-scanner-now-place.

Xie, C. and Yang, C. J. C. 2020. A review on plant high-throughput phenotyping traits using UAV-based sensors. *Computers and Electronics in Agriculture* 178, 105731.

Xiong, Y., Ge, Y., Grimstad, L. and From, P. J. 2020. An autonomous strawberry-harvesting robot: design, development, integration, and field evaluation. *Journal of Field Robotics* 37(2), 202-224.

Xiu, W., Zhijun, M., Wei, M. and Jianbin, J. 2010. The application of GPS for weed investigation in winter wheat field. World Automation Congress.

Yang, W., Duan, L., Chen, G., Xiong, L. and Liu, Q. 2013. Plant phenomics and high-throughput phenotyping: accelerating rice functional genomics using multidisciplinary technologies. *Current Opinion in Plant Biology* 16(2), 180-187.

Yao, L., Wang, Q., Yang, J., Zhang, Y., Zhu, Y., Cao, W. and Ni, J. 2019. UAV-borne dual-band sensor method for monitoring physiological crop status. *Sensors* 19(4), 816.

Young, S. N., Kayacan, E. and Peschel, J. M. 2019. Design and field evaluation of a ground robot for high-throughput phenotyping of energy sorghum. *Precision Agriculture* 20(4), 697-722.

Yuan, W., Li, J., Bhatta, M., Shi, Y., Baenziger, P. S. and Ge, Y. 2018. Wheat height estimation using LiDAR in comparison to ultrasonic sensor and UAS. *Sensors* 18(11), 3731.

Zhang, C., Marzougui, A. and Sankaran, S. 2020. High-resolution satellite imagery applications in crop phenotyping: an overview. *Computers and Electronics in Agriculture* 175, 105584.

Zhang, J., Liu, X., Liang, Y., Cao, Q., Tian, Y., Zhu, Y., Cao, W. and Liu, X. 2019. Using a portable active sensor to monitor growth parameters and predict grain yield of winter wheat. *Sensors* 19(5), 1108.

Zhang, Z., Noguchi, N., Ishii, K., Yang, L. and Zhang, C. 2013. Development of a robot combine harvester for wheat and paddy harvesting. *IFAC Proceedings Volumes* 46(4), 45-48.

Zhao, Y., Gong, L., Huang, Y. and Liu, C. 2016. A review of key techniques of vision-based control for harvesting robot. *Computers and Electronics in Agriculture* 127, 311-323.

Zhou, J., Reynolds, D., Websdale, D., Le Cornu, T., Gonzalez-Navarro, O., Lister, C., Orford, S., Laycock, S., Finlayson, G. and Stitt, T. 2017. CropQuant: an automated and scalable field phenotyping platform for crop monitoring and trait measurements to facilitate breeding and digital agriculture. *bioRxiv*, 161547.

Zhou, J., Tardieu, F., Pridmore, T., Doonan, J., Reynolds, D., Hall, N., Griffiths, S., Cheng, T., Zhu, Y. and Jiang, D. 2018. Plant phenomics: history, present status and challenges. *Journal of Nanjing Agricultural University* 41.

Zhu, Y., Yao, X., Tian, Y., Liu, X. and Cao, W. 2008. Analysis of common canopy vegetation indices for indicating leaf nitrogen accumulations in wheat and rice. *International Journal of Applied Earth Observation and Geoinformation* 10(1), 1-10.

Chapter 7

Advances in high-throughput crop phenotyping using unmanned aerial vehicles (UAVs)

Helge Aasen, Institute of Agricultural Sciences, ETH Zurich and Remote Sensing Team, Division of Agroecology and Environment, Agroscope, Switzerland; and Lukas Roth, Institute of Agricultural Sciences, ETH Zurich, Switzerland

1 Introduction

Over 10 years ago, unmanned aerial vehicles (UAVs) were seen as bringing about a new era in agriculture (Zarco-Tejada, 2008). Seen from today's perspective, the biggest impact in the application of UAVs in agriculture can be seen in high-throughput field phenotyping. Field phenotyping refers to a quantitative description of a plant's phenotype – that is, its anatomical, ontogenetical, physiological and biochemical properties – in its natural environment (Walter et al., 2015). In the context of breeding, where hundreds or even thousands of different genotypes need to be screened for their effect on plant traits and performance, high-throughput field phenotyping allows the timely and rapid screening of multiple traits in the early stages of breeding. This offers the potential to reduce the duration of breeding cycles and avoid loss of potentially important alleles due to linkage drag (Araus and Cairns, 2014; Furbank and Tester, 2011; Rebetzke et al., 2019).

Since UAV systems have matured as remote sensing platforms (Aasen et al., 2018), almost all the 'big players' in the area of field phenotyping (research groups, companies and other organizations) have started to use UAVs for

http://dx.doi.org/10.19103/AS.2022.0102.09

measurement (see reviews by Araus et al., 2018; Cendrero-Mateo et al., 2017; Hund et al., 2019a; Rebetzke et al., 2019; Tardieu et al., 2017). This has led to increasing adaption of remote sensing technologies by agronomists, reflected in an exponential increase in publications (Fig. 1), which has helped significantly in relieving the phenotyping bottleneck.

This is a welcome trend since the potential of remote sensing has long been underexplored in agricultural sciences. However, using UAV technologies also comes with challenges since not every agronomist has a background in remote sensing or image processing. Additionally, some companies can underplay the challenges in flying a particular UAV-mounted sensor for reliable trait extraction. This chapter discusses approaches to UAV remote sensing and data analysis for high-throughput field phenotyping and ecophysiological research.

2 Remote sensing tools: unmanned aerial vehicles and flight protocols

2.1 UAVs

A typical field phenotyping set-up may involve a couple of hundred to several thousand plots. These consist of patches of crop canopies with uniform conditions, ranging in size from single plants/rows to machine harvestable yield plots of a few metres. Each plot needs to be screened. Unmanned aerial vehicle remote sensing combines the potential of capturing highly spatially resolved data with

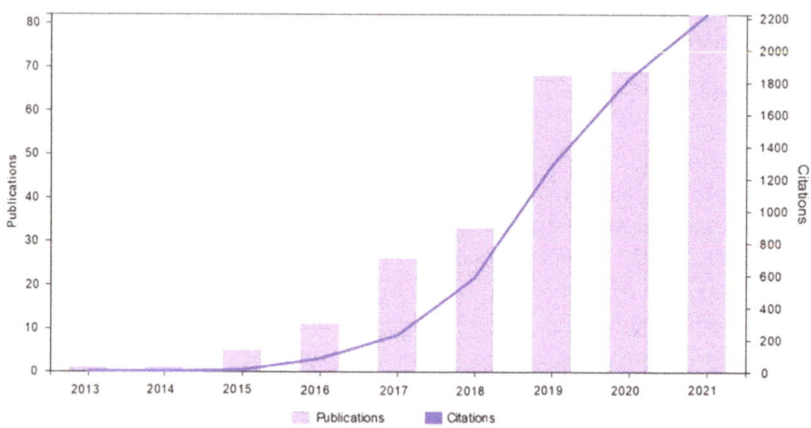

Figure 1 Number of publications and citations with the topics field phenotyping and UAVs. Please note that only publications and citations until August 2021 are shown. Ts = (UAV* or UAS* or 'unmanned aerial vehicle*' or 'unmanned aerial system*' or 'unmanned aircraft system*' or 'RPAS' or 'remotely piloted aircraft system') and (phenotyping or 'field phenotyping' or 'field-phenotyping') and Articles or Review Articles or Early Access (Document Types) and Articles or Review Articles or Book Chapters (Document Types).

high throughput and a wide choice of sensors. In contrast to UAVs, handheld measurement tools are lower cost, easy to set up and highly flexible (Aasen et al., 2020; Perich et al., 2020). They have been supplemented by phenotyping stations (e.g. where arrays of sensors are mounted on gantries which scan plots from above) such as the 'Field Phenotyping Platform' (FIP) at ETH Zürich in Switzerland (Kirchgessner et al., 2017), the 'Field Scanalyzer' in Rothamsted in the UK (Virlet et al., 2017) and similar stations elsewhere (see e.g. Hund et al., 2019b). These technologies arguably have the best ground sampling distance (GSD) in optimizing image resolution. However, sampling large field experiments with such devices is in most cases impractical due to the high running costs and the length of time needed to sample a whole experiment with potentially hundreds of plots. Changes in environmental conditions during the sampling period can also distort the results, for example, in the case of thermal and spectral measurements (Deery et al., 2016; Perich et al., 2020; Sagan et al., 2019) (Fig. 2).

In contrast to phenotyping stations, UAVs have relatively low set-up and running costs. Additionally, the size of the coverable area is only limited by the flight time of the carrier system and can be extended by combining imagery from multiple flights (Perich et al., 2020) or by adjusting the flight height at the expense of lowering the GSD. While fixed-wing UAVs are more efficient when it

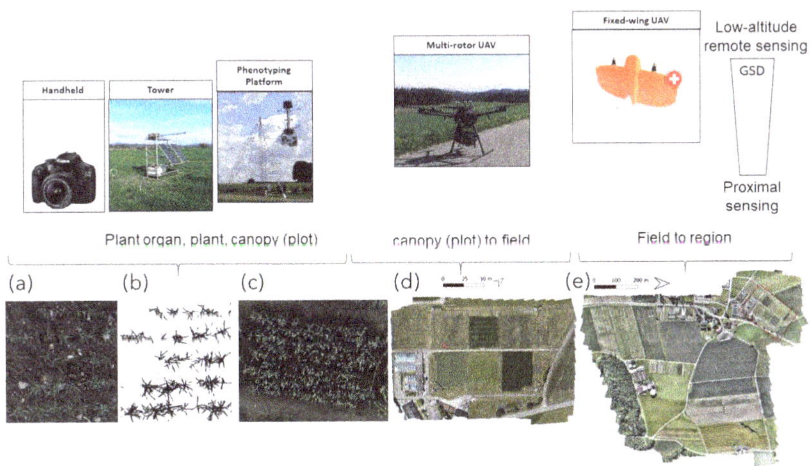

Figure 2 UAV sensing systems in comparison with other field phenotyping approaches regarding coverage and ground sampling distance (GSD). (a) and (b) are images taken with the field phenotyping platform (FIP) with 0.0005 m GSD and a segmentation of leaves using the method of (Zenkl et al., 'in review'). (c) is the area of a plot extracted from a multi-rotor UAV flying at 30 m above ground altitude (AGA) with 0.003 m GSD. (d) is an orthomosaic with an extent of 2 ha and a GSD of 0.008 m generated from a multi-rotor UAV flying at 80 m AGA. The red rectangle marks the area of (c). (e) is an orthomosaic with an extent of approximately 60 ha and a GSD of 0.013 m generated from a fixed-wing UAV flown at 110 m AGA. The red rectangle marks the area of (d).

comes to aerial coverage, multi-rotor systems are more flexible when it comes to flight trajectory. Multi-rotor UAVs also have greater payload potential than fixed-wing UAVs, making it possible to mount a wide range of spectral, thermal, RGB or LiDAR sensors (Aasen et al., 2018; Yang et al., 2017; Lucieer et al., 'in preparation').

Unmanned aerial vehicles can measure traits of interest with relative ease across several hectares in a couple of minutes. The exact coverage over time depends on the GSD necessary to extract the trait of interest. Frequent measurements can be made with no disturbance of the soil or plants. Measurements can also potentially be undertaken by non-experts since flight paths and data capture can be programmed. Unmanned aerial vehicles also allow measurements in multiple locations. Key to repeatable and reliable trait extraction are well-established flight (cf. Section 2.2) and trait extraction protocols (cf. Section 3).

2.2 Flight protocols

Defining UAV measurement protocols can be complex. It requires a good understanding of the sensing system, data processing and analysis (Aasen et al., 2018). Moreover, optimizing measurement results and flight time requires additional infrastructure such as georeferencing ground control points (GCPs) or radiometric ground control points (RCPs). Finally, the flight plan and sensing equipment (flight height, flight speed, angular composition of data, focal length, image resolution, exposure time) need to match the requirements given for the trait of interest (ground sampling distance and motion blur). There are several tools available to help with configuring flight parameters and equipping the area of interest, for example, with GCPs or RCPs (Roth et al., 2018b).

3 Major plant traits that can be extracted using unmanned aerial vehicle remote sensing

The following section discusses some important traits for field phenotyping. Topics are arranged across the growing period from germination to senescence. The discussion ranges from basic feature extraction techniques to their application in an ecophysiological and breeding context.

3.1 Canopy cover and leaf area index

There are different terms to describe plant canopies. Canopy cover describes how well a plant canopy covers the soil. A physiologically more relevant term is leaf area index (LAI) which describes total leaf coverage per unit of ground surface area. The LAI can reach values well above one (e.g. around six for wheat at the end of the vegetative stage) if multiple layers of leaves in a canopy cover a given ground area. It is challenging to measure the LAI nondestructively as soon as multiple leaf layers overlap. Even in the early season, where there are few overlaps, it is not straightforward to estimate LAI. Leaves are often not planar and grow at a certain inclination angle, leading to a complex 3D geometry. This makes it difficult to estimate a correct leaf area; oblique viewing geometries (e.g. towards the edge of an image) may introduce biases due to perspective distortion.[1] Keeping these constraints in mind, two major approaches are used to estimate leaf area:

1 image-based leaf-from-soil segmentation; and
2 absorption-based proxy measurements.

These are discussed in the following sections.

3.1.1 Leaf area estimation by image segmentation

Image segmentation-based leaf area estimation is based on differentiating between 'green' leaf pixels and 'brown' soil pixels using some kind of classification programme. The ratio of 'green' to total pixels is expressed as the canopy cover. These ratios are also referred to as fractional cover or green cover. Besides the constraints already mentioned, the precision of this method of measurement relies on the ability to differentiate 'green' from 'brown' pixels in the image. The GSD and motion blur need to be tuned to reach the necessary precision (Hu et al., 2019; Roth et al., 2018b). This is relative to the organ (leaf) size and is thus crop specific (cf. Fig. 3). Moreover, shadows and overexposure of reflecting leaf parts can compromise the precision of classification.

Decision tree-based segmentation techniques are often used to segment 'green' plant material from the background (e.g. soil and stones). Examples of these classifiers are Random Forest classifiers (in their simplest form using Otsu thresholding to create a binary image) (Torres-Sánchez et al., 2014). These segmentation techniques are used in combination with different colour space transformations to make classification more robust (Roth et al., 2018a, 2020). The most recent segmentation approaches also include information about neighbouring pixels for more precise classification. A recent study by Zenkl et al. ('in review') has compared different segmentation approaches for very-high-resolution RGB images (0.003 m) of the *Eschikon Wheat Segmentation Dataset*. The study concluded that a deep learning-based approach outperformed other machine learning methods using Support Vector Machines and/or

1 The curious reader may grab his/her rubber boots, go to a crop field and look at the same spot from right above the spot and a few metres aside from it.

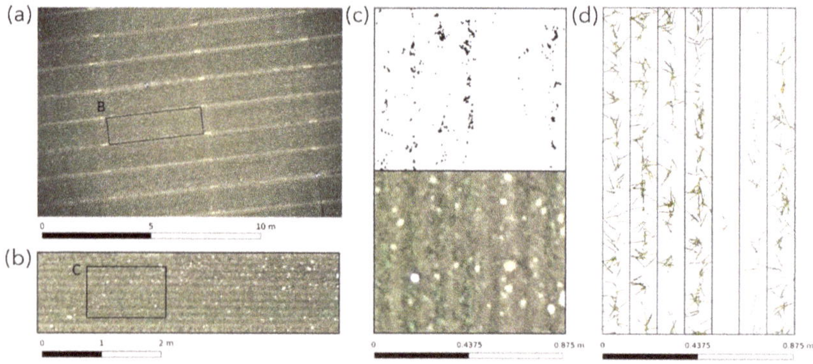

Figure 3 Example of UAV image of wheat (a, b, c) with a GSD of 3 mm per pixel that covers an area of approximately 13 × 9 m in comparison to an image captured with the field phenotyping platform (FIP) of ETH Zurich with a GSD of 0.3 mm per pixel that covers an area of approximately 0.9 × 1.2 m (d), both taken at 12 November 2018 one month after sowing. Illustrated are the original RGB images (a), the shape of the yield plot (b), a subsample of the plot where a Random Forest segmentation approach was applied (Roth et al., 2020) (c) and the same subsample in a FIP image where a deep learning semantic segmentation was applied (Zenkl et al., 'in review') (d).

Random Forest. Most methods are able to assign degrees of uncertainty to the prediction, which means these uncertainties can be taken into account in subsequent processing steps (Roth et al., 2021). However, performing such studies requires well-annotated datasets such as the *Eschikon Wheat Segmentation Dataset* (Zenkl et al., 'in review').

Another 'trick' that can be used to improve leaf area estimation by image segmentation is to make full use of information from multiple overlapping images captured from different angles. Such images are often acquired sequentially by UAVs flying with a constant velocity over a canopy and taking images at equidistant time intervals. Objects on the ground are then captured from slightly different angles in subsequent images. This so-called multiview imaging combines complementary information in these images to improve prediction of leaf area (Roth et al., 2018a, 2020). Nevertheless, all segmentation-based methods reach their limit of accuracy as soon as full canopy closure is reached.

3.1.2 Absorption-based leaf area estimation

A different approach is to use the absorption of plant matter (the change in reflection from a plant surface) to measure the plant material in an area of interest. This approach is particularly useful when the GSD is not sufficient to discriminate plants from the soil background.

This approach has a long history in remote sensing. Several methods and algorithms can be used to generate models for plant trait retrieval from spectral data. Parametric regression methods, such as vegetation indices (VIs), of which the best known is the normalized difference vegetation index (NDVI) (Rouse, Jr. et al., 1974), combine information from several wavelength bands. Linear nonparametric regression methods [sometimes called chemometric techniques; Atzberger et al. (2010)], such as principle component analysis and partial least square regression, take into account the full spectrum. More recently, nonlinear nonparametric methods, often referred to as machine learning regression algorithms (MLRAs), are increasingly being applied in remote sensing. Examples include decision trees (e.g. Random Forest), artificial neural networks and kernel-based regression (e.g. Gaussian process regression) methods. Their main advantage is that they can capture nonlinear relationships without needing to know the underlying data distribution, and hence without assuming a particular probability density distribution. Thus they are perfectly suited to spectroscopic data and have been shown to outperform other methods [for more detailed information, please refer to the reviews of Verrelst et al. (2015, 2019)].

The general advantage of spectral data-based approaches is that they can increase throughput significantly since absorption-based approximation of plant traits does not depend on very high GSD to segment objects. Consequently, they make it possible to fly UAVs higher and faster. They can also measure the amount of plant matter even when multiple layers of leaves overlap. The disadvantage, however, is that, if absolute values from one day to another or one field to another are to be compared, radiometrically and spectrally calibrated devices need to be used. A review on the current state of the art in UAV spectral remote sensing is provided by Aasen et al. (2018). When only relative differences at one location are needed, uncalibrated systems can be used as long as the image-capturing settings and conditions remain constant (Rasmussen et al., 2016).

Structural properties of the canopy such as leaf angles also have an impact on the apparent spectral reflectance of a canopy and may bias the retrieval of plant traits (Wan et al., 2021; e.g. Zou et al., 2018; Zou and Mõttus, 2015). The empirical relationship between spectral reflectance indices and plant traits is not always comparable across years and phenological stages (Aasen et al., 2014; Gnyp et al., 2013). To the authors' knowledge, the robustness of measurements across different genotypes has not yet been assessed. Caution must be exercised when using empirical models for plant trait retrieval from spectral data - particularly in the context of breeding, where structural properties of new varieties are unknown. Inverting physically based radiative transfer models may be a way forward since this procedure builds on mechanistic relationships between spectral formation and biophysical and structural plant traits (Verrelst et al., 2015).

3.2 Plant emergence, density and tillering

An important driver of yield is the number of shoots that bear harvestable organs. This number is limited by two parameters:

1. by the number of plants per area (which itself is the product of sowing density and germination rate); and
2. by the branching rate (often called tillering rate).

Estimating the number of plants per area as well as the number of tillers per area makes it possible to determine the germination rate, tillering rate and related dynamics parameters. Corresponding high-throughput approaches are thus of significant interest in UAV-based field phenotyping.

Monitoring the development of plants and tillers is usually done early in the season when canopies establish. Coverage-based and counting approaches have been used to assess the early growth phase.

Coverage-based approaches use some kind of metric to assess the greenness within an area of interest and are therefore closely related to canopy cover and leaf area approaches (cf. section 2.2). They consequently measure the totality of the canopy and are unable to distinguish tillers from main shoots. They can only estimate the number of tillers and not the number of plants (Phillips et al., 2004; Scotford and Miller, 2004). In addition, coverage-based estimation approaches are based on empirically determined parameters and may therefore be biased by genotype and environment variances that are unrelated to tiller density. On the positive side, coverage-based approaches require a relatively low GSD. To overcome the limitations of pure coverage-based approaches, while working with low GSDs, Roth et al. (2020) used a multiview approach to harvest the information contained in multiple overlapping images to improve the prediction of plant and tiller count estimation.

Counting approaches overcome the limitations of coverage-based approaches by identifying and counting visible plant organs such as leaves. Consequently, those approaches require GSDs that make it possible to clearly separate tiny plant organs in early growth stages from the soil and from each other. In the case of wheat, this typically demands GSD < 0.5 mm per pixel (Roth et al., 2020). Both handcrafted feature extraction and machine learning approaches have proved to be useful to estimate plant densities (Liu et al., 2017, 2018). Transferred to UAV data, those approaches require collecting images while flying at very low heights (Jin et al., 2017) and consequently at low speed to avoid motion blur, which significantly decreases their suitability in large-scale breeding experiments.

3.3 Plant height, growth and lodging

Plant height has become a standard trait extracted by UAV remote sensing. Hoffmeister et al. (2010) introduced the concept of crop surface models (CSMs) later refined by Tilly et al. (2014) to track plant growth. Crop surface models has been adapted for use in UAV remote sensing by Bendig et al. (2013, 2014, 2015). Lodging - which is characterized by a depression in plant height - can be detected using UAV sensing. Plant growth information can then be used, for example, to investigate the interaction of environmental variables, genetic information and stem elongation (Kronenberg et al., 2017, 2021).

Laser scanning and image-based approaches are available for use in UAV systems to extract plant height. Image-based approaches have become popular since they can be used with cheaper and lighter equipment. With photogrammetric algorithms, overlapping 2D images can be aligned and the 3D geometry of an object can be reconstructed using the central projection imaging model (Leuhmann et al., 2014). Particularly important for UAV remote sensing has been the development of the Structure from Motion (SfM) technique within the field of computer vision and its combination with digital photogrammetry (Colomina and Molina, 2014; Eltner and Schneider, 2015). Advances in image matching in digital photogrammetry (Gruen, 2012), orientation (Remondino et al., 2012) and the process of dense 3D point cloud generation allow the reconstruction of surfaces in very high resolution (Haala, 2013).

The drawback of SfM is that, in contrast to active remote sensing systems such as laser scanners, only points visible in at least two images can be reconstructed (Fig. 4). In practice, SfM can barely penetrate the canopy (Harwin and Lucieer, 2012). However, since structural traits are gaining increasing attention, UAV laser-scanning-based approaches hold great potential for field phenotyping applications. With the availability of 2D spectral images, spectral and 3D data can be captured at the same time (Aasen et al., 2015).

3.4 Biomass

Biomass is an important trait, for example, in the context of the harvest index. However, because the measurement is both labour-intensive and destructive, it is often only measured once a season if at all. With remote sensing-based approaches, biomass can now be estimated nondestructively multiple times a season. This makes it possible to investigate dynamic growing patterns such as response to temperature or other environmental parameters.

Structural and spectral approaches have shown great potential in estimating biomass throughout the season (Aasen and Bareth, 2018; Bendig

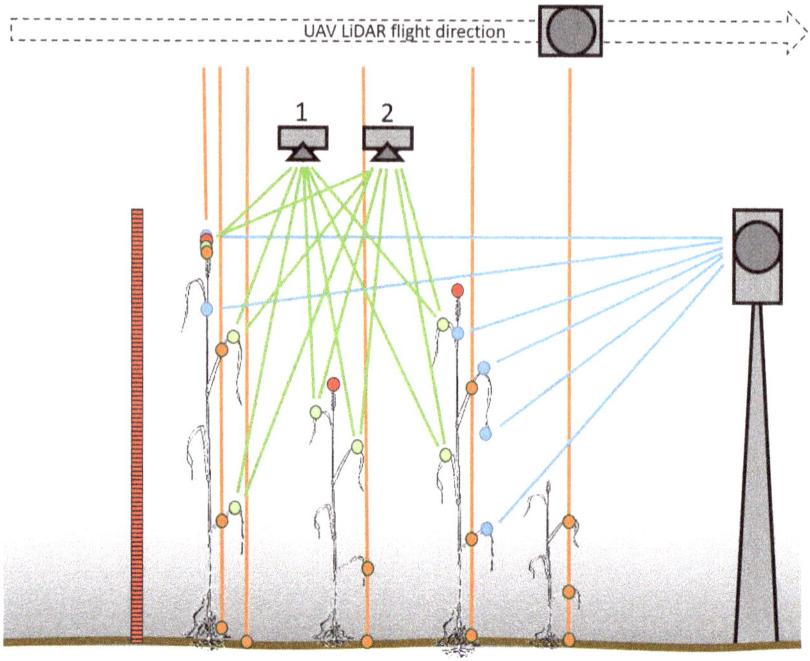

Figure 4 Schematics of plant height estimated with a ruler (red), (single-return) terrestrial laser scanner (blue), UAV (multi-return) laser scanning moving across the area (orange) and structure from motion with two camera positions (green). The active laser scanning systems can penetrate the shade (grey area). Adapted from Aasen (2016).

et al., 2015; Tilly et al., 2015). Structural approaches are mainly based on empirical relationships between biomass and plant height. From a theoretical perspective, one can be sceptical about the robustness of this approach across genotypes and seasons since the relationship of height to biomass may vary between genotypes and due to environmental influences. Extracting the mean plant height per plot represents a massive reduction in dimensionality if working with 3D point clouds as a source. Approaches that include other canopy-related information, for example, canopy roughness (Herrero-Huerta et al., 2020), may lead to improved biomass estimations based on structural data.

Biomass estimation based on spectral data is very common in remote sensing and is now frequently undertaken in field phenotyping applications. Empirical relationships are established between the absorption of an area covered by plants - often measured by means of vegetation indices - and biomass (Aasen et al., 2014; Bendig et al., 2015; Erdle et al., 2013; Fu et al., 2014; Gnyp et al., 2014; Hansen and Schjoerring, 2003; Prabhakara et al., 2015; Tilly et al., 2015). Light absorption is closely related to the amount of absorbing plant matter (cf. Section 3.1.2). From a theoretical point of view, this relationship is more robust than the relationship of plant height to biomass. However, as with LAI, the retrieval of biomass data based on empirical relationships with spectral data may not be robust across genotypes, phenological stages and different years. The use of inversion of physiology-based radiative transfer models in combination with nonparametric machine learning may be a way forward (cf. Section 3.1.2).

3.5 Flowering, fruiting and yield estimation

Flowering and fruiting can be easy to detect for some crops - such as flowering in rapeseed - or very challenging - as in the case of flowering in wheat. Direct and indirect approaches have been used to determine flowering and fruiting.

An indirect method is to monitor the change in colouration of the canopy that indicates that reproduction organs are appearing. A direct approach is to directly detect these organs with image processing techniques. There have been several attempts at direct measurement using traditional, handcrafted features (Guo et al., 2015). With the advent of deep learning, this approach has been largely automated. Today convolutional neural networks can be used to segment plants from soil even in challenging conditions (Zenkl et al., 'in review'), detect and even count flower heads (Gallmann et al., 2022) and wheat ears (David et al., 2020). The publications by David et al. (2020, 2021) are particularly interesting since they represent a worldwide initiative by many institutions to compile a dataset called the *Global Wheat Head Dataset* that now serves as a benchmark dataset for the global machine learning community.

However, for the direct detection approach to work, the GSD needs to be relatively low. This may reduce throughput due to the necessity to lower the flight altitude in order to achieve the necessary resolution. Luckily camera technology is developing fast and current very high-resolution camera systems with 100 megapixels are making it possible to take images at higher altitudes (Lucieer et al., 'in preparation').

Differences in flowering between genotypes can also happen within only a few days, requiring a high frequency of sensing. This might be feasible for permanently manned research stations, but may be particularly suitable for remote field solutions such as PhenoCam which automatically capture images at regular intervals with a camera mounted on a tower (Aasen et al., 2020).

3.6 Senescence

Senescence - the gradual deterioration of all physiological processes - is a developmental process which overlaps with the reproductive phase in annual crop plants. Senescence might reduce crop yield when it is induced prematurely under adverse environmental conditions. On the other hand, a number of studies found that, among so-called 'stay-green' phenotypes, some are functional stay-green plants with increased productivity due to prolonged carbon assimilation (Gregersen et al., 2013).

Since the process of senescence is related to the decolourization of leaf tissue, it is traditionally rated as the ratio of decolourized individual leaves (e.g. flag leaf in wheat) over the whole canopy (Pask et al., 2012). Digital methods follow this approach and quantify canopy colour. (Burkart et al., 2018) used the green-red vegetation index to extract canopy greenness from UAV RGB images across two vegetation periods and could identify different growth stages, including ripening/senescence. Other approaches build on spectral data (Anderegg et al., 2020) and may be used in flying platforms in the near future.

3.7 Canopy temperature

Plants interact with the surrounding environment through carbon, water and energy-exchange processes, maintaining an equilibrium that permits them to grow and adapt to variable growing conditions. Due to the transpiration cooling effect of water leaving the stomata, this energy exchange, among other processes, affects the temperature of foliage. Based on this relationship, many researchers have taken canopy temperature as a proxy for the water status of plants and associated traits such as yield (Li et al., 2019; Lopes and Reynolds, 2010; Rebetzke et al., 2013; Reynolds et al., 1994).

Since field measurements are time consuming and often result in low repeatability if environmental conditions change during the measurement (Deery et al., 2016), airborne thermography (including with UAVs) has gained in popularity. It allows the assessment of canopy temperature in large breeding or genetic experiments in a short time, with improved repeatability compared to manual plot-by-plot measurements (Deery et al., 2016, 2019; Gago et al., 2015; Liebisch et al., 2015; Perich et al., 2020).

However, the relationship between transpiration and canopy temperature is very complex since it depends on the (micro-) environmental conditions around the plant, for example, ambient temperature, wind and vapour pressure (Costa et al., 2013; Maes and Steppe, 2012). This poses significant challenges to translate measurements of canopy temperature to physiologically meaningful results, particularly in areas with potentially fluctuating environmental conditions (e.g. temperate zone countries such as Switzerland). Several approaches to normalizing this effect and establishing a relationship between canopy temperature and transpiration have been proposed. Experimental approaches include wet and dry reference surfaces situated within the image, which is challenging to achieve in UAV remote sensing (Jones, 1999; Maes and Steppe, 2012). Other approaches normalize canopy temperature with air temperature or model reference values based on meteorological data (Berni et al., 2009, 2009; Idso et al., 1981; Jackson et al., 1981; Maes and Steppe, 2012; Zarco-Tejada et al., 2013, 2018). While these existing approaches have been shown to work with airborne and UAV observations in arid and semi-arid regions, to the authors' knowledge they have not been validated for crop genotype screening in temperate climates. In environments where environmental conditions fluctuate sometimes within minutes, the temporal offset between image acquisitions may introduce bias

In cases where the canopy is not fully closed (e.g. wheat), a pixel often does not contain a pure signal from the foliage, but rather an integrated signal from both soil and vegetation, when the canopy is measured by a UAV sensor at a right angle to that surface (NADIR). The soil background interferes with the signal from the foliage (Deery and Jones, 2021). The degree to which it does so depends on shoot biomass, phenology, morphology and structural parameters of the canopy, such as plant height, LAI, ground cover and leaf and spike orientation [reviewed by Prashar and Jones (2014)]. A study by Hund et al. (unpublished) showed that in breeding experiments where genotypes with different heights are placed next to each other, canopy height is the most important driver for canopy temperature. This finding has significant implications for thermal field phenotyping. In such settings, the interpretation of canopy temperature should be handled with care until this effect can be normalized by means of statistical approaches or experimental designs.

4 Conclusion and future trends

The adaption of remote sensing tools by plant scientists has opened up new perspectives and provided the base for high-throughput field phenotyping. Some kind of remote sensing technology, method or philosophy is now part of almost any field phenotyping programme – whether using proximal handheld, robotic or flying platforms [or a combination of those (Pretto et al., 2021)] or the adaption of sophisticated machine learning approaches to make sense of the big data that results from applying remote sensing systems.

However, the rapid adaption of these tools needs to be followed by a deeper understanding of the data captured and the information gathered to be able to truly interpret the information about the extracted traits. This chapter has reviewed recent trends in high-throughput field phenotyping from the perspective of remote sensing. In doing so we have not only highlighted promising approaches but provided some background on remote sensing theory to cast light into the shadows that hide potential pitfalls.

But we should not stop here. The remote sensing community is becoming increasingly aware of the needs of plant scientists and more and more remote sensing researchers are discovering the plant to plot scale as a fruitful field (Machwitz et al., 2021). Fostering this development and bringing plant scientists, remote sensing specialists, ecologists and breeders to the same (dinner) table to exchange ideas and initiate collaborations would allow us to unlock the full potential of current remote sensing technologies by identifying common interests and enabling transfer of knowledge. Moreover, this would also allow the transfer of insights gained by applying remote sensing technology to the plant to plot scale back to the landscape scale by improving satellite remote sensing approaches. Finally, in the whole field phenotyping community, there is a risk that research is driven by technology rather than applications. This exchange should therefore go both ways: we call for plant scientists to help guide the research of remote sensing specialists and the whole field phenotyping community into directions that yield improvements towards a more sustainable agriculture.

5 Authors' contributions

H. A. conceived and wrote major parts of the text. L. R. contributed partially to the text, revised the text and created Fig. 3.

6 Acknowledgements

We acknowledge Quirina Merz for providing the orthomosaic of the fixed-wing UAV in Fig. 2. We acknowledge the funding of the Swiss Science Foundation for the projects 'PhenomEn' (grant number IZCOZ0_198091) and 'PhotoInd' (grant number CRSK-3_195591), the Swiss Innovation Agency Innosuisse for the project 'Trait spotting' (grant number KTI P-Nr 27059.2 PFLS-LS), and the Swiss National Science Foundation (SNSF), within the framework of the National Research Programme 'Sustainable Economy: resource-friendly, future-oriented, innovative' (NRP 73), in the InnoFarm project (grant number 407340_172433). Also, we thank Radek Zenkl for his contributions to Fig. 2. Finally, we thank the whole Crop Science Group of ETH Zürich with Professor Achim Walter and the Remote Sensing and GIS Group at the University of Cologne with Professor Georg Bareth. Gaining the knowledge summarized in this chapter can only stem from great group efforts.

7 References

Aasen, H. 2016. *The acquisition of hyperspectral digital surface models of crops from UAV snapshot cameras* (Dissertation). University of Cologne, Cologne.

Aasen, H. and Bareth, G. 2018. Ground and UAV sensing approaches for spectral and 3D crop trait estimation. In: Thenkabail, P., Lyon, J. G. and Huete, A. (Eds). *Hyperspectral*

Remote Sensing of Vegetation (Vol. II): Biophysical and Biochemical Characterization and Plant Species Studies. Chapman and Hall/CRC, Milton.

Aasen, H., Burkart, A., Bolten, A. and Bareth, G. 2015. Generating 3D hyperspectral information with lightweight UAV snapshot cameras for vegetation monitoring: From camera calibration to quality assurance. *ISPRS J. Photogramm.* 108, 245-259. https://doi.org/10.1016/j.isprsjprs.2015.08.002.

Aasen, H., Gnyp, M. L., Miao, Y. and Bareth, G. 2014. Automated hyperspectral vegetation index retrieval from multiple correlation matrices with HyperCor. *Photogramm. Eng. Remote Sens.* 80(8), 785-795. https://doi.org/10.14358/PERS.80.8.785.

Aasen, H., Honkavaara, E., Lucieer, A. and Zarco-Tejada, P. 2018. Quantitative remote sensing at ultra-high resolution with UAV spectroscopy: A review of sensor technology, measurement procedures, and data correction workflows. *Remote Sens.* 10(7), 1091. https://doi.org/10.3390/rs10071091.

Aasen, H., Kirchgessner, N., Walter, A. and Liebisch, F. 2020. PhenoCams for field phenotyping: Using very high temporal resolution digital repeated photography to investigate interactions of growth, phenology, and harvest traits. *Front. Plant Sci.* 11, 593. https://doi.org/10.3389/fpls.2020.00593.

Adão, T., Hruška, J., Pádua, L., Bessa, J., Peres, E., Morais, R. and Sousa, J. 2017. Hyperspectral imaging: A review on UAV-based sensors, data processing and applications for agriculture and forestry. *Remote Sens.* 9(11), 1110. https://doi.org/10.3390/rs9111110.

Anderegg, J., Yu, K., Aasen, H., Walter, A., Liebisch, F. and Hund, A. 2020. Spectral vegetation indices to track senescence dynamics in diverse wheat germplasm. *Front. Plant Sci.* 10, 1749. https://doi.org/10.3389/fpls.2019.01749.

Anderson, K. and Gaston, K. J. 2013. Lightweight unmanned aerial vehicles will revolutionize spatial ecology. *Front. Ecol. Environ.* 11(3), 138-146. https://doi.org/10.1890/120150.

Araus, J. L. and Cairns, J. E. 2014. Field high-throughput phenotyping: The new crop breeding frontier. *Trends Plant Sci.* 19(1), 52-61. https://doi.org/10.1016/j.tplants.2013.09.008.

Araus, J. L., Kefauver, S. C., Zaman-Allah, M., Olsen, M. S. and Cairns, J. E. 2018. Translating high-throughput phenotyping into genetic gain. *Trends Plant Sci.* 23(5), 451-466. https://doi.org/10.1016/j.tplants.2018.02.001.

Atzberger, C., Guérif, M., Baret, F. and Werner, W. 2010. Comparative analysis of three chemometric techniques for the spectroradiometric assessment of canopy chlorophyll content in winter wheat. *Comput. Electron. Agric.* 73(2), 165-173. https://doi.org/10.1016/j.compag.2010.05.006.

Bendig, J., Bolten, A. and Bareth, G. 2013. Monitoring des Pflanzenwachstums mit Hilfe multitemporaler und hoch auflösender Oberflächenmodelle von Getreidebeständen auf Basis von Bildern aus UAV-Befliegungen. *pfg* 2013(6), 551-562. https://doi.org/10.1127/1432-8364/2013/0200.

Bendig, J., Bolten, A., Bennertz, S., Broscheit, J., Eichfuss, S. and Bareth, G. 2014. Estimating biomass of barley using crop surface models (CSMs) derived from UAV-based RGB imaging. *Remote Sens.* 6(11), 10395-10412. https://doi.org/10.3390/rs61110395.

Bendig, J., Yu, K., Aasen, H., Bolten, A., Bennertz, S., Broscheit, J., Gnyp, M. L. and Bareth, G. 2015. Combining UAV-based plant height from crop surface models, visible, and near infrared vegetation indices for biomass monitoring in barley. *Int. J. Appl. Earth Obs. Geoinf.* 39, 79-87. https://doi.org/10.1016/j.jag.2015.02.012.

Berni, J. A. J., Zarco-Tejada, P. J., Sepulcre-Cantó, G., Fereres, E. and Villalobos, F. 2009. Mapping canopy conductance and CWSI in olive orchards using high resolution thermal remote sensing imagery. *Remote Sens. Environ.* 113(11), 2380–2388. https://doi.org/10.1016/j.rse.2009.06.018.

Burkart, A., Hecht, V. L., Kraska, T. and Rascher, U. 2018. Phenological analysis of unmanned aerial vehicle based time series of barley imagery with high temporal resolution. *Precis. Agric.* 19(1), 134–146. https://doi.org/10.1007/s11119-017-9504-y.

Cendrero-Mateo, M. P., Muller, O., Albrecht, H., Burkart, A., Gatzke, S., Janssen, B., Keller, B., Körber, N., Kraska, T., Matsubara, S., Jinquan, L., Müller-Linow, M., Pieruschka, R., Pinto, F., Rischbeck, P., Schickling, A., Steier, A., Watt, M., Schurr, U. and Rascher, U. 2017. Field phenotyping: Concepts and examples to quantify dynamic plant traits across scales in the field. In: Chabbi, A. and Loescher, H. W. (Eds). *Terrestrial Ecosystem Research Infrastructures: Challenges and Opportunities, Abad Chabbi.* Abad Chabbi, Henry W. Loescher; CRC Press, Boca Raton, pp. 53–80.

Colomina, I. and Molina, P. 2014. Unmanned aerial systems for photogrammetry and remote sensing: A review. *ISPRS J. Photogramm.* 92, 79–97. https://doi.org/10.1016/j.isprsjprs.2014.02.013.

Costa, J. M., Grant, O. M. and Chaves, M. M. 2013. Thermography to explore plant-environment interactions. *J. Exp. Bot.* 64(13), 3937–3949. https://doi.org/10.1093/jxb/ert029.

David, E., Madec, S., Sadeghi-Tehran, P., Aasen, H., Zheng, B., Liu, S., Kirchgessner, N., Ishikawa, G., Nagasawa, K., Badhon, M. A., Pozniak, C., de Solan, B., Hund, A., Chapman, S. C., Baret, F., Stavness, I. and Guo, W. 2020. Global wheat head detection (GWHD) dataset: A large and diverse dataset of high-resolution RGB-labelled images to develop and benchmark wheat head detection methods. *Plant Phenom.* 2020, 3521852. https://doi.org/10.34133/2020/3521852.

David, E., Serouart, M., Smith, D., Madec, S., Velumani, K., Liu, S., Wang, X., Espinosa, F. P., Shafiee, S., Tahir, I. S. A., Tsujimoto, H., Nasuda, S., Zheng, B., Kichgessner, N., Aasen, H., Hund, A., Sadhegi-Tehran, P., Nagasawa, K., Ishikawa, G., Dandrifosse, S., Carlier, A., Mercatoris, B., Kuroki, K., Wang, H., Ishii, M., Badhon, M. A., Pozniak, C., LeBauer, D. S., Lilimo, M., Poland, J., Chapman, S., de Solan, B., Baret, F., Stavness, I. and Guo, W. 2021. Global Wheat Head Dataset 2021: More diversity to improve the benchmarking of wheat head localization methods. *CS*, ArXiv210507660.

Deery, D. M. and Jones, H. G. 2021. Field phenomics: Will it enable crop improvement? *Plant Phenom.* 2021, 9871989. https://doi.org/10.34133/2021/9871989.

Deery, D. M., Rebetzke, G. J., Jimenez-Berni, J. A., Bovill, W. D., James, R. A., Condon, A. G., Furbank, R. T., Chapman, S. C. and Fischer, R. A. 2019. Evaluation of the phenotypic repeatability of canopy temperature in wheat using continuous-terrestrial and airborne measurements. *Front. Plant Sci.* 10, 875. https://doi.org/10.3389/fpls.2019.00875.

Deery, D. M., Rebetzke, G. J., Jimenez-Berni, J. A., James, R. A., Condon, A. G., Bovill, W. D., Hutchinson, P., Scarrow, J., Davy, R. and Furbank, R. T. 2016. Methodology for high-throughput field phenotyping of canopy temperature using airborne thermography. *Front. Plant Sci.* 7, 1808. https://doi.org/10.3389/fpls.2016.01808.

Eltner, A. and Schneider, D. 2015. Analysis of different methods for 3D reconstruction of natural surfaces from parallel-axes UAV images. *Photogram. Rec.* 30(151), 279–299. https://doi.org/10.1111/phor.12115.

Erdle, K., Mistele, B. and Schmidhalter, U. 2013. Spectral high-throughput assessments of phenotypic differences in biomass and nitrogen partitioning during grain filling of wheat under high yielding Western European conditions. *Field Crops Res.* 141, 16–26. https://doi.org/10.1016/j.fcr.2012.10.018.

Fu, Y., Yang, G., Wang, J., Song, X. and Feng, H. 2014. Winter wheat biomass estimation based on spectral indices, band depth analysis and partial least squares regression using hyperspectral measurements. *Comput. Electron. Agric.* 100, 51–59. https://doi.org/10.1016/j.compag.2013.10.010.

Furbank, R. T. and Tester, M. 2011. Phenomics – Technologies to relieve the phenotyping bottleneck. *Trends Plant Sci.* 16(12), 635–644. https://doi.org/10.1016/j.tplants.2011.09.005.

Gago, J., Douthe, C., Coopman, R. E., Gallego, P. P., Ribas-Carbo, M., Flexas, J., Escalona, J. and Medrano, H. 2015. UAVs challenge to assess water stress for sustainable agriculture. *Agric. Water Manag.* 153, 9–19. https://doi.org/10.1016/j.agwat.2015.01.020.

Gallmann, J., Schüpbach, B., Jacot, K., Albrecht, M., Winizki, J. and Aasen, H. 2022. Flower mapping in grasslands with drones and deep learning. *Front. Plant Sci.* 12, 774965. doi:10.3389/fpls.2021.774965

Gnyp, M. L., Miao, Y., Yuan, F., Yu, K., Yao, Y., Huang, S. and Bareth, G. 2014. *Derivative Analysis to Improve Rice Biomass Estimation at Early Growth Stages*. https://doi.org/10.5880/TR32DB.KGA94.7.

Gnyp, M. L., Yu, K., Aasen, H., Yao, Y., Huang, S., Miao, Y. and Bareth, G. 2013. Analysis of crop reflectance for estimating biomass in rice canopies at different phenological stages. *pfg* 2013(4), 351–365. https://doi.org/10.1127/1432-8364/2013/0182.

Gregersen, P. L., Culetic, A., Boschian, L. and Krupinska, K. 2013. Plant senescence and crop productivity. *Plant Mol. Biol.* 82(6), 603–622. https://doi.org/10.1007/s11103-013-0013-8.

Gruen, A. 2012. Development and status of image matching in photogrammetry: Development and status of image matching in photogrammetry. *Photogramm. Rec.* 27(137), 36–57. https://doi.org/10.1111/j.1477-9730.2011.00671.x.

Guo, W., Fukatsu, T. and Ninomiya, S. 2015. Automated characterization of flowering dynamics in rice using field-acquired time-series RGB images. *Plant Methods* 11, 7. https://doi.org/10.1186/s13007-015-0047-9.

Haala, N. 2013. *The Landscape of Dense Image Matching Algorithms*. Presented at the Photogrammetric Week'13, pp. 271–281.

Hansen, P. M. and Schjoerring, J. K. 2003. Reflectance measurement of canopy biomass and nitrogen status in wheat crops using normalized difference vegetation indices and partial least squares regression. *Remote Sens. Environ.* 86(4), 542–553. https://doi.org/10.1016/S0034-4257(03)00131-7.

Harwin, S. and Lucieer, A. 2012. Assessing the accuracy of georeferenced point clouds produced via multi-view stereopsis from unmanned aerial vehicle (UAV) imagery. *Remote Sens.* 4(6), 1573–1599. https://doi.org/10.3390/rs4061573.

Herrero-Huerta, M., Bucksch, A., Puttonen, E. and Rainey, K. M. 2020. Canopy roughness: A new phenotypic trait to estimate aboveground biomass from unmanned aerial system. *Plant Phenom.* 2020, 6735967. https://doi.org/10.34133/2020/6735967.

Hoffmeister, D., Bolten, A., Curdt, C., Waldhoff, G. and Bareth, G. 2010. High-resolution crop surface models (CSM) and crop volume models (CVM) on field level by terrestrial laser scanning. In: *Proceedings Volume 7840, Sixth International Symposium on*

Digital Earth: Models, Algorithms, and Virtual Reality. International Society for Optics and Photonics, Beijing, China, p. 6.

Hu, P., Guo, W., Chapman, S. C., Guo, Y. and Zheng, B. 2019. Pixel size of aerial imagery constrains the applications of unmanned aerial vehicle in crop breeding. *ISPRS J. Photogramm.* 154, 1–9. https://doi.org/10.1016/j.isprsjprs.2019.05.008.

Hund, A., Kislinger, F. and Aasen, H. in-preparation. Plant height drives canopy temperature in breeding experiments.

Hund, A., Kronenberg, L., Anderegg, J., Yu, K. and Walter, A. 2019a. Non-invasive phenotyping of cereal growth and development characteristics in the field. In: Ordon, F. and Friedt, W. (Eds). *Advances in Crop Breeding Techniques*. Burleigh Dodds.

Hund, A., Kronenberg, L., Anderegg, J., Yu, K. and Walter, A. 2019b. Non-invasive field phenotyping of cereal development. *Burleigh Dodds Series in Agricultural Science*, 249–292. https://doi.org/10.19103/AS.2019.0051.13.

Idso, S. B., Jackson, R. D., Pinter, P. J., Reginato, R. J. and Hatfield, J. L. 1981. Normalizing the stress-degree-day parameter for environmental variability. *Agric. Meteorol.* 24, 45–55. https://doi.org/10.1016/0002-1571(81)90032-7.

Jackson, R. D., Idso, S. B., Reginato, R. J. and Pinter, P. J. 1981. Canopy temperature as a crop water stress indicator. *Water Resour. Res.* 17(4), 1133–1138. https://doi.org/10.1029/WR017i004p01133.

Jin, X., Liu, S., Baret, F., Hemerlé, M. and Comar, A. 2017. Estimates of plant density of wheat crops at emergence from very low altitude UAV imagery. *Remote Sens. Environ.* 198, 105–114. https://doi.org/10.1016/j.rse.2017.06.007.

Jones, H. G. 1999. Use of thermography for quantitative studies of spatial and temporal variation of stomatal conductance over leaf surfaces. *Plant Cell Environ.* 22(9), 1043–1055. https://doi.org/10.1046/j.1365-3040.1999.00468.x.

Kirchgessner, N., Liebisch, F., Yu, K., Pfeifer, J., Friedli, M., Hund, A. and Walter, A. 2017. The ETH field phenotyping platform FIP: A cable-suspended multi-sensor system. *Funct. Plant Biol.* 44(1), 154–168. https://doi.org/10.1071/FP16165.

Kronenberg, L., Yates, S., Boer, M. P., Kirchgessner, N., Walter, A. and Hund, A. 2021. Temperature response of wheat affects final height and the timing of stem elongation under field conditions. *J. Exp. Bot.* 72(2), 700–717. https://doi.org/10.1093/jxb/eraa471.

Kronenberg, L., Yu, K., Walter, A. and Hund, A. 2017. Monitoring the dynamics of wheat stem elongation: Genotypes differ at critical stages. *Euphytica* 213(7). https://doi.org/10.1007/s10681-017-1940-2.

Li, X., Ingvordsen, C. H., Weiss, M., Rebetzke, G. J., Condon, A. G., James, R. A. and Richards, R. A. 2019. Deeper roots associated with cooler canopies, higher normalized difference vegetation index, and greater yield in three wheat populations grown on stored soil water. *J. Exp. Bot.* 70(18), 4963–4974. https://doi.org/10.1093/jxb/erz232.

Liebisch, F., Kirchgessner, N., Schneider, D., Walter, A. and Hund, A. 2015. Remote, aerial phenotyping of maize traits with a mobile multi-sensor approach. *Plant Methods* 11, 9. https://doi.org/10.1186/s13007-015-0048-8.

Liu, S., Baret, F., Andrieu, B., Burger, P. and Hemmerlé, M. 2017. Estimation of wheat plant density at early stages using high resolution imagery. *Front. Plant Sci.* 8, 739. https://doi.org/10.3389/fpls.2017.00739.

Liu, T., Yang, T., Li, C., Li, R., Wu, W., Zhong, X., Sun, C. and Guo, W. 2018. A method to calculate the number of wheat seedlings in the 1st to the 3rd leaf growth stages. *Plant Methods* 14, 101. https://doi.org/10.1186/s13007-018-0369-5.

Lopes, M. S. and Reynolds, M. P. 2010. Partitioning of assimilates to deeper roots is associated with cooler canopies and increased yield under drought in wheat. *Funct. Plant Biol.* 37(2), 147. https://doi.org/10.1071/FP09121.

Lucieer, A. and Aasen, H. in preparation. Advances in remote sensing of vegetation with drones - Review and outlook.

Luehmann, H. P., Pressly, E. D., Detering, L., Wang, C., Pierce, R., Woodard, P. K., Gropler, R. J., Hawker, C. J. and Liu, Y. 2014. PET/CT imaging of chemokine receptor CCR5 in vascular injury model using targeted nanoparticle. J. *Nucl. Med.* 55(4), 629-634. https://doi.org/10.2967/jnumed.113.132001.

Machwitz, M., Pieruschka, R., Berger, K., Schlerf, M., Aasen, H., Fahrner, S., Jimenez-Berni, J. A., Baret, F. and Rascher, U. 2021. Bridging the gap between remote sensing and plant phenotyping - Challenges and opportunities for the next generation of sustainable agriculture. *Front. Plant Sci.* 12, 749374. doi:10.3389/fpls.2021.749374.

Maes, W. H. and Steppe, K. 2019. Perspectives for remote sensing with unmanned aerial vehicles in precision agriculture. *Trends Plant Sci.* 24(2), 152-164. https://doi.org/10.1016/j.tplants.2018.11.007.

Maes, W. H. and Steppe, K. 2012. Estimating evapotranspiration and drought stress with ground-based thermal remote sensing in agriculture: A review. *J. Exp. Bot.* 63(13), 4671-4712. https://doi.org/10.1093/jxb/ers165.

McCabe, M. F., Rodell, M., Alsdorf, D. E., Miralles, D. G., Uijlenhoet, R., Wagner, W., Lucieer, A., Houborg, R., Verhoest, N. E. C., Franz, T. E., Shi, J., Gao, H. and Wood, E. F. 2017. The future of Earth observation in hydrology. *Hydrol. Earth Syst. Sci.* 21(7), 3879-3914. https://doi.org/10.5194/hess-21-3879-2017.

Pádua, L., Vanko, J., Hruška, J., Adão, T., Sousa, J. J., Peres, E. and Morais, R. 2017. UAS, sensors, and data processing in agroforestry: A review towards practical applications. *Int. J. Remote Sens.* 38(8-10), 2349-2391. https://doi.org/10.1080/01431161.2017.1297548.

Pajares, G. 2015. Overview and current status of remote sensing applications based on unmanned aerial vehicles (UAVs). *Photogramm. Eng. Remote Sens.* 81(4), 281-330. https://doi.org/10.14358/PERS.81.4.281.

Pask, A., Pietragalla, J., Mullan, D. and Raynolds, M. (Eds). 2012. Physiological breeding II. In: *A Field Guide to Wheat Phenotyping*. CIMMYT, Mexico.

Perich, G., Hund, A., Anderegg, J., Roth, L., Boer, M. P., Walter, A., Liebisch, F. and Aasen, H. 2020. Assessment of multi-image unmanned aerial vehicle based high-throughput field phenotyping of canopy temperature. *Front. Plant Sci.* 11, 150. https://doi.org/10.3389/fpls.2020.00150.

Phillips, S. B., Keahey, D. A., Warren, J. G. and Mullins, G. L. 2004. Estimating winter wheat tiller density using spectral reflectance sensors for early-spring, variable-rate nitrogen applications. *Agron. J.* 96(3), 591-600. https://doi.org/10.2134/agronj2004.0591.

Prabhakara, K., Hively, W. D. and McCarty, G. W. 2015. Evaluating the relationship between biomass, percent groundcover and remote sensing indices across six winter cover crop fields in Maryland, United States. *Int. J. Appl. Earth Obs. Geoinf.* 39, 88-102. https://doi.org/10.1016/j.jag.2015.03.002.

Prashar, A. and Jones, H. 2014. Infra-red thermography as a high-throughput tool for field phenotyping. *Agronomy* 4(3), 397-417. https://doi.org/10.3390/agronomy4030397.

Pretto, A., Aravecchia, S., Burgard, W., Chebrolu, N., Dornhege, C., Falck, T., Fleckenstein, F. V., Fontenla, A., Imperoli, M., Khanna, R., Liebisch, F., Lottes, P., Milioto, A., Nardi,

D., Nardi, S., Pfeifer, J., Popovic, M., Potena, C., Pradalier, C., Rothacker-Feder, E., Sa, I., Schaefer, A., Siegwart, R., Stachniss, C., Walter, A., Winterhalter, W., Wu, X. and Nieto, J. 2021. Building an aerial-ground robotics system for precision farming: An adaptable solution. *IEEE Robot. Automat. Mag.* 28(3), 29–49. https://doi.org/10.1109/MRA.2020.3012492.

Rasmussen, J., Ntakos, G., Nielsen, J., Svensgaard, J., Poulsen, R. N. and Christensen, S. 2016. Are vegetation indices derived from consumer-grade cameras mounted on UAVs sufficiently reliable for assessing experimental plots? *Eur. J. Agron.* 74, 75–92. https://doi.org/10.1016/j.eja.2015.11.026.

Rebetzke, G. J., Jimenez-Berni, J., Fischer, R. A., Deery, D. M. and Smith, D. J. 2019. Review: High-throughput phenotyping to enhance the use of crop genetic resources. *Plant Sci.* 282, 40–48. https://doi.org/10.1016/j.plantsci.2018.06.017.

Rebetzke, G. J., Rattey, A. R., Farquhar, G. D., Richards, R. A. and Condon, A. T. G. 2013. Genomic regions for canopy temperature and their genetic association with stomatal conductance and grain yield in wheat. *Funct. Plant Biol.* 40(1), 14–33. https://doi.org/10.1071/FP12184.

Remondino, F., Del Pizzo, S., Kersten, T. P. and Troisi, S. 2012. Low-cost and open-source solutions for automated image orientation – A critical overview. In: Ioannides, M., Fritsch, D., Leissner, J., Davies, R., Remondino, F. and Caffo, R. (Eds). *Progress in Cultural Heritage Preservation*. Springer, Berlin Heidelberg, Berlin, Heidelberg, pp. 40–54.

Reynolds, M., Balota, M., Delgado, M., Amani, I. and Fischer, R. 1994. Physiological and morphological traits associated With spring wheat yield Under hot, irrigated conditions. *Funct. Plant Biol.* 21(6), 717. https://doi.org/10.1071/PP9940717.

Roth, L., Aasen, H., Walter, A. and Liebisch, F. 2018a. Extracting leaf area index using viewing geometry effects–A new perspective on high-resolution unmanned aerial system photography. *ISPRS J. Photogramm.* 141, 161–175. https://doi.org/10.1016/j.isprsjprs.2018.04.012.

Roth, L., Hund, A. and Aasen, H. 2018b. PhenoFly planning tool: Flight planning for high-resolution optical remote sensing with unmanned areal systems. *Plant Methods* 14, 116. https://doi.org/10.1186/s13007-018-0376-6.

Roth, L., Camenzind, M., Aasen, H., Kronenberg, L., Barendregt, C., Camp, K. H., Walter, A., Kirchgessner, N. and Hund, A. 2020. Repeated multiview imaging for estimating seedling tiller counts of wheat genotypes using drones. *Plant Phenom.* 2020, 3729715. https://doi.org/10.34133/2020/3729715.

Roth, L., Rodríguez-Álvarez, M. X., van Eeuwijk, F., Piepho, H.-P., and Hund, A. (2021). Phenomics data processing: A plot-level model for repeated measurements to extract the timing of key stages and quantities at defined time points. *Field Crops Research*, 274 (108314). https://doi.org/10.1016/j.fcr.2021.108314.

Rouse, J. W., Jr., Haas, R. H., Schell, J. A. and Deering, D. W. 1974. Monitoring vegetation systems in the Great Plains with Erts. In: *Third Earth Resources Technology Satellite-1 Symposium- Volume I: Technical Presentations*. National Astronautics and Space Administration, Washington, DC, pp. 309–317.

Sagan, V., Maimaitijiang, M., Sidike, P., Eblimit, K., Peterson, K. T., Hartling, S., Esposito, F., Khanal, K., Newcomb, M., Pauli, D., Ward, R., Fritschi, F., Shakoor, N. and Mockler, T. 2019. UAV-based high resolution thermal imaging for vegetation monitoring, and plant phenotyping using ICI 8640 P, FLIR Vue Pro R 640, and thermoMap cameras. *Remote Sens.* 11(3). https://doi.org/10.3390/rs11030330.

Sanchez-Azofeifa, A., Antonio Guzmán, J., Campos, C. A., Castro, S., Garcia-Millan, V., Nightingale, J. and Rankine, C. 2017. Twenty-first century remote sensing technologies are revolutionizing the study of tropical forests. *Biotropica* 49(5), 604-619. https://doi.org/10.1111/btp.12454.

Scotford, I. M. and Miller, P. C. H. 2004. Estimating tiller density and leaf area index of winter wheat using spectral reflectance and ultrasonic sensing techniques. *Biosyst. Eng.* 89(4), 395-408. https://doi.org/10.1016/j.biosystemseng.2004.08.019.

Tardieu, F., Cabrera-Bosquet, L., Pridmore, T. and Bennett, M. 2017. Plant phenomics, from sensors to knowledge. *Curr. Biol.* 27(15), R770-R783. https://doi.org/10.1016/j.cub.2017.05.055.

Tilly, N., Aasen, H. and Bareth, G. 2015. Fusion of plant height and vegetation indices for the estimation of barley biomass. *Remote Sens.* 7(9), 11449-11480. https://doi.org/10.3390/rs70911449.

Tilly, N., Hoffmeister, D., Cao, Q., Huang, S., Lenz-Wiedemann, V., Miao, Y. and Bareth, G. 2014. Multitemporal crop surface models: Accurate plant height measurement and biomass estimation with terrestrial laser scanning in paddy rice. *J. Appl. Remote Sens.* 8(1), 083671. https://doi.org/10.1117/1.JRS.8.083671.

Torres-Sánchez, J., Peña, J. M., de Castro, A. I. and López-Granados, F. 2014. Multi-temporal mapping of the vegetation fraction in early-season wheat fields using images from UAV. *Comput. Electron. Agric.* 103, 104-113. https://doi.org/10.1016/j.compag.2014.02.009.

Verrelst, J., Camps-Valls, G., Muñoz-Marí, J., Rivera, J. P., Veroustraete, F., Clevers, J. G. P. W. and Moreno, J. 2015. Optical remote sensing and the retrieval of terrestrial vegetation bio-geophysical properties - A review. *ISPRS J. Photogramm.* 108, 273-290. https://doi.org/10.1016/j.isprsjprs.2015.05.005.

Verrelst, J., Malenovský, Z., Van der Tol, C., Camps-Valls, G., Gastellu-Etchegorry, J.-P., Lewis, P., North, P. and Moreno, J. 2019. Quantifying vegetation biophysical variables from imaging spectroscopy data: A review on retrieval methods. *Surv. Geophys.* 40(3), 589-629. https://doi.org/10.1007/s10712-018-9478-y.

Virlet, N., Sabermanesh, K., Sadeghi-Tehran, P. and Hawkesford, M. J. 2017. Field scanalyzer: An automated robotic field phenotyping platform for detailed crop monitoring. *Funct. Plant Biol.* 44(1), 143-153. https://doi.org/10.1071/FP16163.

Walter, A., Liebisch, F. and Hund, A. 2015. Plant phenotyping: From bean weighing to image analysis. *Plant Methods* 11, 14. https://doi.org/10.1186/s13007 015 0056-8.

Wan, L., Zhu, J., Du, X., Zhang, J., Han, X., Zhou, W., Li, X., Liu, J., Liang, F., He, Y. and Cen, H. 2021. A model for phenotyping crop fractional vegetation cover using imagery from unmanned aerial vehicles. *J. Exp. Bot.* 72(13), 4691-4707. https://doi.org/10.1093/jxb/erab194.

Warner, T. A. and Cracknell, A. P. 2017. Unmanned aerial vehicles for environmental applications. *Int. J. Remote Sens.* 38(8-10), 2029-2036. https://doi.org/10.1080/01431161.2017.1301705.

Yang, G., Liu, J., Zhao, C., Li, Z., Huang, Y., Yu, H., Xu, B., Yang, X., Zhu, D., Zhang, X., Zhang, R., Feng, H., Zhao, X., Li, Z., Li, H. and Yang, H. 2017. Unmanned aerial vehicle remote sensing for field-based crop phenotyping: Current status and perspectives. *Front. Plant Sci.* 8, 1111. https://doi.org/10.3389/fpls.2017.01111.

Zarco-Tejada, P. J. 2008. A new era in remote sensing of crops with unmanned robots. *SPIE Newsroom*. https://doi.org/10.1117/2.1200812.1438.

Zarco-Tejada, P. J., Camino, C., Beck, P. S. A., Calderon, R., Hornero, A., Hernández-Clemente, R., Kattenborn, T., Montes-Borrego, M., Susca, L., Morelli, M., Gonzalez-Dugo, V., North, P. R. J., Landa, B. B., Boscia, D., Saponari, M. and Navas-Cortes, J. A. 2018. Previsual symptoms of Xylella fastidiosa infection revealed in spectral plant-trait alterations. *Nat. Plants* 4(7), 432–439. https://doi.org/10.1038/s41477-018-0189-7.

Zarco-Tejada, P. J., González-Dugo, V., Williams, L. E., Suárez, L., Berni, J. A. J., Goldhamer, D. and Fereres, E. 2013. A PRI-based water stress index combining structural and chlorophyll effects: Assessment using diurnal narrow-band airborne imagery and the CWSI thermal index. *Remote Sens. Environ.* 138, 38–50. https://doi.org/10.1016/j.rse.2013.07.024.

Zenkl, R., Timofte, R., Kirchgessner, N., van Gool, L., Walter, A. and Aasen, H. in review. *Outdoor Wheat Segmentation With Deep Learning.* https://www.frontiersin.org/articles/10.3389/fpls.2021.774068/full.

Zhang, C. and Kovacs, J. M. 2012. The application of small unmanned aerial systems for precision agriculture: A review. *Precis. Agric.* 13(6), 693–712. https://doi.org/10.1007/s11119-012-9274-5.

Zou, X., Haikarainen, I., Haikarainen, I., Mäkelä, P., Mõttus, M. and Pellikka, P. 2018. Effects of crop leaf angle on LAI-sensitive narrow-band vegetation indices derived from imaging spectroscopy. *Appl. Sci.* 8(9), 1435. https://doi.org/10.3390/app8091435.

Zou, X. and Mõttus, M. 2015. Retrieving crop leaf tilt angle from imaging spectroscopy data. *Agric. For. Meteorol.* 205, 73–82. https://doi.org/10.1016/j.agrformet.2015.02.016.

Part 4

Data analysis

Chapter 8

Meeting computer vision and machine learning challenges in crop phenotyping

Hanno Scharr, Institute of Bio- and Geosciences: Plant Sciences (IBG-2) and Institute for Advanced Simulation: Data Analytics and Machine Learning (IAS-8), Forschungszentrum Jülich, Germany; and Sotirios A. Tsaftaris, The University of Edinburgh and Alan Turing Institute, UK

1 Introduction

Even without any subsequent processing, imaging data can aid in visualization and documentation in plant physiological experiments, breeding, or precision agriculture applications. However, 'image processing' or 'computer vision' (CV) techniques are needed to analyze images for obtaining further useful information such as measurements (plant size, colour, degree of pest/disease infestation, etc.) or other qualitative values (e.g. whether plants are healthy or stressed). This further step has been described as performing image-based 'registration of physiological parameters' (Demidchik et al., 2020). In plant phenotyping, such measurements and values are combined with omics and environmental data to draw conclusions about plant physiology and its underlying mechanisms (Fiorani and Schurr, 2013). There is a plethora of different imaging techniques addressing a wide range of physiological parameters (see, e.g. Fiorani and Schurr, 2013; Demidchik et al., 2020; Li et al., 2020 for systematic overviews). Some combined analytical tools also exist (see Lobet et al., 2013, and Mochida et al., 2019).

http://dx.doi.org/10.19103/AS.2022.0102.11

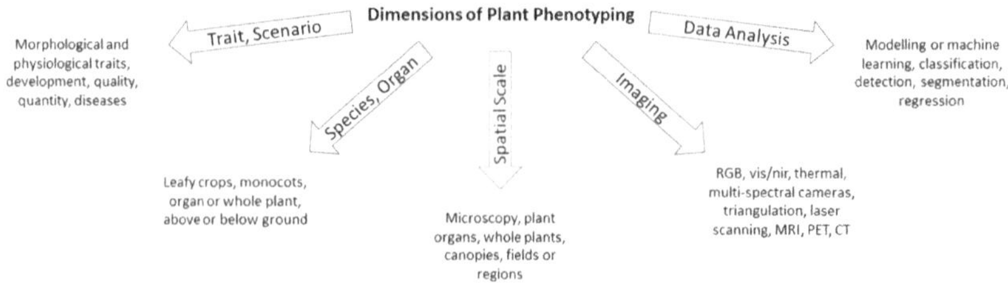

Figure 1 Dimensions of plant phenotyping.

Non-invasive, image-based methods are increasingly utilized in plant phenotyping. The resulting images need to be analyzed in a high-throughput, robust, accurate, and reliable manner (Minervini et al., 2015b). The problems later in the paragraph differ from the usual tasks addressed by the CV community because of the unique challenges involved. Plants are complex, self-changing systems whose complexity increases over time. Typical phenotyping problems include measuring the size, shape, 3D surface structure, architecture, and other structural traits of plants and their organs (leaves, fruit, roots, etc.). Phenotyping often requires quantitative description of plant populations. Key problems include reliable detection and multi-label segmentation of many similar objects or the reconstruction of reflecting, almost featureless and overlapping, surfaces. Quantitative description of the growth of these complex, deforming objects is vital and requires suitable tracking, optical flow, and/or scene flow estimation methods. Tracked plant objects change their appearance over time. In some cases, images may be acquired under controlled conditions to limit variability, but they are increasingly likely to be taken in more challenging natural environments like greenhouses, in the field, or by remote sensing drones or satellites.

In this chapter, we do not aim to give a complete overview of applications, tools, and challenges in CV. We aim to pick out key themes and highlight recent review papers describing the current landscape of image-based plant phenotyping or phenomics and some of its key dimensions (Fig. 1). It is necessary to start with relevant dimensions of interest (trait or scenario, plant type, organ, and scale). This then makes it possible to select suitable imaging sensors and CV tools as well as address challenges in their use.

2 Key dimensions to consider in computer vision applications in plant phenotyping

The key dimensions to consider are as follows:

1 Trait or scenario: Typically, a distinction is made between morphological and physiological traits, development or post-harvest properties, and pest

infestations and diseases. Jiang and Li (2020) distinguish between plant stress, plant development, and post-harvest quality. Six typical agricultural application scenarios highlighted in Tian et al. (2020) are as follows:

- growth monitoring of crops;
- detection of crop diseases, insect pests, and weeds;
- automated crop harvesting;
- classification and inspection of agricultural products;
- automated farm management; and
- monitoring of farmland information with unmanned aerial vehicles (UAV).

In contrast, Patricio and Rieder (2018) use categories such as diseases, quality, phenotyping, and phenology.

2 Plant type and organ: CV needs to evaluate the properties of dicots or leafy crops (Arabidopsis, tobacco, and salad), monocots (maize, rice, wheat, and barley) or trees, and their fruits (apples and avocado) which differ significantly. A list of approx. 20 examples of properties for different species can be found in the study by Paul et al. (2020). Li et al. (2020) classify plants by organ or whole plant, above and below the ground. Patricio and Rieder (2018) focus on grain crops (rice, soybean, wheat, maize).

3 Scale: Spatial scale ranges from microscopic analysis of cells to measurements of plant organs and whole plants, up to plant canopies, airborne or even satellite remote sensing of fields or regions (Minervini et al., 2015b). Plant organ and whole plant studies are at the core of plant phenotyping, while cell-scale studies are often characterized as plant physiology, and field-scale applications are more often classified under precision agriculture.

These dimensions then raise a range of technical issues, that is, hardware constraints (imaging technique and devices, carrier systems) and data analysis needs (classic solutions or machine learning (ML), data availability). Technical issues include the following:

4 Imaging technique: Depending on the trait, plant type, organ, and the required scale, different imaging methods may be required. For example, the structure of root systems in soil in smaller pots can be measured mainly non-invasively by magnetic resonance imaging (MRI) or computed tomography (CT). Roots of seedlings in transparent agar can be imaged using red, green, blue (RGB) or near infra-red (NIR) cameras while, at the field scale, there are no non-invasive imaging methods available. Fiorani and Schurr (2013), Mochida et al. (2019),

and Demidchik et al. (2020) summarize the most common techniques and devices, for example, RGB, Vis/NIR, thermal or multi-/hyperspectral cameras, stereo and multi-camera systems, 3D by time-of-flight, as well as laser scanning, triangulation, volumetric imaging by MRI, PET, and CT. Fiorani and Schurr (2013) also relate imaging techniques to examples of applications as well as identify their pros and cons. Demidchik et al. (2020) list the most common commercial carrier systems for either plant-to-sensor or sensor-to-plant solutions (Fiorani and Schurr, 2013). Li et al. (2020) provide examples of different high-throughput plant phenotyping platforms.

5 Data analysis: Methods and algorithms for image analysis are currently categorized between 'classic' solutions, based on model assumptions made by experts, and machine learning solutions, based on available data. Fiorani and Schurr (2013) list seven 'classic' software packages for root morphometric and growth dynamic parameters based on 2D imaging of roots grown in a variety of growth media. Li et al. (2020) classify approx. 30 different 2D plant organ image segmentation methods according to the type of underlying method. ML-based solutions are usually categorized as classification, detection, segmentation, and regression methods. Evaluating qualitative status (e.g., quality, health) typically uses a classification method, while localizing objects and (sometimes) counting is done via detection or segmentation if analysed at pixel level. Measuring or estimating quantitative status (size, shape, colour, number) would typically be considered a task for regression methods. A detailed and comprehensive review of neural networks in plant phenotyping is provided by Jiang and Li (2020), including a representative list of the most used deep neural network (DNN) architectures, public code resources, and a summary of DNN-based data analysis approaches in imaging-based plant phenotyping.

Over the last decade, the CV revolution, which has driven rapid progress in plant phenotyping, has been facilitated by DNNs and other ML techniques. In addition to the new and improved ML-based plant phenotyping solutions under development, there have also been advances in underlying methods. Techniques are data-hungry, and appropriate training and testing data are indispensable for developing and evaluating novel approaches. Most solutions are mostly trained offline in a supervised fashion. This requires large amounts of training data, usually many thousands of images, together with the desired 'ground truth' output (to validate the accuracy of classification). Such output can be bounding boxes tightly drawn around the object to be detected. Accurate segmentation pixel-accurate labels may also be needed if tasks remain unsolved by ML. By definition, these are usually generated or cleaned up manually to offset the limitations of automated methods (see Fig. 2). Labelling can be done by citizen scientists, provided tasks are clearly described and do not need expert knowledge (see Giuffrida et al., 2018a).

Solutions trained on a certain set of data are specific to that data. This means it is unclear how well these solutions generalize for different data. This means current solutions suffer from high specificity, that is, they only work for the particular range of data they have been trained for. Detecting data that is not in a sample distribution is an active field of research and makes it possible to adapt available solutions to novel scenarios, that is, to perform a domain shift (to enhance the performance of an algorithm based on a training set in analysing novel data sets). A well-known example is agricultural robotics where methods developed on one set of agricultural fields may not work well on other fields. Solutions such as active learning strategies have been proposed to minimize the need for additional training data on new fields (see, e.g. Sheikh et al., 2020). The key to a good solution is the availability of suitable training data that well describe the application task to be addressed. In developing effective ML solutions, ML experts and application experts, practitioners, or scientists can and should work together symbiotically (see, e.g. Tsaftaris and Scharr, 2019). In the following sections, we describe how this symbiosis can be leveraged to advance CV and crop phenotyping. Together, practitioners and scientists can actively produce better data analysis solutions to address specific plant phenotyping or agricultural problems.

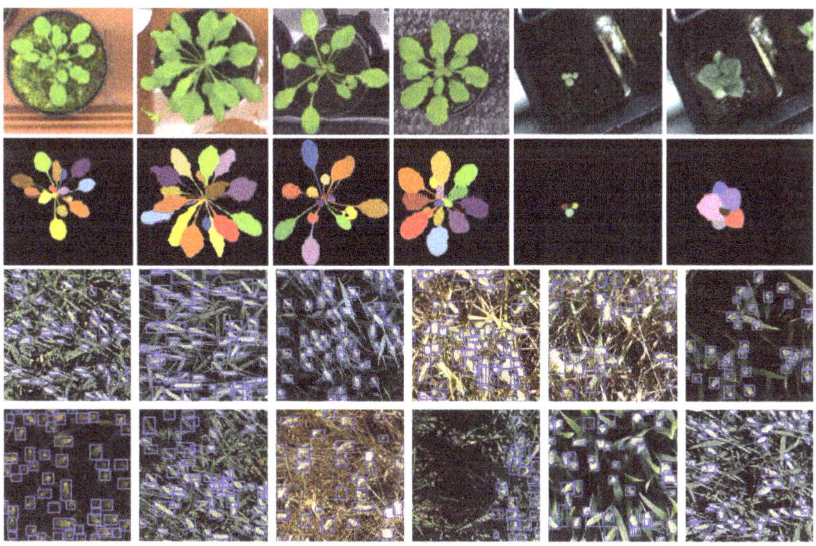

Figure 2 Images from typical training data sets for (top) 'instance segmentation' (Scharr et al., 2014) and (bottom) 'object detection' (David et al., 2020).

3 Creating synergies between research communities: the Computer Vision Problems in Plant Phenotyping (CVPPP) Workshop

In 2014 (and still, to some extent, today) CV and ML researchers were largely unaware of plant phenotyping as a potential area of application for CV techniques. In addition, appropriate data for agricultural and other plant-related image-based problems were lacking, making image processing a persistent bottleneck in plant phenotyping and other agricultural applications (see Minervini et al., 2015b).

Image-based crop phenotyping poses many significant challenges for CV and ML. Image data and the questions it can help answer are the provinces of the crop specialists while CV techniques and solutions are the domain of CV researchers. Bringing crop specialists and CV experts together and introducing this fascinating and important field to CV/ML researchers were therefore our primary goals when organizing the 'Computer Vision Problems in Plant Phenotyping' (CVPPP) Workshop series.

We observed that stand-alone workshops on specific applications mostly attract experts already interested in that area but rarely bring in *new* people. We therefore planned a workshop series in conjunction with major international CV conferences. In order to attract both crop scientists and CV experts, the CVPPP workshop series focused on the challenges and state of the art in CV for plant phenotyping. This not only allows CV/ML researchers new to crop science to explore this new field but also provides an incentive to authors to present their work since such workshops are highly prestigious publication venues. To ensure sufficient quality, double-blind peer review of workshop papers is overseen by a programme committee of well-established experts. Secondly, accepted papers are published as part of the official workshop proceedings of the main CV conference since this is a priority for CV/ML researchers. In addition, to broaden access, we also decided to accept extended abstracts to the workshop. Abstracts undergo peer review; they are not published in the official proceedings but on the relevant workshop website (see online Tsaftaris et al., 2020). We also organized a special issue (Scharr et al., 2016a) in a well-known CV journal including extended versions of selected workshop contributions.

Workshops addressing developments in CV for particular applications are not unusual. Publications typically present some novel algorithm or other CV solution. However, up until 2014, there was no publication channel to present work on unsolved crop phenotyping challenges. We therefore proposed 'data set papers' as a novel type of submission allowing crop experts to discuss key unsolved problems (see, e.g. Uchiyama et al., 2017; Skovsen et al., 2019; Mortensen et al., 2019; Scharr et al., 2020; for recent examples). Such papers introduce challenging CV problems in plant phenotyping, accompanied by benchmark data sets and suitable performance evaluation methods. This approach means that imaging scenarios and specific CV problems can be described from a plant science and imaging perspective. Such papers need to include benchmark data, that is, data sets required to develop ML solutions. These publications may also suggest a possible baseline solution for further discussion and resolution.

As an example of how to showcase a CV problem in plant phenotyping, we started the 'Leaf Segmentation Challenge' (LSC, 2014) (see online Tsaftaris and Scharr, 2014) and 'Leaf Counting Challenge' (LCC, 2015) as described in the next section.

4 Data challenges to accelerate progress in computer vision techniques: leaf counting and segmentation

To address the lack of awareness of plant phenotyping and agricultural CV/ML challenges, we encouraged the collection of plant data for several phenotyping tasks. These are described in Scharr et al. (2014) and Minervini et al. (2015a) and on a dedicated webpage (see online Tsaftaris et al., 2016). Our objectives in selecting the 'right' challenges in collecting data were not only to assess how well existing ML algorithms could solve major phenotyping tasks but also to expand the community of scientists addressing phenotyping applications. The challenges addressed two 'hot' CV/ML problems that concern many CV researchers:

- leaf segmentation as a 'segmentation' problem; and
- leaf counting as an 'object counting' problem.

They both pose specific challenges for CV algorithms.

Looking at the first, there are different types of segmentation tasks:

- semantic segmentation;
- instance segmentation; and
- more recently, plenoptic segmentation.

Semantic segmentation aims at assigning an appropriate class label to every pixel in an image, for example, leaf, soil, pot, table, etc. This means every leaf in an image is assigned to the same class and thus gets the same label. In our context, this would boil down to the much simpler task of 'foreground/background' segmentation, mainly an advanced version of finding green pixels in an image. We did not ask for this because it was too simplistic to be of interest to CV researchers. Instance segmentation and plenoptic segmentation require giving every instance of a class a different label, for example, leaf 1, leaf 2, leaf 3, etc. In instance segmentation, this is done only for a single object class. In plenoptic segmentation, this is done for a wide variety of classes simultaneously. In 2014, instance segmentation was feasible but still very challenging, suggesting the right degree of difficulty to make the problem interesting and attractive to CV researchers.

Counting objects using an effective instance segmentation solution is simple. However, training an instance segmentation solution requires labelled images such as the ones shown in Fig. 2. Such labelled images are expensive to obtain for unsolved problems. However, counting can be learned with extremely simple labels, that is, the number of visible leaves in an image rather than the pixel-accurate labelling. We therefore presented counting objects as a separate challenge in 2015. Counting has also been addressed in other CV contexts, for example, counting pedestrians in images or counting cells in microscopy images, and is still considered an interesting problem.

The challenges were not an immediate success. It took some effort to promote them, for example, in 2015, we offered participants the opportunity to co-author a collation study summarizing results to date (see Scharr et al., 2016b). Through continuous promotion over a few years, the data sets became very popular, even reaching 'standard data set' status for instance segmentation and object counting. Most importantly, we saw tremendous progress in solving these two tasks. We believe promoting the challenges was vital for their success. Simply making data available without suitable promotion does not work since dozens of challenges are made available every year (see online van Ginneken et al., 2012; Kaggle, 2019; Codalab, 2012). There are also thousands of data sets available (see online, e.g. Fisher, 2019), even if only some address plant-related tasks. In order to be successful, a challenge needs to be highlighted, be easy to work on, and yield immediate results of value to CV researchers.

In the following discussion, we offer advice on making problems interesting and corresponding data useful for the CV/ML community. In order to make data challenges most attractive and sustainable, challenges need to be designed such that they follow FAIR+ principles (**F**easiblity, **A**menability, **I**ncentives, **R**eproduciblity, and **C**onvenience, see also Tsaftaris and Scharr, 2019). These are discussed in more detail below:

- Feasibility highlights a task that is challenging but solvable. In general, standard approaches achieve above chance predictions, but more sophisticated methods are expected to achieve more precise results.
- Amenability means describing the domain-specific problem in a way that is easily understandable for CV/ML experts who are generally not crop scientists.

- Incentives can be rewards (monetary or not) given for the most accurate, fastest, most concise, or other 'best' solution or means to offer rapid, visible publication routes for CV/ML researchers. Incentives can be financial, but full papers at international conferences or offering a special issue in a suitable journal are also attractive.
- Reproducibility in each case is given through 'ground truth' which involves a well-defined splitting of training, validation and test data, as well as standardized evaluation procedures. This means performance can be measured by defined metrics and compared between different solutions. Ideally, results are collected on a leader board on a suitable internet website.
- Convenience: Data should be made available in standard formats together with scripts for easy reading and annotation.

Providing and promoting an attractive data set is an achievable goal without the need to be an ML expert or invest in significant resources. However, it requires some work and effort. It may therefore be worth considering collaborating with existing challenge initiatives and contributing data to support them. In the following section, we list recent challenges and plant data sets from our own and other sources that may offer opportunities for synergies.

5 Recent agriculture-related computer vision challenges

5.1 The challenges

The creation of data-driven challenges to advance the state of the art in CV is not a new idea. In this section, we provide a description of CV challenges related to plants and agriculture. The methodology we followed was to first use searches on Google and Google Scholar. We found papers that mentioned challenges and we traced back the original reference. In addition, we used common databases where information on relevant challenges is sometimes recorded. It is quite possible that some challenges may have been missed and we apologize to those concerned. The latter highlights the need for a central location to collect such information.

5.1.1 The leaf segmentation and counting challenges *(LSC, LCC, 2014 onwards; see online Tsaftaris et al., 2016)*

The competition was organized for the first time in 2014 as part of the inaugural CVPPP workshop, and these challenges aim to develop algorithms that can segment or count each leaf in images of Arabidopsis and tobacco plants. The competition initially operated via email submissions and now partly runs on

Codalab. As of now (April 2022), the accompanying data set (Minervini et al., 2015a) has been downloaded 7000 times (see online Tsaftaris et al., 2016, 'impact'). As reported in Tsaftaris and Scharr (2019), the availability of this challenge and data set has led to an almost fourfold reduction in counting error and a 20% improvement in leaf segmentation. Currently, solutions are dominated by deep-learning multi-instance segmentation approaches.

5.1.2 PlantCLEF *(2011 onwards; see online Sanderson and Clough, 2003)*

This competition has been one of the longest-running competitions largely targeted at image retrieval and image classification. Organized under the Cross-Language Education and Function (CLEF) umbrella, this competition has evolved to consider different images of plants (and leaves). To the best of our knowledge, the competition started to focus on plants in 2011, as reported by Goëau et al. (2011), with the goal of identifying tree species from leaf images.

5.1.3 The global wheat challenge *(2020; see online David et al, 2020)*

This challenge was organized for the first time in 2020 using the global wheat head data set (David et al, 2020) to develop algorithms that automatically count wheat ears in images. The competition ran on the Kaggle platform from 4 May to 4 August 2020 and was held in conjunction with CVPPP 2020. The data set, provided by an international consortium, had 190 000 images of wheat ears available. The first prize is US$8000, the second is US$4000, and the third US$3000, sponsored by the Global Institute for Food Security at the University of Saskatchewan, Canada. More than 2200 teams participated, with the winning solution from a team in Hanoi (Vietnam). The winning solution used a state-of-the-art model for object detection (Tan et al., 2020) with a considerable number of augmentations (DungNB, 2020).

5.1.4 The wheat rust challenge *(2020; see online Zindi, 2020a)*

Wheat rust is a devastating plant disease that affects many African crops, reducing yields, affecting the livelihoods of farmers, and decreasing food security across the continent. The disease is difficult to monitor over a large scale, making it difficult to control and eradicate. The objective of this challenge is to build an ML algorithm to correctly classify if a plant is healthy or has stem or leaf rust. The challenge was held in conjunction with the Computer Vision for Agriculture (CV4GC) Workshop (Cv4gc, 2020). About 900 images were made available to train algorithms on a classification task. The winning solution by Picek and Valan relied on Inception V4 and Inception-ResNet V2 with significant augmentations during training and testing (Picek, 2020).

5.1.5 Crop detection from satellite imagery *(2020; see online Zindi, 2020b)*

The objective of this competition was to create an ML model to classify fields by crop type from images collected by the Sentinel-2 satellite during the growing season. The fields in this training set were across western Kenya, and the images were collected by PlantVillage (PlantVillage, 2021). The challenge was held in conjunction with the Computer Vision for Agriculture (CV4GC) Workshop (Cv4gc, 2020). The data set contained more than 4000 field images using 12 bands of observations from the Sentinel-2 L2A satellite. The winning solution by Karim Magdy Amer used a fully convolutional neural network with considerable augmentations (Amer, 2020).

5.1.6 Agriculture-Vision *(2020 and 2021; see Chiu et al., 2020a and 2020b)*

The data set used in this challenge is a subset of the Agriculture-Vision dataset (Chiu et al., 2020b). The challenge data set contained about 21 000 aerial farmland images captured throughout 2019 across the United States using RGB and NIR sensors. The task was to obtain a per-pixel semantic segmentation to distinguish farmland and waterways (including annotations on field anomalies). The winner of the 2020 challenge was Team Hyunseong: Hyunseong Park, Junhee Kim, Sungho Kim (Agency for Defense Development, South Korea). The method used was Residual DenseNet with Expert Network for Semantic Segmentation. This challenge was held in 2020 and repeated in 2021 in conjunction with a workshop (Agriculture-Vision, 2021).

5.1.7 The Plant Pathology 2020 challenge *(see online Plant Pathology, 2020)*

This challenge invited models to address a fine-grained classification problem of diseases on plant leaves, based on a bespoke data set created for this purpose (Thapa et al., 2020). It was held in conjunction with a workshop on finegrained recognition at the IEEE/CVF Conference on Computer Vision and Pattern Recognition (CVPR) (see online Mac Aodha et al., 2020a). The competition had 1300 teams, and the winning solution (by a team from China) used a recent ResNet architecture (Hu et al., 2020) with simple data augmentation (see online Mac Aodha et al., 2020b).

5.1.8 Grass clover data set semantic segmentation and biomass composition challenges *(2019 onwards; see online Skovsen, 2019)*

This challenge came directly from a CVPPP workshop and the 'data set papers' category. Presented first as a data set at CVPPP 2019 (Skovsen et al., 2019), this challenge invites participants to segment images to identify grass clover and estimate biomass. Overall, the organizers provide both synthetic and real images (about 30 000) with only few of the real images labelled at pixel level. At present, six participants have participated in these recurring challenges, but unfortunately, we cannot source information about the solutions used. A sister challenge on oil radish is also available (see online Mortensen, 2019), again inspired by a data set paper at CVPPP 2019 (Mortensen et al., 2019).

We have attempted to catalogue as many recent competitions as possible, from which we can see several trends. For example, classification competitions are more common and competitions that offer rewards are more popular. Labelling images with classes is much easier than competitions that require, for example, instance segmentation such as the Leaf Segmentation Challenge. Financial rewards naturally attract community attention and widen participation which is very welcome. We observe that solutions tend to rely on fully supervised methods and use augmentations to boost performance and edge higher on leader boards. Some of these 'augmentation' tricks, however, tend to produce methods that do not generalize well in the 'wild' (i.e. when applied to new data from the lab or the field). However, some tricks such as augmentation are necessary to offset data sets that can be small in size. Moving forward, we are confident that competitions will be a well-established mechanism for advancing the state of the art in CV for agricultural applications. Unfortunately, for those readers interested in this field, there is no single point of reference to consult. We advise readers to follow a similar methodology as the one discussed in the beginning of this section to keep up to date with developments.

5.2 Finding data sets

The previous section described challenges applied to existing data sets. These data sets can be the source of future challenges. The concluding section suggests that future challenges may benefit from agglomerating similar data sets. The examples are the LSC and LCC challenges from 2017 onwards, which included an additional Arabidopsis data set by Bell and Dee (2016). In this section, we provide sources where typically data sets and accompanying papers are listed:

- Lobet et al. (2013) offer also a curated list of plant data sets.
- Lu and Young (2020) provide a recent survey of data sets. Unfortunately, it is not comprehensive since several of the data sets available in the curated lists described previously are not included.
- Hasan et al. (2021) list several data sets on crops or weeds.
- Fisher (2019) provides a curated list of data sets in several CV domains, including agriculture.
- Kaggle (2019) can be consulted using the tag 'plants' to retrieve a repository of relevant data sets.
- Tensorflow (2015) offers a collection of plant data sets in their catalogue.

However, because none of the above sources are considered definitive, there remains a need for a careful review of the literature to identify new data sets. We hope that the authors of the papers will ensure that their data sets are appropriately listed so that they are easily discoverable. In addition, we hope to see the establishment of a single, universally accepted source where all relevant data sets are listed.

6 Summary

Overall, we are enthusiastic about the progress made over the last few years. Thinking ahead, we want to conclude this chapter by offering some general observations and suggestions on how the communities can come together and ensure future progress.

While initially challenges were geared towards developing vision solutions where more fine-grained and detailed information (useful for understanding plant function) can be extracted, we have recently seen challenges where less fine-grained output is required. More fine-grained information typically leads to segmentation and tracking and is useful, for example, in monitoring plant growth. However, this is a laborious and time-consuming process that results in competitions based on small(er) data sets.

Recent challenges contain thousands of images but with coarser annotation (e.g. presence or not of disease). Several of these challenges are geared towards robotic and remote sensing (satellite or drones) applications at a larger scale in terms of imaging resolution (e.g. level of canopy). This contrasts with past competitions which sought to design vision tasks for extracting information at the organ level (e.g. leaf). This evolution has both positive and negative implications in advancing the state of the art in this area. On the positive side, if a biological problem aligns with classification-level tasks, for example, a breeder looking for a general response in a field, this new development will meet such needs. On the negative side, such classification-level tasks may not lead to rapid advancements for more fine-grained and biologically motivated problems, for example, leaf shape in response to stress.

We suggest that plant scientists interested in fine-grained applications openly release their data and invite other communities to work on them. We encourage work across disciplines to identify the right way of posing problems that can attract sustained interest from the CV community.

As noted earlier, setting up a challenge does not only mean releasing a data set on an open server. It requires data that is sufficiently large (in number), easy to access, easy to understand, and, of course, with appropriate image-level annotation and description. If the solution needs to work for data from different labs, data variability is essential.

We suggest working together with other plant or crop scientists so that data sets become rich and variable.

An interesting CV problem becomes even more interesting when we discuss issues about generalization and ability to adapt to new settings, as we have demonstrated in Giuffrida et al. (2018b). It can be useful to agglomerate data sources and improve performance (Dobrescu et al., 2017, 2020). Based on examples such as autonomous driving or general CV (Ranftl et al., 2020, the more data sets we can combine the better. Given that collecting plant image data is hard (particularly when originating from controlled experiments), one dataset will never be enough unless it is huge!

We suggest identifying common vision problems and then creating data sets to address a similar problem.

Finally, it is very important to sustain interest to make a true impact. Competitions should offer monetary incentives where possible but, at the same time, should be sustainable to ensure continued progress. 'One off' competitions may not lead to solutions that can translate to practice. After all, the incentive of competition participants is to win the challenge and it is quite possible that their solutions are not easy to translate (e.g. too complex) or generalize in the 'wild' (i.e. due to overfitting the training and testing sets from the competition).

It is important that competitions are designed to reflect as much as possible the translation requirements of users of the solutions.

All of the above recommendations pose the question: What are the new (sustainable) analysis challenges that we should focus on? Clearly, this requires a collaboration between plant scientists, computer scientists, and engineers as well as industry (including breeders). Despite the many events or working groups that aim to bring us together, a 'moon-shot' challenge (inspired by the recent advances we see in reaching Mars) has not yet surfaced. This suggests repurposing the question: If we do not know what challenges may emerge, how do we design mechanisms that can identify new challenges for the future? We do not know the answer, but we invite readers to approach us with suggestions.

To conclude, we are certain that considerable advances can be made by exploiting competitions and challenges to advance the state of the art and bring communities together. We can and should work together to shape how progress can be made in the future.

7 Where to look for further information

Section 5 describes the currently most relevant publications and websites for finding challenges and datasets and provides good starting points for looking for further information. Computer vision for crop phenotyping and agricultural applications is often addressed in special events like workshops in conjunction with major computer vision conferences. Computer vision conference lists which can be searched for 'agriculture' exist, see for example:

- http://conferences.visionbib.com.
- https://conferenceindex.org/conferences/computer-vision.

Organizations which have in the past addressed events bringing computer vision and plant experts together are for example:

- The international plant phenotyping network (IPPN, https://www.plant-phenotyping.org/).
- The European plant phenotyping network (EPPN, https://www.plant-phenotyping-network.eu/).
- The PhenomUK plant phenotyping network in the UK (PhenomUK, https://www.phenomuk.net).
- National plant phenotyping networks as listed e.g. here https://emphasis.plant-phenotyping.eu/phenotyping-landscape/national-communities.

Examples of major projects, research centers, or scientific institutions addressing crop phenotyping in conjunction with computer vision (and usually also robotics):

- The German cluster of excellence PhenoRob (https://www.phenorob.de/).
- The University of Saskatchewan plant phenotyping and imaging research center (P2IRC, https://p2irc.usask.ca/).

- The Netherlands plant eco-phenotyping centre (NPEC, https://www.npec.nl/).
- The Australian plant phenomics facility (APPF, https://www.plantphenomics.org.au).
- The Jülich plant phenotyping centre (JPPC, https://www.fz-juelich.de/jppc).
- The EPSRC Centre for Doctoral Training in Agri-Food Robotics in the UK (AgriFoRwArdS, https://agriforwards-cdt.blogs.lincoln.ac.uk).

However, when looking for up-to-date information an internet search using a combination of the search phrases 'computer vision', 'challenge', 'workshop', 'agriculture', 'phenotyping', and/or 'crop' using ones favorite search engine yields many useful links for further reading.

8 References

Agriculture-Vision. (2021). Agriculture-vision: Challenges and opportunities for computer vision in agriculture. Available at: https://www.agriculture-vision.com/agriculture-vision-2021 (accessed 20 April 2021).

Amer, M. K. (2020). Meet Karim Magdy Amer, winner of the Radiant Earth Crop Detection Challenge and deep learning computer vision expert. Available at: https://zindi.africa/learning/meet-karim-magdy-amer-winner-of-the-radiant-earth-crop-detection-challenge-and-deep-learning-computer-vision-expert (accessed 19 April 2021).

Bell, J. and Dee, H. M. (2016). *Aberystwyth Leaf Evaluation Dataset*. https://doi.org/10.5281/ZENODO.168158.

Chiu, M. T., Xu, X., Wang, K., Hobbs, J., Hovakimyan, N., Huang, T. S. and Shi, H. (2020a). The 1st agriculture-vision challenge: Methods and results. In: *Proceedings of the IEEE/CVF Conference on Computer Vision and Pattern Recognition (CVPR) Workshops*. Available at: https://github.com/SHI-Labs/Agriculture-Vision (accessed 19 April 2021).

Chiu, M. T., Xu, X., Wei, Y., Huang, Z., Schwing, A. G., Brunner, R., Khachatrian, H., Karapetyan, H., Dozier, I., Rose, G., Wilson, D., Tudor, A., Hovakimyan, N., Huang, T. S. and Shi, H. (2020b). Agriculture-vision: A large aerial image database for agricultural pattern analysis. In: *Proceedings of the IEEE/CVF Conference on Computer Vision and Pattern Recognition (CVPR)*. Available at: https://github.com/SHI-Labs/Agriculture-Vision (accessed 19 April 2021).

Codalab. (2012). CodaLab competitions. Available at: https://competitions.codalab.org/ (accessed 19 April 2021).

Cv4gc. (2020). Computer vision for global challenges. Available at: http://cv4gc.org (accessed 19 April 2021).

David, E., Madec, S., Sadeghi-Tehran, P., Aasen, H., Zheng, B., Liu, S., Kirchgessner, N., Ishikawa, G., Nagasawa, K., Badhon, M. A., Pozniak, C., de Solan, B., Hund, A., Chapman, S. C., Baret, F., Stavness, I., and Guo, W. (2020). Global wheat head detection (GWHD)dataset: A large and diverse dataset of high-resolution RGB-labelled images to develop and benchmark wheat head detection methods. *Plant Phenomics* 2020, 3521852. https://doi.org/10.34133/2020/3521852. Available at: http://www.global-wheat.com/2020-challenge (accessed 19 April 2021).

Demidchik, V. V., Shashko, A. Y., Bandarenka, U. Y., Smolikova, G. N., Przhevalskaya, D. A., Charnysh, M. A., Pozhvanov, G. A., Barkosvkyi, A. V., Smolich, I. I., Sokolik, A. I., Yu, M. and Medvedev, S. S. (2020). Plant phenomics: Fundamental bases, software and hardware platforms, and machine learning. *Russian Journal of Plant Physiology* 67(3), 397–412. https://doi.org/10.1134/S1021443720030061.

Dobrescu, A., Giuffrida, M. V. and Tsaftaris, S. A. (2017). Leveraging multiple datasets for deep leaf counting. In: *Proceedings of the IEEE International Conference on Computer Vision (ICCV)*, pp. 2072–2079.

Dobrescu, A., Giuffrida, M. V. and Tsaftaris, S. A. (2020). Doing more with less: A multitask deep learning approach in plant phenotyping. *Frontiers in Plant Science* 11, 141. https://doi.org/10.3389/fpls.2020.00141.

DungNB. (2020). Kaggle Discussion 1st place solution [MIT-Compliant]. Available at: https://www.kaggle.com/c/global-wheat-detection/discussion/172418 (accessed 19 April 2021).

Fiorani, F. and Schurr, U. (2013). Future scenarios for plant phenotyping. *Annual Review of Plant Biology* 64(1), 267–291.

Fisher, R. (2019). Cv online: Image databases. Available at: https://homepages.inf.ed.ac.uk/rbf/CVonline/Imagedbase.htm (accessed 19 April 2021).

Giuffrida, M. V., Chen, F., Scharr, H. and Tsaftaris, S. A. (2018a). Citizen crowds and experts: Observer variability in image-based plant phenotyping. *Plant Methods* 14, 12. https://doi.org/10.1186/s13007-018-0278-7.

Giuffrida, M. V., Doerner, P. and Tsaftaris, S. A. (2018b). Pheno-deep counter: A unified and versatile deep learning architecture for leaf counting. *Plant Journal: For Cell and Molecular Biology* 96(4), 880–890. https://doi.org/10.1111/tpj.14064.

Goëau, H., Joly, P. B. A., Boujemaa, N., Barthelemy, D., Molino, J.-F., Birnbaum, P., Mouysset, E. and Picard, M. (2011).'The CLEF 2011 plant images classification task. In: *CLEF 2011 Working Notes*, Amsterdam, The Netherlands. Available at: https://www.imageclef.org/2011/Plants (accessed 19 April 2021).

Hasan, A. S. M. M., Sohel, F., Diepeveen, D., Laga, H. and Jones, M. G. K. (2021). A survey of deep learning techniques for weed detection from images. *Computers and Electronics in Agriculture* 184, 106067. https://doi.org/10.1016/j.compag.2021.106067.

Hu, J., Shen, L., Albanie, S., Sun, G. and Wu, E. (2020). Squeeze-and-excitation networks. *IEEE Transactions on Pattern Analysis and Machine Intelligence* 42(8), 2011–2023. https://doi.org/10.1109/tpami.2019.2913372.

Jiang, Y. and Li, C. (2020). Convolutional neural networks for image-based high-throughput plant phenotyping: A review. *Plant Phenomics* 2020, Article ID 4152816, 22 pages. https://doi.org/10.34133/2020/4152816.

Kaggle. (2019). Competitions. Available at: https://www.kaggle.com/competitions (accessed 19 April 2021).

Li, Z., Guo, R., Li, M., Chen, Y. and Li, G. (2020). A review of computer vision technologies for plant phenotyping. *Computers and Electronics in Agriculture* 176, 105672. https://doi.org/10.1016/j.compag.2020.105672.

Lobet, G., Draye, X. and Périlleux, C. (2013). An online database for plant image analysis software tools. *Plant Methods* 9, 38. https://doi.org/10.1186/1746-4811-9-38.

Available at: https://www.quantitative-plant.org/software (accessed 19 April 2021) and Available at: https://www.quantitative-plant.org/dataset (accessed 19 April 2021).

Lu, Y. and Young, S. (2020). A survey of public datasets for computer vision tasks in precision agriculture. *Computers and Electronics in Agriculture* 178.

Mac Aodha, O., Farrell, R., Maji, S., Kaeser-Chen, C. and Van Horn, G. (2020a). Competitions. In: *The Seventh Workshop on Fine-Grained Visual Categorization (FGVC7)*. Available at: https://sites.google.com/view/fgvc7/competitions (accessed 21 April 2021).

Mac Aodha, O., Farrell, R., Maji, S., Kaeser-Chen, C. and Van Horn, G. (2020b). 1st place solution (single model). In: *The Seventh Workshop on Fine-Grained Visual Categorization (FGVC7)*. Available at: https://www.kaggle.com/c/plant-pathology-2020-fgvc7/discussion/154056 (accessed 21 April 2021).

Minervini, M., Fischbach, A., Scharr, H. and Tsaftaris, S. A. (2015a). Finely-grained annotated datasets for image-based plant phenotyping. *Pattern Recognition Letters*, 1–10. https://doi.org/10.1016/j.patrec.2015.10.013.

Minervini, M., Scharr, H. and Tsaftaris, S. A. (2015b). Image analysis: The new bottleneck in plant phenotyping [applications corner]. *IEEE Signal Processing Magazine* 32(4), 126–131. https://doi.org/10.1109/msp.2015.2405111.

Mochida, K., Koda, S., Inoue, K., Hirayama, T., Tanaka, S., Nishii, R. and Melgani, F. (2019). Computer vision-based phenotyping for improvement of plant productivity: A machine learning perspective. *GigaScience* 8(1, January), giy153. https://doi.org/10.1093/gigascience/giy153.

Mortensen, A. K. (2019). Oil radish dataset semantic segmentation and yield estimation challenges. Available at: https://competitions.codalab.org/competitions/20981#learn_the_details (accessed 21 April 2021).

Mortensen, A. K., Skovsen, S., Karstoft, H. and Gislum, R. (2019). The oil radish growth dataset for semantic segmentation and yield estimation. In: *Proceedings of the IEEE/CVF Conference on Computer Vision and Pattern Recognition (CVPR) Workshops* 2019, pp. 2703–2710.

Patrício, D. I. and Rieder, R. (2018). Computer vision and artificial intelligence in precision agriculture for grain crops: A systematic review. *Computers and Electronics in Agriculture* 153, 69–81. https://doi.org/10.1016/j.compag.2018.08.001.

Paul, A., Ghosh, S., Das, A. K., Goswami, S., Das Choudhury, S. and Sen, S. (2020). A review on agricultural advancement based on computer vision and machine learning. In: Mandal, J. and Bhattacharya, D. (Eds) *Emerging Technology in Modelling and Graphics. Advances in Intelligent Systems and Computing* (vol. 937). Singapore: Springer. https://doi.org/10.1007/978-981-13-7403-6_50.

Picek, L. (2020). CGIAR computer vision for crop disease challenge presentation. Available at: https://docs.google.com/presentation/d/1slv9KHiSmlQvGcZgXxevUTCkfBNMcR13XAvA1yxUwTU/edit#slide=id.p (accessed 19 April 2021).

Plant Pathology. (2020). Plant pathology FGVC7: Identify the category of foliar diseases in apple trees. Available at: https://www.kaggle.com/c/plant-pathology-2020-fgvc7 (accessed 21 April 2021).

PlantVillage. (2021). PlantVillage. Available at: https://plantvillage.psu.edu (accessed 19 April 2021).

Ranftl, R., Lasinger, K., Hafner, D., Schindler, K. and Koltun, V. (2020). Towards robust monocular depth estimation: Mixing datasets for zero-shot cross-dataset transfer. *IEEE Transactions on Pattern Analysis and Machine Intelligence* 44(3), 1623–1637.

Sanderson, M. and Clough, P. (2003). ImageCLEF - Cross-language image retrieval evaluations. Available at: https://www.imageclef.org (accessed 19 April 2021).

Scharr, H., Bruns, B., Fischbach, A., Roussel, J., Scholtes, L. and Stein, J. (2020). Germination detection of seedlings in soil: A system, dataset and challenge. In: Bartoli, A. and Fusiello, A. (Eds) *Lecture Notes in Computer Science* (vol. 12540) Computer Vision - ECCV 2020 Workshops. Cham: Springer. https://doi.org/10.1007/978-3-030-65414-6_25.

Scharr, H., Dee, H., French, A. P. and Tsaftaris, S. A. (2016a). Special issue on computer vision and image analysis in plant phenotyping. *Machine Vision and Applications* 27(5), 607–609. https://doi.org/10.1007/s00138-016-0787-1.

Scharr, H., Minervini, M., French, A. P., Klukas, C., Kramer, D. M., Liu, X., Luengo, I., Pape, J. M., Polder, G., Vukadinovic, D., Yin, X. and Tsaftaris, S. A. (2016b). Leaf segmentation in plant phenotyping: A collation study. *Machine Vision and Applications* 27(4), 585-606. https://doi.org/10.1007/s00138-015-0737-3.

Scharr, H., Minervini, M., Fischbach, A. and Tsaftaris, S. A. (2014). Annotated image datasets of rosette plants. Technical Report FZJ-2014-03837. Forschungszentrum Jülich. Available at: http://hdl.handle.net/2128/5848.

Sheikh, R., Milioto, A., Lottes, P., Stachniss, C., Bennewitz, M. and Schultz, T. (2020). *Gradient and Log-Based Active Learning for Semantic Segmentation of Crop and Weed for Agricultural Robots* (vol. 2020), Paris, France: IEEE International Conference on Robotics and Automation (ICRA), pp. 1350–1356. https://doi.org/10.1109/icra40945.2020.9196722.

Skovsen, S. (2019). Grass clover dataset semantic segmentation and biomass composition challenges. Available at: https://competitions.codalab.org/competitions/21122#learn_the_details (accessed 21 April 2021).

Skovsen, S., Dyrmann, M., Mortensen, A. K., Laursen, M. S., Gislum, R., Eriksen, J., Farkhani, S., Karstoft, H. and Jorgensen, R. N. (2019). The Grass Clover image dataset for semantic and hierarchical species understanding in agriculture. In: *Proceedings of the IEEE/CVF Conference on Computer Vision and Pattern Recognition (CVPR) Workshops* (vol. 2019).

Tan, M., Pang, R. and Le, Q. V. (2020). EfficientDet: Scalable and efficient object detection (vol. 2020). IEEE/CVF Conference on Computer Vision and Pattern Recognition (CVPR), Seattle, WA, USA, pp. 10778–10787. https://doi.org/10.1109/cvpr42600.2020.01079.

Tensorflow. (2015). Catalog. Available at: https://www.tensorflow.org/datasets/catalog/plant_leaves (accessed 19 April 2021).

Thapa, R., Zhang, K., Snavely, N., Belongie, S. and Khan, A. (2020). The plant pathology challenge 2020 data set to classify foliar disease of apples. *Applications in Plant Sciences* 8(9), e11390. https://doi.org/10.1002/aps3.11390.

Tian, H., Wang, T., Liu, Y., Qiao, X. and Li, Y. (2020). Computer vision technology in agricultural automation – A review. *Information Processing in Agriculture* 7(1), 1–19. https://doi.org/10.1016/j.inpa.2019.09.006.

Tsaftaris, S. A., Minervini, M. and Scharr, H. (2016). Plant phenotyping datasets. Available at: https://www.plant-phenotyping.org/datasets-home (accessed 19 April 2021).

Tsaftaris, S. A. and Scharr, H. (2014). Leaf segmentation challenge (LSC). Available at: https://www.plant-phenotyping.org/CVPPP2014-challenge (accessed 19 April 2021).

Tsaftaris, S. A. and Scharr, H. (2019). Sharing the right data right: A symbiosis with machine learning. *Trends in Plant Science* 24(2), 99–102.

Tsaftaris, S. A., Scharr, H. and Pridmore, T. (2020). 2020, Glasgow, UK, online event (vol. 28)(August), ECCV 2020 Workshop on Computer Vision Problems in Plant Phenotyping (CVPPP 2020): About the workshop series. Available at: https://www.plant-phenotyping.org/CVPPP2020-About (accessed 19 April 2021).

Uchiyama, H., Sakurai, S., Mishima, M., Arita, D., Okayasu, T., Shimada, A. and Taniguchi, R. (2017). *An Easy-to-Setup 3D Phenotyping Platform for KOMATSUNA Dataset* (vol. 2017). IEEE International Conference on Computer Vision Workshops (ICCVW), Venice, Italy, 2017, pp. 2038–2045. https://doi.org/10.1109/ICCVW.2017.239.

van Ginneken, B., Kerkstra, S. and Meakin, J. (2012). Grand challenges website. Available at: https://grand-challenge.org/challenges/ (accessed 19 April 2021).

Zindi, C. (2020a). ICLR Workshop Challenge #1: CGIAR computer vision for crop disease. Available at: https://zindi.africa/competitions/iclr-workshop-challenge-1-cgiar-computer-vision-for-crop-disease (accessed 19 April 2021).

Zindi, R. (2020b). ICLR Workshop Challenge #1: CGIAR computer vision for crop disease. Available at: https://zindi.africa/competitions/iclr-workshop-challenge-2-radiant-earth-computer-vision-for-crop-recognition (accessed 19 April 2021).

Chapter 9

Digital phenotyping and genotype-to-phenotype (G2P) models to predict complex traits in cereal crops

Nicolas Virlet, Rothamsted Research, UK; Danilo H. Lyra, Biometrics and Breeding Research, BASF, Belgium; and Malcolm J. Hawkesford, Rothamsted Research, UK

1 Introduction

With the development of Breeding 4.0, new genotyping and phenotyping tools are needed to help the breeding process to increase the productivity of genotypes (Van Eeuwijk et al., 2019, Wallace et al., 2018). This includes a trend to integrate multiple layers of genomics, high-throughput plant phenotyping (HTPP), and large-scale envirotyping to improve the prediction of complex traits (Crossa et al., 2021, Cooper et al., 2014). Whole genome-enabled prediction, referred to as genomic prediction (GP) or genomic selection (GS), is the main approach to integrate these new tools into breeding programs for supporting the delivery of high- and sustainable-yielding cultivars. The main goal of GS is to predict complex traits based on marker information, increasing the precision of selection by generating a genomic estimated breeding value for candidates. Therefore, GS is potentially superior to phenotypic selection for increasing genetic gains per unit time and shortening the length of the breeding cycle (Crossa et al., 2017). Recently, the requirements of breeders are increasingly shifting towards accommodating HTPP data and environmental information into the multi-environment trial analysis (Araus et al., 2018). However, it is

http://dx.doi.org/10.19103/AS.2022.0102.12

uncertain whether the increasing number of phenotypes measured and the depth of information from the new phenotyping platforms will improve the prediction of quantitative traits.

Improvements in yield have been achieved mainly by evaluating grain yield as the sole selection criterion. However, grain yield in cereals is an integrative trait resulting from the genetic ability of a plant to grow, to capture resources, and to transfer them into the grain in a specific environment. Determination of grain yield involves multiple processes, and it is therefore under multi-gene control with complex interactions with the environment. Reynolds and Langridge (2016) suggested that to accelerate yield improvement, physiological traits need to be considered for breeding in addition to the advanced genomics methods (Reynolds et al., 2012). Furthermore, since a comprehensive genetic basis explaining cultivar-level differences in performance does not yet exist for any crop, physiological breeding currently relies heavily on phenomics.

Digital phenotyping tools enable rapid screening of populations in field conditions in a non-invasive and non-destructive approach. These tools make use of remote sensing and close-range technologies (from airplane to handheld device) for screening experimental trials with hundreds or thousands of plots in a fast, non-invasive, and non-destructive way (Araus et al., 2018). Digital phenotyping enables repeat measurements over time and thus provides valuable dynamic information on crop growth as well as crop physiology and phenology, to enable an understanding of the whole life-cycle growth performance of the genetic materials in relation to the end product, that is grain yield. In addition, conducting experiments at multiple locations should help to deliver optimal final products targeting specific environments with optimal yield and performance for farmers. However, while a wealth of information is now available from digital phenotyping, not everything is necessarily needed or even relevant for the breeding community. One important aspect to consider is the cost and the throughput of the technologies being deployed for breeding trials. Reynolds et al. (2012) emphasized that breeders are interested in reducing the size of their test populations to achieve the required genotype at a reasonable probability and cost, even if the methods might be seen as imperfect from a physiological and theoretical point of view.

Digital phenotyping involving the application of remote sensing technologies and methods at spatially close proximal scales enables access to the different components of plant growth with unprecedented accuracy (in terms of temporal, spatial, and spectral resolution). Quantifying the morphological, phenological, and physiological traits at plot scale is now possible even if some approaches still require refinement to replace traditional approaches. The use of optical sensors offers the possibility to assess biomass, nitrogen, and chlorophyll estimation along the plant development cycle, as well as physiological parameters linked to photosynthesis and light use efficiency (LUE) of the crop. Thermal infrared technology measuring leaf and canopy temperature provides information on plant water transpiration status and water use. Lidar technology gives access to plant stature and morphology. Visible imaging, using simple camera technology, offers the greatest versatility, providing high pixel resolution and recording information akin to the human eye. This is extremely valuable as traditionally breeders rely tremendously on visual observation, the 'breeder's eye'.

High-throughput plant phenotyping records numerous traits with high spatiotemporal resolution at a level far beyond what humans can assimilate. As a result, a new method of selecting phenotypes has recently been proposed, known as 'phenomic selection' (Rincent et al., 2018). In this approach, data such as hyperspectral information are used the same way as molecular markers to make inferences about relatedness. In other words, phenotypic variables (referred to as secondary traits) replace genetic information in a traditional GP approach. Integration of both types of information (genetic plus phenomic) as predictors in genotype-to-phenotype (G2P) models has been used to improve the predictive ability for yield in wheat (Krause et al., 2019). To that end, the complexity of the new HTPP data raises many questions about how to efficiently integrate genetic and non-genetic components to improve the accuracy of prediction for complex traits.

Applications of HTPP data for predictive G2P models may be grouped into three categories. The first strategy uses the phenotypic data generated from the platforms as the target phenotype (Lyra et al., 2020, Watanabe et al., 2017). The second strategy uses thousands of reflectance data points (hyperspectral bands or derived vegetation indices (VIs)), captured at many stages of crop development as secondary phenotypes (genotype-specific covariables), with G2P models to improve the prediction of primary (target) traits, that is grain yield (Rutkoski et al., 2016, Sun et al., 2017). A third strategy is modelling plant growth from longitudinal traits (e.g., light interception, biomass accumulation, canopy height), allowing the selection of high-yielding cultivars at the early stages of development (Moreira et al., 2020). Several authors have applied different statistical models to handle this data complexity, including multivariate analysis (Rutkoski et al., 2016, Sun et al., 2017), factorial regression (Van Eeuwijk et al., 2019), functional regression (Montesinos-Lopez et al., 2017b), multi-kernel regression (Krause et al., 2019), deep learning (Cuevas et al., 2019), regularized selection indices (Lopez-Cruz et al., 2020), mega-scale linear mixed model (MegaLMM) (Runcie et al., 2021), and crop growth models (Van Eeuwijk et al., 2019). Furthermore, with the recent developments of precision envirotyping platforms, the integration of new phenotypic and envirotypic information will become a new standard.

Envirotyping is the process of collecting environmental factors (e.g. soil and climate information) in multi-environment trials (MET), intending to characterize the variation of the phenotypic performance of genotypes over, for example environmental gradients (envirotype) (Cooper et al., 2014, Xu, 2016). Another promising concept that could be applied in crop breeding for a more optimized variety selection is 'enviromics', which is the integration of envirotypes with multiple enviromic markers which correspond to environmental variables that may interact with the genetic background (Resende et al., 2021). Envirotyping information has been used to explore envirotype-to-phenotype dynamics and has been incorporated in G2P models to improve the prediction of complex traits (Costa-Neto et al., 2020, 2021a, b, Millet et al., 2019, Porker et al., 2020).

In this chapter, recent literature on incorporating new phenotyping data into predictive G2P models is explored, along with an introduction to the use of envirotyping information. The chapter is presented in two sections. First, the main 'traits' that are currently used and/or should be useful for breeding are introduced. Second, the most common statistical methods used to integrate markers, environmental information, and HTPP data are described.

2 Digital phenotyping as a tool to support breeding programs

Yield is considered a primary trait and is the goal in most breeding programs. For decades, breeding has been focused only on selecting genetic material based on the yield itself at harvest. However, as mentioned in the study by Van Eeuwijk et al. (2019), secondary phenotypes have proved useful as a covariate in prediction models for yield. These secondary phenotypes are defined as basic or intermediate traits (Bustos-Korts et al., 2019). The basic traits correspond to response mechanism/sensitivities to environmental conditions (e.g. sensitivity to photoperiod, water uptake capacity, radiation use efficiency), whilst the intermediate traits result from the integration of several processes over time (e.g. biomass accumulation, flowering time, grain number).

The introduction of secondary traits in prediction models is called physiological breeding (Reynolds and Langridge, 2016). However, the application may be difficult to implement depending on the secondary phenotypes considered. For example, measuring height or scoring for heading or anthesis time is straightforward (despite the fact it is time-consuming), while other phenotypes such as biomass or photosynthesis are more constraining in terms of requiring destructive measurement or demands to collect data at a large scale in a short amount of time. Pask et al. (2012) summarized most of the secondary phenotypes used for breeding as well as the practical methods to collect them. The authors also brought to light some of the remote sensing methods applied at the plot scale, for example using spectroradiometers and thermal infrared sensors, as non-destructive and non-invasive technologies to produce proxies of biomass and transpiration traits, respectively. Although a vast amount of data are now available from the multiple technologies for phenotyping, such as visual, multi-/hyperspectral imaging, thermal infrared, and fluorescence sensors, as well as lidar technologies, not all this information will be relevant for breeders. Digital phenotyping must balance the time of acquisition with the relevance of the proxies and the cost to deploy the technologies.

In this section, we will review the secondary phenotypes measured by high-throughput digital phenotyping tools that are potentially useful for breeding. The secondary phenotypes will be considered in terms of the level of integration and complexity regarding their use as single/multi-indirect traits for yield prediction, covariates for yield prediction, or for integrating over time using growth curves to access new parameters summarizing growth events.

2.1 Simple secondary traits

This section summarizes secondary phenotypes considered as simple variables. In this case, 'simple' means that the variable is the output from a sensor and is not the result of the combination of different metrics or sensors. The secondary phenotypes are sub-divided according to the morphological, physiological, and phenological classification of the traits (Violle et al., 2007).

2.1.1 Morphological traits

Morphological traits such as height and canopy cover are the most straightforward variables to collect using either lidar technologies and/or visible cameras (Red-Green-Blue (RGB)). Height can be easily calculated not only from different lidar systems (Deery et al., 2014, Friedli et al., 2016, Virlet et al., 2016) but also from the structure from motion (SfM) principle when RGB cameras are coupled with an appropriate vector, such as unmanned aerial vehicles (UAV) or tractor-based systems (Holman et al., 2016, Jay et al., 2015). The output of these two approaches is called a point cloud, with each point having *x*, *y*, and *z* coordinates. Height is the easiest parameter derived from the point cloud. Recent studies have shown capabilities to estimate the volume of the canopy, the above-ground biomass (AGB), the leaf area index (LAI), or the stem diameter by processing the point clouds using voxelization (Che et al., 2020, Hosoi et al., 2013, Jimenez-Berni et al., 2018, Salas Fernandez et al., 2017, Xiao et al., 2020).

With sufficient spatial resolution, processing the point cloud enables access to other morphological parameters such as the leaf area and angle (Mantilla-Perez and Salas Fernandez, 2017), as well as the volume of the reproductive organs in cereal crops. It is also possible to extract the fraction vegetation cover (FVC) from the point clouds as in the study by Duan et al. (2016). However, FVC is generally computed from RGB images as the ratio of crop green pixel to the total number of pixels contained in the region of interest. Numerous approaches have been developed to segment the green plant pixel from the background as the ambient illumination is a limiting factor for the efficiency of the segmentation (Casadesús et al., 2007, Guo et al., 2013, Hamuda et al., 2016, Sadeghi-Tehran et al., 2017a). Those methods range from simple threshold approaches to deep learning methods overcoming the illumination issues by training the algorithm with a wide range of illumination conditions. There are multiple applications of monitoring FVC such as enabling quantification of canopy development from emergence to maturity (Borra-Serrano et al., 2020, Sadeghi-Tehran et al., 2017a, Varela et al., 2021).

In recent years, there has been a growing interest in using RGB images to quantify the number of ears/panicles for the major grain crops (Duan et al., 2015, Fernandez-Gallego et al., 2018, Lu et al., 2017, Xiong et al., 2017, Chandra et al., 2020, Ghosal et al., 2019, Liu et al., 2020, Madec et al., 2019, Sadeghi-Tehran et al., 2019, Velumani et al., 2020, Zhou et al., 2019). Many of these approaches rely on machine learning methods and require the acquisition of solid training data sets to encompass the various stages of maturation of the reproductive organs and the genetic variation. To facilitate such data set acquisition, a group of researchers from nine different institutions across seven countries and three continents developed the Global Wheat Head Detection data set that can be used to benchmark methods proposed in the computer vision community (David et al., 2020).

2.1.2 Physiological traits

Digital phenotyping also enables the measurement of physiological traits related to carbon and nitrogen metabolism. Physiological traits are mainly assessed using the spectral properties of the plant, whether in the optical domain or in the thermal infrared. Multi- and hyperspectral technologies (signal point or utilizing imaging systems) are used to derive spectral reflectance indices (SRIs), which are mainly related to biomass accumulation and component traits (light interception and light/radiation use efficiency). SRIs are obtained by the arithmetic combination of two or more wavelengths designed to highlight a particular property of vegetation. Each SRI is designed to accentuate a particular vegetation property (https://www.l3harrisgeospatial.com/docs/vegetationindices.html).

More than 100 indices have been developed over the past 50 years. Most of the indices developed are related to the biomass accumulation and the light interception such as the normalized differences vegetation index (NDVI) and the developed derivatives to overcome NDVI limitations (Corti et al., 2018, Cabrera-Bosquet et al., 2011, Gutierrez et al., 2012, Haboudane et al., 2004, Yue et al., 2017). Chlorophyll and nitrogen leaf/canopy content as well as other pigments have been extensively investigated over recent decades using

Published by Burleigh Dodds Science Publishing Limited, 2022.

vegetation reflectance properties (Berger et al., 2020, Blackburn, 2007, Boegh et al., 2002, Camino et al., 2018a, Cammarano et al., 2014, Fitzgerald et al., 2010, Gitelson and Solovchenko, 2018, Jay et al., 2015). Post-harvest traits related to nitrogen have been also explored extensively in the literature (Erdle et al., 2013, Frels et al., 2018, Pavuluri et al., 2015, Prey et al., 2020, 2019, Prey and Schmidhalter, 2020, 2019). The above-mentioned agronomical traits reflect the process of biomass accumulation, which can be monitored and assessed using different VIs such as NDVI and NDVI-like indices. However, those indices do not report short-term variation that might occur with changes in ambient conditions (Dobrowski et al., 2005). Among the SRIs, the photochemical/physiological reflectance index (PRI), a normalized reflectance index that uses the 531 nm and 570 nm wavelengths, is generally used as a direct method to assess LUE over short periods (Gamon et al., 2016). PRI has been shown to negatively correlate with non-photochemical quenching (NPQ) and the de-epoxidation state of the xanthophyll cycle (Evain et al., 2004, Peguero-Pina et al., 2008, Porcar-Castell et al., 2012, Rascher et al., 2007) and positively correlate with steady-state fluorescence, F', and the photosystem II (PSII) operating efficiency (F_q'/F_m') under differing irrigation regimes in controlled or natural environmental conditions (Evain et al., 2004, Peguero-Pina et al., 2008). The use of active sensors such as the laser-induced fluorescence transient should be useful to assess photosynthetic dynamics of the crop and replace gas-exchange measurements. However, this technology still needs refinement to be applied practically on a large scale (Wyber et al., 2017, 2018).

The canopy water status may be investigated using SRIs of specific wavelengths sensitive to water content. The most used of these indices are the water band index and normalized water band indices computed from the wavelengths from the 850–970 nm range (Prasad et al., 2007). However, as shown in Winterhalter et al. (2011), other SRIs have been investigated using various wavelengths in the visible-near infrared domain and wavelengths of the shortwave infrared domain. In addition to the canopy water content, canopy transpiration may be assessed using thermal infrared technologies (Jones et al., 2009), which measure the canopy temperature and its variation due to the control of stomatal opening/closure regulating water loss in response to changes in environmental conditions (Gonzalez-Dugo et al., 2014, Kelly et al., 2019, Maes and Steppe, 2012, Munns et al., 2010).

2.1.3 Phenological traits

Crop phenology is usually scored by using Zadoks, Feekes-Large, or the Haun systems. The remote sensing community has been using VIs to estimate the stage of development of various crops using variations in the VIs during growth (Duncan et al., 2015, Piao et al., 2019, Yamasaki et al., 2017). Recent advances in the phenotyping community allowing collection of images at a closer range to the crops have enabled an unprecedented spatial resolution. The use of RGB cameras, coupled with appropriate algorithms, has enabled detection and quantification of some of the important growth stages such as heading and flowering stages in cereal crops.

The dynamic of ears per image may be used to estimate heading dates (Velumani et al., 2020, Wang et al., 2019). Flowering information has been collected in field conditions for various annual and perennial crops such as *Lesquerella*, apple tree, grapevine, rice, wheat, and cotton (Aquino et al., 2015, Guo et al., 2015, Hocevar et al., 2014, Sadeghi-Tehran et al., 2017b, Thorp and Dierig, 2011, Xu et al., 2018). Different approaches have been used for the flower segmentation using threshold-based algorithms (Hocevar et al., 2014, Thorp and Dierig, 2011), for segmentation and quantification using mathematical morphology and pyramidal algorithm (Aquino et al., 2015), for identification of flowering within RGB images using k-means clustering and support vector machine (SVM) (Guo et al., 2015, Sadeghi-Tehran et al., 2017b), and for identification of flowers in three-dimensional (3D) images from aerial pictures using convolutional neural networks (Xu et al., 2018).

The senescence period has also been investigated using RGB, multi-, and hyperspectral sensors. The temporal data of the senescence enable to derive parameters such as the onset, midpoint, and senescence duration, and the topic will be detailed in Section 2.3 (Anderegg et al., 2020, Borra-Serrano et al., 2020, Christopher et al., 2014, Kipp et al., 2014, Montazeaud et al., 2016).

2.2 Multi-sensor secondary traits modelling

To increase the efficiency of agronomic trait prediction (e.g. biomass and physiological traits), many approaches may be used such as (i) arithmetic combination of secondary traits to try to mimic as much as possible the agronomic trait, (ii) multi-linear/non-linear regression model to select the secondary phenotypes enabling better prediction, and (iii) multivariate analysis which summarizes secondary phenotypes information to predict the agronomic traits of interest. In this Section 2.2, only statistical methods used to predict traits without including genomic information are described.

2.2.1 Arithmetic combination of secondary phenotypes

AGB has been predicted using lidar outputs such as height and volume as well as using spectral VIs and FVC. The accuracy of the predictions varies depending on the crop, the type of trial, treatment, and stage of development. Height itself is not an indicator of biomass. The use of SRIs as a biomass predictor, as well as the FVC derived from the RGB camera, is often limited by the excess (saturating)

Published by Burleigh Dodds Science Publishing Limited, 2022.

density of the cover, and height itself might not be sufficient enough as a biomass indicator. It has been shown that the combination of SRI or FVC (or both) with height improves biomass prediction. Yue et al. (2017) showed that the incorporation of the height data by multiplication or division of the eight VIs tested improved the biomass prediction in all cases. Maimaitijiang and Sidike (2019) developed a canopy volume index based on the parameters extracted from the crop height model from the orthomosaic image obtained by UAV and combine them with different SRIs to improve the biomass prediction with success. Another approach was taken by Jin et al. (2019) to predict biomass by estimating the volume of the plot. The authors combined height at anthesis time when it is considered to be at its maximum, the post-harvest stem diameter, and the number extracted from RGB images capturing the residual stems standing straight after harvest. Lu et al. (2021) also used a combination of VI and height measurement to estimate the canopy nitrogen content (CNC). They showed that the combination of height, VI, and FVC increased the CNC prediction.

SRIs have been used extensively to determine the canopy nitrogen (N) status in various crops. However, as mentioned by Fitzgerald et al. (2010), measuring the N status by SRI was found to be challenging across the season, as the ground coverage and the canopy structure are both changing along with crop growth, as well as there being an N dilution effect. The canopy chlorophyll content index (CCCI) has been developed to differentiate N from water stress in irrigated cotton and separate the confounding effects of canopy density (Barnes et al., 2000, Clarke et al., 2001, Rodriguez et al., 2006). Fitzgerald et al. (2010) have developed a proxy of canopy nutrition index (CNI, also called nutrition nitrogen index), allowing the monitoring of crop N status at different growth stages without being affected by the N dilution effect. The CNI is built on the relationship between dry biomass and the % N of the vegetation. Based on the same principle, the CCCI, a planar domain VI measuring plant biophysical parameters in a mixed soil/plant pixel, can be built by analysing the relationship between one chlorophyll and one biomass-related SRIs plotted in a two-dimensional (2D) space (Clarke et al., 2001, Pancorbo et al., 2021). Another index, the water deficit index (WDI) has been developed in the 1990s to overcome the confounding N and water stress effect on the SRI. The WDI is also a planar domain index: it is based on the relationship between a VI such as the NDVI and the variable $T_c - T_a$ from a thermal infrared sensor and by building 2D spaces, for the estimation of the evapotranspiration of the crops with less effect on the canopy density is possible (Moran et al., 1994, Pancorbo et al., 2021, Virlet et al., 2014).

2.2.2 Multi-linear/non-linear regression

In Section 2.2.1, most studies generally focused on single regression analysis based on a simple or a complex index to estimate the trait of interest. However, the single regression approaches might not be sufficient to estimate integrative traits such as biomass and grain yield in cereals. Here, different approaches to integrate parameters using multiple regression analysis (linear or non-linear) are reviewed. The first part covers research works predicting traits based on a single date of data collection using multiple traits, and the second section covers studies with data collection on multiple days, irrespective of whether the date was used as an independent variable in the models.

Multi-linear regression has been applied to estimate grain yield on maize, barley, and soft red wheat (Gracia-Romero et al., 2017, Kefauver et al., 2017, Pavuluri et al., 2015). On maize and barely, the best model for improving grain yield prediction from UAV imaging was a combination of two to five indices derived from RGB and/or multispectral imaging (Gracia-Romero et al., 2017, Kefauver et al., 2017). At the ground level, Pavuluri et al. (2015) showed models with three SRIs gave the best predictions for grain yield. On sorghum, Li et al. (2018) used simple and multiple exponential regression analyses to estimate fresh and dry biomass. In this case, a simple exponential regression model based on canopy height gave in most of the cases the best biomass prediction, with some improvements of the prediction by combining VIs. Vargas et al. (2019) used the Lasso algorithm and found that the best prediction for AGB in peas was a combination of VIs, FVC, and canopy volume. In the study by Santini et al. (2019), a combination of VIs and thermal infrared data gave the best prediction for the stem volume of adult trees of *Pinus halepensis*.

Multiple regression has also been applied to nutrient content prediction. Gracia-Romero et al. (2017) used this approach to predict phosphorus (P) content in maize and improve P content prediction compared to simple regression analysis. However, the model prediction was still low. Pavuluri et al. (2015) had more success estimating grain N uptake, yield, and protein nitrogen use efficiency using models built from data collected at different growth stages (stem elongation, booting, heading, and grain filling). The predictions were better by combining the 2 years of measurement for each growth stage.

Bendig et al. (2015) looked at biomass monitoring prediction in barley based on a combination of plant height and VIs using data collected prior and post-heading date. Models based on single and multiple exponential or linear regressions were tested using either the full data set or only the pre-heading data set. Cross-validation showed that the combination of VIs with plant height performed better than using the VIs alone, whatever the use of multiple linear or non-linear regression. In a different approach, Hoyos-Villegas et al. (2014) combined canopy height, FVC, AGB, photosynthesis, and rate of growth collected at 13 time points into multiple regression models to predict grain yield, AGB, photosynthesis rate, and rate of growth. The approach taken by the authors provided interesting model predictions for each of the parameters and showed that multi-sensor data produce highly successful predictions. The main drawback was the upscaling feasibility to large populations for breeding purposes due to the amount of data to collect, especially for the gas exchange.

Magney et al. (2016) used NDVI data and the parameters from the growth and senescence curves (growth rate and duration of NDVI for tillering, stem elongation, heading, and maturation) to predict grain yield, total biomass, protein content, and grain N yield using multi-linear regression models. The best results were obtained by the models using two or three phenological stages. This study also showed that using NDVI-derived phenological metrics from the early season (tillering and stem extension) substantially improved early prediction of yield and biomass, as compared to daily NDVI data, whereas protein and grain N were primarily driven by metrics associated with the reproductive development of the crop (heading and ripening). More recently, Buchaillot et al. (2019) used multiple regression analysis to predict grain yield in maize based on agronomic data coupled with (i) field sensors, (ii) ground imagery, and (iii) UAV imagery. Most of the models improved the grain yield prediction using four to nine parameters. This study also showed that the simple use of the date of anthesis, the duration from anthesis to the silking stage, the chlorophyll index given by the SPAD-502 at the vegetative and reproductive stage gave the best prediction for grain yield prediction (r^2 = 0.61). The model using UAV image data showed promising results (r^2 = 0.60) but used nine parameters instead of the four used in the previously mentioned model.

Camino et al. (2018b) used multiple regression analysis to assess the contribution of sun-induced fluorescence (SIF) indices for nitrogen prediction in wheat. The performance of the built models with and without SIF was compared with the performance of models built with plant traits estimated by the PROSPECT-SAILH model inversion and with standard approaches based on single SRIs. The use of SIF indices increased the performance of the models based on the output of the inversion models (r^2 = 0.68–0.77 without SIF, r^2 = 0.92–0.93 with SIF) and also outperformed the prediction for the chlorophyll content by the narrow-band indices (r^2 = 0.25–0.57).

2.2.3 Partial least square regression

Partial least square regression (PLSR), a multivariate analysis, has been proved useful for dealing with large numbers of parameters and volumes of data. The use of PLSR in plant phenotyping is increasing and shows promising results to estimate and predict different plant parameters. During the last decade, it has been used to estimate grain yield on barley, maize, and wheat (Barmeier and Schmidhalter, 2017, Elsayed et al., 2018, 2017, 2015, Garriga et al., 2017, Rischbeck et al., 2016, Weber et al., 2012), biomass at tillering and anthesis time on barley, maize, and wheat (El-Hendawy et al., 2019a, Elsayed et al., 2018,

Montes et al., 2011), leaf nitrogen and chlorophyll content (Ecarnot et al., 2013, Silva-Perez et al., 2018, Vigneau et al., 2011), canopy water status on barley and wheat (El-Hendawy et al., 2019a, b, Elsayed et al., 2015, 2017), physiological and photosynthesis parameters on cotton, blueberry, tobacco (Fu et al., 2019, 2020, Lobos et al., 2019, Meacham-Hensold et al., 2020, Silva-Perez et al., 2018, Thorp et al., 2015). Most of those studies are using the reflectance data as input for the PLSR. However, some authors have been using PLSR with SRIs directly. Elsayed et al. (2015) used SRIs from different sensors to estimate grain yield. The performance of the PLSR model improved grain yield prediction compared to the single regression model. El-Hendawy et al. (2019a) and El-Hendawy et al. (2019b) built a PLSR model based on SRIs and a selection of the most influential wavelengths. While in the El-Hendawy et al. (2019a) study, both approaches gave similar results, the El-Hendawy et al. (2019b) study showed the PLSR model with higher values of r^2 and lower values of the root mean squared error (RMSE) between observed and predicted values of the measured parameters was the model based on SRIs. Similarly, Fu et al. (2020) compared the performance of the PLSR model based on reflectance data and based on SRIs to estimate two photosynthetic parameters (V_{cmax} and J_{max}). For V_{cmax}, the PLSR model based on SRIs gave better prediction than the model based on the wavelength, while for J_{max}, both approaches gave similar predictions.

PLSR has also been used to integrate variables from different sensors. Montes et al. (2011) used this method to predict early biomass in maize based on height and reflectance data, while Rischbeck et al. (2016) used the method to predict grain yield in barley. Elsayed et al. (2017) used PLSR with different parameters to predict grain yield on wheat. Three PLSR models were built, one based on selected SRIs, one data fusion model based on SRIs and thermal infrared data, and one based on SRIs, thermal infrared data as well as relative water content and canopy water content. In the three last studies, the use of data fusion for PLSR improved the prediction showing the strong potential of the PLSR in plant phenotyping to integrate data from different sensors and cameras.

Some authors have compared the PLSR model to other approaches such as machine learning algorithms, multi-linear regression models, or even to the predicted parameters from the inversion radiative transfer model PROSAIL. In the study by Lobos et al. (2019), the PLSR model performed better than the multi-linear regression model to estimate fluorescence parameters in blueberry based on reflectance data, while in the study by El-Hendawy et al. (2019b), the two approaches gave similar results for yield prediction. Garriga et al. (2017) showed that for most of the traits and conditions tested, the estimations provided by the ridge regression and modified SVM were the same or better than those provided by the SRIs. They also found that among the classification methods, the PLS-discriminant analysis showed the best performance, and unlike the SRI and regression models, most traits were relatively well classified within a specific hydric condition. In the study by Montes et al. (2011), the SVM regression model to estimate biomass at the early stage performed slightly better than the PLSR. Fu et al. (2019) developed a framework combining six machine learning algorithms, including artificial neural network (ANN), SVM, least absolute shrinkage and selection operator, random forest (RF), Gaussian process and PLSR to optimize high-throughput analysis of the two photosynthetic variables, V_{cmax} and J_{max}, based on reflectance data. The PLSR model was the second-best model with r^2 and RMSE close to the best model. Thorp et al. (2015) used PLSR to estimate four phenotypes: leaf water content (C_w), specific leaf mass (C_m), leaf chlorophyll a and b content (C_{ab}), and LAI. The results showed the PSLR model was the most robust to estimate the four investigated parameters compared to linear regression, based on SRIs and the PROSAIL model inversion.

2.3 Integrative secondary traits

Different vectors (platforms) have been used to look at plant growth parameters (canopy height, FVC, and SRIs) such as handheld devices or telescopic poles (Anderegg et al., 2020, Cairns et al., 2012, Christopher et al., 2014, Grieder et al., 2015, Montazeaud et al., 2016, Velumani et al., 2020), tractor-based systems (Comar et al., 2012, Kipp et al., 2014), ground platforms (Aasen et al., 2020, Beauchêne et al., 2019, Kronenberg et al., 2017), or UAV systems (Borra-Serrano et al., 2020, Burkart et al., 2018, Han et al., 2018, Holman et al., 2019, Holman et al., 2016, Varela et al., 2021). A number of these studies have collected time-series data to visually identify (i) the beginning of stem elongation period using canopy height data (Holman et al., 2016), (ii) growth stages on barley based on the growth profile (Burkart et al., 2018), (iii) rapid growth expansion and senescence dates and durations on soybean (Aasen et al., 2020), and (iv) growth pattern among maize breeding material (Han et al., 2018).

Fitted curves maximize the use of time series data, quantifying and summarizing the different curve phases corresponding to the stage of the crop development: slow and rapid canopy development, a plateau, and the senescence phase. Fitted curve parameters enable the comparison of the growth and senescence pattern of multiple genetic materials for genetic studies.

Grieder et al. (2015) used spline functions to smooth the FVC data and compute daily relative growth rate (RGR) to look at the influence of the temperature on early growth data for 29 random lines from wheat germplasm. Varela et al. (2021) also used spline fitting curves on FVC, height, and SRIs data. The author aimed to predict AGB on sorghum using the daily data and their corresponding RGR derived from the spline fitting using an RF algorithm and to understand the relationships between growth dynamics, temporal resolution, and end-of-season AGB in 869 diverse accessions of a highly productive photoperiod sensitive sorghum.

Cairns et al. (2012) used NDVI time series data collected over a growing season of a large set of tropical and subtropical maize inbred lines and single cross hybrids to look at their grain performance. NDVI data were used to compute the area under the curve (AUC) as an integrative indicator of growth and senescence patterns. The AUC was divided into sections corresponding to the different stages of development to investigate the link with grain yield. Comar et al. (2012) used a bi-linear model for investigating the SRIs time series to derive integrative parameters to compare the six wheat varieties growing with different nitrogen levels and planted at different densities. More recently, Borra-Serrano et al. (2020) used a sigmoidal model (Gompertz) and Beta functions to fit FVC and canopy height data. The authors derived parameters including maximum absolute growth rate, early vigour, maximum height, and senescence for a collection of soybean genotypes and were integrated into multi-linear regression model to estimate seed yield.

Some studies have only focused on the senescence period. Christopher et al. (2014), Kipp et al. (2014), and Montazeaud et al. (2016) used NDVI and other spectral data to model the dynamics of the senescence to identify stay-green phenotype using logistic, PLSR, and linear models, respectively. From the fitted curves, parameters such as the onset and mid-point of senescence and the senescence rate have been used to characterize the stay-green phenotype. Anderegg et al. (2020) have investigated different SRIs and modelling approaches to fit the data collected during the senescence period. PLS and cubist regressions have been used on the SRIs, and the derived parameters were assessed by prediction of the grain protein content and grain yield.

The last few years have seen an increase in interest to quantify the number of reproductive organs in cereal crops, as mentioned earlier. Wang et al. (2019) and Velumani et al. (2020) have been using a series of daily RGB images to quantify the number of ears. The authors fitted a logistic curve on the ears data and extracted the inflection point corresponding at half of the total ears to estimate the heading date.

3 Genotype-to-phenotyping (G2P) models: integrating data from phenomics and envirotyping in predictive breeding

3.1 Implementing complex G2P models in breeding programs

It is well known that increasing productivity largely depends on genotype (G) and environment (E). With the newly developed technologies, the G component can now be accessed through multi-omics layers (Scossa et al.,

2021), the phenotype (P) component can be precisely measured using HTPP tools (Araus et al., 2018), and the E factors can be evaluated using envirotyping techniques (Xu, 2016). In this section, how to integrate these new phenotyping data into G2P models to improve the prediction of complex traits is discussed. The utilization of HTPP data for genetic dissection (gene/trait discovery) of quantitative traits has been discussed previously (Moreira et al., 2020, Zhang et al., 2021, Mir et al., 2019, Brown et al., 2014, Li and Sillanpää, 2015).

G2P models are considered here as different types of statistical models (e.g., linear, mixed, factorial regression, crop growth) that allow the identification of genetic and environmental factors driving phenotypic variation (Van Eeuwijk et al., 2019, Cooper et al., 2021). G2P models have a wide range of applications in breeding, mainly exploiting the genotype × environment interactions (GEI). Environmental effects and GEI are estimated through METs by evaluating genotypes in many locations and years. Traditionally, breeders have been using the final trait value for selection, which is the cumulative effect of many factors including genotype, environment, and interactions among them. Consequently, a simple linear model can be used to untangle the phenotypic variance quantified as $V_P = V_G + V_E + V_{GE} + V_\varepsilon$, where V_G stands for genotypic variance, V_E for environmental variance, V_{GE} for GEI variance, and $V\varepsilon$ represents the random residual. Note that there are many strategies to handle GEI and the most common approaches are the mixed models, the additive main effects and multiplicative interaction models, the genotype plus genotype-by-environment models, the analysis of variances, and the stability measures (Van Eeuwijk et al., 2016, Malosetti et al., 2013).

Recently, the term 'envirotyping' has been introduced in the context of GEI, and its integration with the new genotyping and phenotyping tools will help support breeding programs speeding up the development of high-yielding cultivars (Cooper et al., 2014, Van Eeuwijk et al., 2019). Envirotyping refers to the process of characterizing environmental factors such as climate, soil, biotic, and crop management (Xu, 2016). Precise characterization of E factors provides the key component for refining field experiments, reducing environmental variance, and thus increasing heritability estimation. Environment-related data have been applied in predictive G2P models resulting in improvements in predictive ability (Pérez-Rodríguez et al., 2015, De Los Campos et al., 2020). For example, Jarquín et al. (2014) and Pérez-Rodríguez et al. (2015) used 68 and 76 environmental covariates in wheat and cotton, respectively. New interest has been going towards 'enviromics' for the exploration of key genetic components linked to environmental factors or development stages (Crossa et al., 2021, Resende et al., 2021). Another hot topic is the establishment of a four-dimensional (4D) profile with G-P-E information incorporated with time series (Xu, 2016), but this process has to be integrated with the new phenotyping data in breeding pipelines.

In recent years, HTPP tools have been used extensively in crop trials generating large amounts of data with high spatial and temporal resolution (Mir et al., 2019, Rebetzke et al., 2019). Many approaches (discussed in Section 2) already have high potential to be integrated into the routine of breeding pipelines such as (i) estimating NDVI to characterize biomass accumulation and canopy greenness in the phenological stages (Rebetzke et al., 2016), (ii) identifying flowering time and the number of spikes in wheat using sequential images (Sadeghi-Tehran et al., 2017b, 2019), (iii) using canopy temperature to identify drought-tolerant genotypes, and (iv) quantification of the severity of infection (Yu et al., 2018). The primary goal in breeding programs is the improvement of yield (commonly referred to as target or primary trait), but characterizing secondary phenotypes could be of great value in selecting yield as they are genetically correlated. Secondary traits can be characterized as intermediate (biomass, flowering time, grain number) or basic (sensitivity to photoperiod, water uptake capacity, radiation use efficiency) phenotypes. Similarly, hundreds of hyperspectral data can be used as secondary traits for predicting yield (Fig. 1). However, it is not clear which secondary traits should be prioritized and how they can be integrated into G2P models (Van Eeuwijk et al., 2019). Furthermore, additional costs for phenotyping secondary traits may not necessarily translate into the improved predictive ability for the primary trait (Bustos-Korts et al., 2019).

Despite grain yield being the primary trait of interest, breeders might also be interested in using other phenotypes collected through phenotyping tools such as canopy height and FVC, to use in combination with secondary traits such as hyperspectral data (Fig. 1). In this case, indirect selection and selection index are two common methods used when seeking improvements of yield in a target population of environments. In indirect selection, a primary trait $\boldsymbol{y}$ is selected indirectly by selecting for another genetically correlated trait $\boldsymbol{x}$ (Fischer and Rebetzke, 2018) that can be quantified using HTPP, for example using spectral reflectance traits to select materials with high yield (Lozada et al., 2020, Kyratzis et al., 2017). In the selection index (SI) context, a primary trait $\boldsymbol{y}$ is selected based on an index calculated from a set of $\boldsymbol{t}$ secondary traits. A linear index is built as follows:

$$I = \sum_{j=1}^{t} \boldsymbol{b}_j \boldsymbol{t}_j ,$$

where $\boldsymbol{b}_j$ and $\boldsymbol{t}_j$ are the weight and phenotypic value of the trait, respectively. The weight represents the relative importance of each trait. Optimization of SI applying hyperspectral data to reduce overfitting and sub-optimal accuracy of indirect selection has been proposed in the recent literature (Lopez-Cruz et al., 2020).

The development of robust G2P frameworks that can accommodate these high-dimensional secondary traits together with genomic and environmental

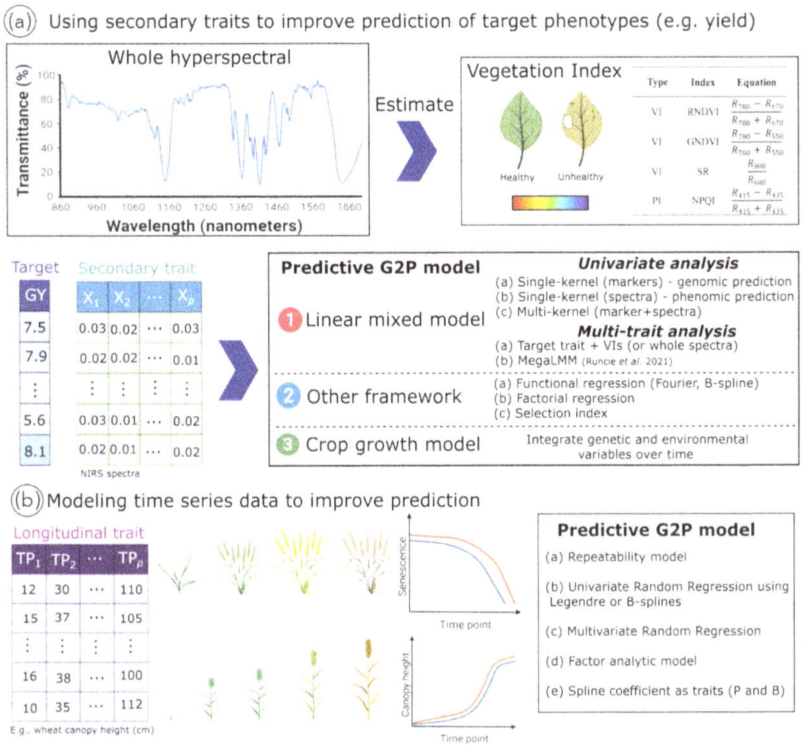

Figure 1 Schematic representation of the integration of HTPP data into genotype-to-phenotype (G2P) predictive models in crop breeding. (a) The first approach is to integrate secondary traits into G2P models to improve the prediction of target or primary phenotypes, that is, grain yield (GY). Secondary phenotypes are represented by whole hyperspectral reflectance data (e.g. near-infrared reflectance spectroscopy) and vegetation (VI) and pigmented (PI) indices. Secondary traits could be incorporated (1) in a linear mixed model (LMM) using hyperspectral data as predictors (single kernel) in an approach called *phenomic prediction* or by using a multivariate linear mixed model (MvLMM), (2) using other frameworks as functional and factorial regression, and (3) by using crop growth models. (b) The second approach is to explore G2P models for longitudinal traits to capture plant functioning at different stages of crop development. The most common predictive G2P models used for such an approach are traditional MvLMM (with specific variance-covariance) and random regression model. TP, time point; MegaLMM, mega-scale linear mixed model.

data is required and has been discussed in recent reviews (Van Eeuwijk et al., 2019, Morota et al., 2019). Primarily, this includes (i) extraction and characterization of HTPP secondary traits and environmental data, (ii) robust experimental design management (e.g. correction for spatial factors), (iii) dynamic modelling (e.g. time point (TP) and environment), and (iv) target trait prediction (using robust approaches as mixed and crop growth models). For a more detailed description of this workflow, see the study by Van Eeuwijk et al. (2019). Briefly, images are pre-processed (alignment, calibration, and segmentation) to extract informative data. This step can be laborious and time-consuming as large amounts of information will be reduced to a small number of variables relevant for phenotypic and genomic prediction. In the second stage, adjustment of genotypic means including spatial variation correction is required. Analyses can be performed per TP (also referred to as time series, longitudinal, and repeated measurements) using a mixed-model approach including first-order auto-regressive structure or 2D P-spline basis to row and column directions. In the third scenario, genotypic means are treated as unique observations and modelled in the function of time (in case the target trait is longitudinal). Parametric (logistic, Gompertz, and exponential growth functions) and spline (P and B) models are used to capture the dynamics of traits, for example, canopy height, AGB, leaf area. Lastly, after estimating genotype-specific parameters, we need to expand this approach to multiple environments. Therefore, a function integrating both factors can effectively adjust for temporal and environmental gradients (Chenu et al., 2009, Cooper et al., 2021).

In summary, G2P models can predict target traits from genotype-specific and environmental inputs. In the future, more measurements using field phenotyping tools in multiple environments will be available; therefore, G2P models must be fine-tuned to incorporate secondary phenotyping information in a breeding pipeline in an efficient way.

3.2 Combining HTPP data into predictive G2P models

The integration of HTPP and genomic data has been a current reality in plant breeding programs (Araus et al., 2018). Many adjustments in standard G2P models have been made to account for the complexity of secondary traits in GS approaches to improve the selection efficiency of the target trait (Table 1). The most used method in whole-genome regression models is the genomic best linear unbiased prediction (GBLUP), which utilizes a genomic relationship matrix to estimate the genetic merit of an individual (Morota and Gianola, 2014). The matrix defines the covariance between individuals based on observed similarity at the genomic level, rather than on expected similarity based on pedigree. Therefore, integrating HTPP data into genomic selection schemes can be classified into two common statistical methods: univariate and multivariate approaches.

3.2.1 Univariate predictive G2P models

The univariate model focuses on predicting one primary trait collected either using manual phenotyping (grain yield) or high-throughput platforms (e.g. canopy height, NDVI, senescence). Several studies used traits collected using UAV as target traits in GP studies, reducing the labour cost compared to traditional

Published by Burleigh Dodds Science Publishing Limited, 2022.

measurements (Watanabe et al., 2017). In a mixed-model perspective, GP can be performed with these mentioned traits using a single- or multi-kernel derived from molecular markers and/or traits derived from image data as described in Section 2. The type of hyperspectral data (whole spectra vs. SRIs) and feature selection (e.g., based on heritability) which should be used in G2P models have been discussed (van Eeuwijk et al., 2019, Tardieu et al., 2017). When using only genetic variants (or different layers like gene expression) to predict the genetic values, it is referred to as *genomic prediction*. Alternatively, *phenomic selection or phenomic prediction* is referred to when using only spectral data or any other data collected through digital phenotyping (Rincent et al., 2018).

Phenomic selection was first introduced by Rincent et al. (2018) using only secondary traits such as near-infrared spectroscopy (NIRs), in the same way as genomic regressors (genomic kernel), allowing predictions of untested wheat and poplar (*Populus nigra* L.) genotypes. The authors observed higher predictive ability compared to a single genomic kernel. Lane et al. (2020) further explored this approach predicting grain yield in maize genotypes using NIR spectra as kernel and observed promising results. In practical terms, phenomic selection might be capturing the phenotypic relationship between individuals from a biological composition basis (Rincent et al., 2018). The reflectance spectrum is a result of numerous wavelengths capturing chemical bonds in the analysed tissue. Interpretation of phenomic prediction results must be done with caution as they do not directly model the genetic component of correlations between primary and secondary traits (Runcie et al., 2021).

Suppose we have three scenarios of predictive G2P models from a mixed-model perspective (ignoring environment-specific information and TP variables) as follows:

$$\hat{\boldsymbol{y}} = \boldsymbol{X}\beta + \boldsymbol{Z_g g} + \varepsilon \left(\text{single-kernel genomic prediction}\right) \tag{1}$$

$$\hat{\boldsymbol{y}} = \boldsymbol{X}\beta + \boldsymbol{Z_t T} + \varepsilon \left(\text{single-kernel phenomic prediction}\right) \tag{2}$$

$$\hat{\boldsymbol{y}} = \boldsymbol{X}\beta + \boldsymbol{Z_g g} + \boldsymbol{Z_t T} + \varepsilon \left(\text{multi-kernel prediction}\right) \tag{3}$$

where $\hat{\boldsymbol{y}}$ is a vector of phenotypic values, $\boldsymbol{\beta}$ is a vector of fixed effects, $\boldsymbol{g}$ is a vector of random additive genetic effects of markers, $\boldsymbol{t}$ is a vector of random secondary trait (e.g., whole hyperspectral, NIR) effects, and $\boldsymbol{\varepsilon}$ is a vector of random residuals. The incidence matrices for $\boldsymbol{\beta}$, $\boldsymbol{g}$, and $\boldsymbol{t}$ are $\boldsymbol{X}$, $\boldsymbol{Z_g}$, and $\boldsymbol{Z_t}$, respectively. The random effects are independent and normally distributed $\boldsymbol{g} \sim N\left(0, \sigma_g^2 \boldsymbol{G_g}\right)$, $\boldsymbol{t} \sim N\left(0, \sigma_t^2 \boldsymbol{G_t}\right)$, and $\varepsilon \sim N\left(0, \sigma_\varepsilon^2 \boldsymbol{I}\right)$, where $\boldsymbol{I}$ was the identity matrix. $\boldsymbol{G_g}$ is a kernel or relationship matrix (covariance matrix describing genomic similarities between pairs of genotypes) using $\boldsymbol{G_g} = \boldsymbol{WW'}/m$, where $\boldsymbol{W}$ is a $n \times m$ matrix of scaled and centred markers from n individuals and m

Table 1 Summary of publications using predictive G2P model incorporating HTPP secondary traits

Goal	Crop	Target trait	Secondary trait	No TP	No spectra/ band	G2P model	Reference
Incorporation of secondary traits to improve prediction of target traits	Wheat	Grain yield	CT, GNDVI, RNDVI	12	-	Single- and multi-trait	Rutkoski et al. (2016)
	Wheat	Grain yield	CT, GNDVI, RNDVI	12	-	Simple repeatability, multi-trait, random regression	Sun et al. (2017)
	Maize	Grain yield	Six spectral VI	5	62	Three single-trait models	Aguate et al. (2017)
	Wheat	19 end-use quality traits	NMR, NIR	1	-	Single-trait, multi-trait	Hayes et al. (2017)
	Wheat	Grain yield	Eight spectral VI, whole hyperspectral	9	250	Seven models including one functional regression	Montesinos-Lopez et al. (2017b)
	Wheat	Heading date, grain yield	NIRS	1	1050[a]	Single kernel using markers or NIRS	Rincent et al. (2018)
	Poplar	Height, circumference, bud flush, bud set	NIRS	1	2001[b]	Single kernel using markers or NIRS	Rincent et al. (2018)
	Wheat	Grain yield	CT, NDVI	4	-	Single- and multi-trait	Crain et al. (2018)
	Soybean	Grain yield, days to maturity	CC	58	-	Nine models exploring interactions	Jarquin et al. (2018)
	Wheat	Grain yield	Whole hyperspectral	13 (2013/2014) 14 (2014/2015)	62	Single and multi-kernel using markers and/or hyperspectral	Krause et al. (2019)

Published by Burleigh Dodds Science Publishing Limited, 2022.

				8 (2015/2016) 14 (2016/2017)			
	Wheat	Grain yield	GNDVI	2-4	-	IBCF approach	Juliana et al. (2019)
	Wheat	Grain yield	NIR	1	1059	Three models including one deep kernel	Cuevas et al. (2019)
	Sugarcane	Fibre and sucrose content	NIR	1	3112	Single- and multi-kernel using markers and/or NIR	Gonçalves et al. (2021)
	Maize	Grain yield	NIRS	1	3076	PLSR, single kernel, functional regression	Lane et al. (2020)
	Wheat	Grain yield	Whole hyperspectral	9	250	Regularized selection indices	Lopez-Cruz et al. (2020)
	Wheat	Grain yield	Whole hyperspectral	14 (2014/2015)	62	MegaLMM	Runcie et al. (2021)
Improve prediction for longitudinal traits	Rice	Projected shoot area	-	20	-	Single-trait, random regression	Campbell et al. (2018)
	Rice	Projected shoot area	-	20	-	Random regression with Legendre polynomials and B-splines	Momen et al. (2019)
	Rice	Projected shoot area	-	20	-	Multi-trait random regression	Baba et al. (2020)
	Wheat	Canopy height	-	26	-	Factor analytic and B-splines	Lyra et al. (2020)

GNDVI and RNDVI, green and red normalized difference vegetation index; CT, canopy temperature; VI, vegetative index; NIR or NIRS, near-infrared reflectance spectroscopy; NMR, nuclear magnetic resonance; CC, canopy coverage; IBCF, item-based collaborative filtering; PLSR, partial least square regression; LMM, linear mixed model; TP, time point; G2P, genotype to phenotype.

[a] NIR spectra obtained on wheat grains.

[b] NIR spectra obtained on poplar wood.

is the total number of markers. $\boldsymbol{G_t}$ is a kernel or covariance matrix describing similarities based on secondary trait information between pairs of genotypes using $\boldsymbol{G_t} = \boldsymbol{TT}'/s$, where $\boldsymbol{T}$ is a $n \times s$ matrix of scaled and centred secondary traits from n individuals and s is the total number of secondary traits.

Several studies have integrated HTPP data in predictive models using the multi-kernel approach mentioned above and expanded further on the integration of GEI and envirotyping (Montesinos-Lopez et al., 2017b, Aguate et al., 2017, Rutkoski et al., 2016, Krause et al., 2019, Gonçalves et al., 2021). For example, Aguate et al. (2017) incorporated wavelengths into a GP model in maize hybrids showing that using wavelength as predictors increased predictive ability compared to VIs. Montesinos-Lopez et al. (2017b) compared the full and a subset of bands when predicting grain yield on wheat concluding that using all wavelengths resulted in higher predictive ability. Recently, Krause et al. (2019) used a multi-kernel GBLUP approach combining marker, pedigree, and hyperspectral reflectance kernels and observed the highest predictive ability compared to single kernels when predicting genotypes. Cuevas et al. (2019) applied a non-linear arc-cosine kernel (deep learning artificial neural networks) in wheat data set demonstrating promising results.

3.2.2 Multivariate predictive G2P models

The multivariate approach focuses on predicting multiple traits as response variables using single kernel from molecular markers. Therefore, phenotypes can be a combination of the targeted trait (mainly yield) and the correlated secondary traits (usually VIs). This approach is expected to improve predictive ability relative to single trait by enabling information to be shared among correlated traits, particularly for traits showing low heritability. Initial implementation of such analyses was published by Rutkoski et al. (2016) followed by Sun et al. (2017) and Crain et al. (2018). These three publications used canopy temperature and NDVI as secondary traits to improve the predictive ability of grain yield in wheat. Rutkoski et al. (2016) showed that the integration of both types of secondary traits into single-kernel GBLUP model increased the predictive ability by 70% compared to the univariate model. Sun et al. (2017) showed that the incorporation of secondary traits in both training and testing sets in a multivariate model yielded the best results in terms of predictive ability. Hayes et al. (2017) observed improvements in predictive ability when incorporating NIR and nuclear magnetic resonance spectra as secondary traits, in a predictive multi-trait framework in wheat. Crain et al. (2018) included grain yield, canopy temperature, and VI measurements in a multi-trait model showing a slight increase (7% gains) in predictive ability when compared to the single-trait model.

Direct implementation of thousands of hyperspectral spectra in a traditional G2P multivariate model would be computationally demanding. Several authors have used feature selection to identify those traits that are relevant (Crain et al., 2018, Rutkoski et al., 2016, Sun et al., 2017). Other authors have replaced the multivariate method with a direct regression on the secondary phenotype data, for example using secondary traits (wavelengths or environmental input) as functional covariates in functional regression (Montesinos-Lopez et al., 2017a, b) and factorial regression (van Eeuwijk et al., 2018) models. A review on functional regression analysis using HTPP data has been written by Montesinos-Lopez et al. (2018). Alternative methods have also been proposed to handle efficiently this high complexity data such as the 'item-based collaborative filtering' (IBCF) (Juliana et al., 2019) and regularized selection indices (Lopez-Cruz et al., 2020). For example, the IBCF approach can be effectively applied in scenarios where trait correlations with the primary target (grain yield) are moderate to high (Juliana et al., 2019).

While these methods mentioned above (e.g. functional regression) are powerful alternatives to the traditional multi-trait model, they do not explicitly account for the high-dimensional genetic correlations between traits. To overcome this issue, Runcie et al. (2021) proposed a multivariate linear mixed model (MegaLMM) taking advantage of the factor-analytic approach. They used the original data set from Krause et al. (2019) and observed higher predictive ability when using MegaLMM compared to the multi-kernel (genomic + phenomic) approach. According to the authors, by directly modelling the genetic covariance between the hyperspectral reflectance traits, MegaLMM should be more efficient, mainly when secondary traits and the target (focal) traits are measured on the same plants.

3.2.3 Predictive G2P models using longitudinal (time-series) traits

Measuring plant development has become feasible with the advancements of HTPP in recent years. A 4D profile can be built based on the genotype, phenotype, and envirotype (3D profile) while the fourth involves the developmental stages (temporal data) (Xu, 2016). For example, collecting time-series data enables the comparison between the canopy height of different accessions at the same growth stage in sorghum (Watanabe et al., 2017) and also allows the selection of high-yielding cultivars at an early plant developmental stage (Sun et al., 2017). Analysing G2P models from different TPs collected in the same environment is straightforward, but it is more difficult when measurements are made across environments. Also, Lyra et al. (2020) pointed out the importance of sampling data in later stages of growth development to capture more genetic variance. Therefore, many G2P frameworks have been adapted to incorporate information from new phenotyping and envirotyping data across developmental stages. A review on using longitudinal data from HTPP in predictive G2P models has been discussed by Moreira et al. (2020).

For longitudinal data, a simple univariate mixed model can be fitted at each TP for traits like FVC, canopy height, senescence, and biomass accumulation. However, to capture the covariance between TPs, a multivariate approach can be used allowing each TP to have a unique variance and covariance between TPs. Other common methods used to model longitudinal data more flexibly are random regression and factorial regression (Baba et al., 2020, Campbell et al., 2018, Momen et al., 2019). Random regression model (RRM) (Schaeffer, 2016, Schaeffer, 2004) is used to estimate covariance functions and to model the shape/trajectory of observations taken over time. This trajectory (covariance across TPs) can be modelled using Legendre polynomials and splines. Sun et al. (2017) applied an RRM using a cubic smoothing spline in wheat aiming to capture the trait development (canopy temperature and NDVI) during the growth stages across multiple environments. Campbell et al. (2018) observed improvements of predictive ability using RRM using Legendre polynomials to model shoot growth trajectories in rice when compared to single-time prediction. Lyra et al. (2020) also observed higher predictive ability compared to single TP using B-spline basis coefficients as phenotypes after adjusting the values using a factor analytic model in 26 TPs for canopy height in wheat.

4 Conclusion

In this chapter, the current methods accessible for monitoring crop canopy development and nutrition, for predicting grain yield and yield components, as well as the various approaches available for integrating data from phenomics and envirotyping in predictive breeding, have been reviewed. Whilst much has been achieved in the last decade, digital phenotyping as a tool to support breeding programs is still in its infancy. A limitation has been the application of digital phenotyping into the breeding process. The temporal and spatial resolution required to look at a trait of interest is likely to be different at earlier or the later stages of the selection programs, requiring tools to scan hundreds to thousands of genetic materials or fined-tuned tools for detailed analysis of the performance of the targeted traits on a sub-sample of a population. Such considerations define the cost-benefit of the technology deployed. Further limitations are the ability to deploy such technologies at different locations and the coordination of the data acquisition in multi-environment studies. Finally, the level of data integration to develop efficient pipelines to integrate digital phenotyping data into G2P models still requires further development. None of these points should be insurmountable with appropriate investment. The last decade has seen an exponential increase in terms of initiatives and studies around digital phenotyping applications to breeding, which should lead to significant breakthroughs in the next decade.

5 Acknowledgements

Rothamsted Research receives grant-aided support from the Biotechnology and Biological Sciences Research Council (BBSRC) through the Designing Future Wheat programme [BB/P016855/1].

6 Where to look for further information

The following articles provide a good overview of the subject:

- Crossa, J., Perez-Rodriguez, P., Cuevas, J., Montesinos-Lopez, O., Jarquin, D., de los Campos, G., Burgueno, J., Gonzalez-Camacho, J. M., Perez-Elizalde, S., Beyene, Y., Dreisigacker, S., Singh, R., Zhang, X. C., Gowda, M., Roorkiwal, M., Rutkoski, J. and Varshney, R. K. 2017. Genomic selection in plant breeding: methods, models, and perspectives. *Trends in Plant Science* 22, 961–975.
- Pask, A. J. D., Pietragalla, J., Mullan, D. M. and Reynolds, MP. (Eds.) 2012. *Physiological Breeding II: A Field Guide to Wheat Phenotyping*. Mexico, D.F.: CIMMYT.
- van Eeuwijk, F. A., Bustos-Korts, D., Millet, E. J., Boer, M. P., Kruijer, W., Thompson, A., Malosetti, M., Iwata, H., Quiroz, R., Kuppe, C., Muller, O., Blazakis, K. N., Yug, K., Tardieu, F. and Chapman, S. C. 2018. Modelling strategies for assessing and increasing the effectiveness of new phenotyping techniques in plant breeding. *Plant Science*. doi: 10.1016/j.plantsci.2018.06.018.
- Reynolds, M. and Langridge, P. 2016. Physiological breeding. *Current Opinion in Plant Biology* 31, 162–171.

Further information about infrastructure and initiatives about phenotyping:

- IPPN (International Plant Phenotyping Network) is an association representing the major plant phenotyping centers and providing relevant information about plant phenotyping: www.plant-phenotyping.org/IPPN_home.
- EMPHASIS is a project aiming to address the limitations of the current phenotyping to fully exploit the genetic and genomic resources available for crop improvement in changing climate: https://emphasis.plant-phenotyping.eu.

Database for spectral reflectance indices:

- L3 Harris Geospatial documentation center is regrouping a vast database of spectral indices: www.l3harrisgeospatial.com/docs/vegetationindices.html.

7 Abbreviations

AGB	Above-ground biomass
ANN	Artificial neural network
AUC	Area under the curve
C_{ab}	Leaf chlorophyll a and b content
CCCI	Canopy chlorophyll content index
C_m	Specific leaf mass
CNC	Canopy nitrogen content
CNI	Canopy nutrition index
CT	Canopy temperature
C_w	Leaf water content
F'	Steady-state fluorescence
F_q'/F_m'	Photosystem II operating efficiency
FVC	Fraction vegetation cover
G	Genotype
G2P	Genotype to phenotype
GBLUP	Genomic best linear unbiased prediction
GEBV	Genomic estimated breeding value
GEI	Genotype by environment interaction
GP	Genomic Prediction
GP	Gaussian process
GS	Genomic selection
HTPP	High-throughput plant phenotyping
IBCF	Item-based collaborative filtering
LAI	Leaf area index
LUE	Light use efficiency
MegaLMM	Mega-scale linear mixed model
MET	Multi-environment trials
N	Nitrogen
NDVI	Normalized difference vegetation index
NIRs	Near-infrared spectroscopy
NPQ	Non-photochemical quenching
P	Phenotype
PLSR	Partial least square regression
PRI	Photochemical/physiological reflectance index
PSII	Photosystem 2
RF	Random Forest
RGB	Red-green-blue
RMM	Random regression model
SfM	Structure from motion
SI	Selection index

SIF	Sun-induced fluorescence
SRI	Spectral reflectance index
SVM	Support vector machine
$T_c - T_a$	Canopy minus air temperature
TP	Time point
UAV	Unmanned aerial vehicle
VE	Environmental variance
V_ε	Random residual variance
V_{GE}	GEI variance
VI	Vegetation index
VP	Phenotypic variance
WDI	Water deficit index

8 References

Aasen, H., Kirchgessner, N., Walter, A. and Liebisch, F. 2020. PhenoCams for field phenotyping: using very high temporal resolution digital repeated photography to investigate interactions of growth, phenology, and harvest traits. *Frontiers in Plant Science* 11, 593.

Aguate, F. M., Trachsel, S., Perez, L. G., Burgueno, J., Crossa, J., Balzarini, M., Gouache, D., Bogard, M. and De Los Campos, G. 2017. Use of hyperspectral image data outperforms vegetation indices in prediction of maize yield. *Crop Science* 57(5), 2517–2524.

Anderegg, J., Yu, K., Aasen, H., Walter, A., Liebisch, F. and Hund, A. 2020. Spectral vegetation indices to track senescence dynamics in diverse wheat germplasm. *Frontiers in Plant Science* 10, 1749.

Aquino, A., Millan, B., Gutierrez, S. and Tardaguila, J. 2015. Grapevine flower estimation by applying artificial vision techniques on images with uncontrolled scene and multi-model analysis. *Computers and Electronics in Agriculture* 119, 92–104.

Araus, J. L., Kefauver, S. C., Zaman-Allah, M., Olsen, M. S. and Cairns, J. E. 2018. Translating high-throughput phenotyping into genetic gain. *Trends in Plant Science* 23(5), 451–466.

Baba, T., Momen, M., Campbell, M. T., Walia, H. and Morota, G. 2020. Multi-trait random regression models increase genomic prediction accuracy for a temporal physiological trait derived from high-throughput phenotyping. *PLoS ONE* 15(2), e0228118.

Barmeier, G. and Schmidhalter, U. 2017. High-throughput field phenotyping of leaves, leaf sheaths, culms and ears of spring barley cultivars at anthesis and dough ripeness. *Frontiers in Plant Science* 8, 1920.

Barnes, E. M., Clarke, T. R., Richards, S. E., Colaizzi, P. D., Haberland, J., Kostrzewski, M., Waller, P., Choi, C., Riley, E. and Thompson, T. 2000. Coincident detection of crop water stress, nitrogen status and canopy density using ground based multispectral data. Proceedings of the 5th International Conference on Precision Agriculture and Other Resource Management July 16-19, 2000, Bloomington, MN USA, 15 p. Madison, WI: Precision Agriculture Center, University of Minnesota, ASA-CSSA-SSSA, c2000.

Beauchêne, K., Leroy, F., Fournier, A., Huet, C., Bonnefoy, M., Lorgeou, J., De Solan, B., Piquemal, B., Thomas, S. and Cohan, J. P. 2019. Management and characterization of abiotic stress via PhénoField®, a high-throughput field phenotyping platform. *Frontiers in Plant Science* 10, 904.

Bendig, J., Yu, K., Aasen, H., Bolten, A., Bennertz, S., Broscheit, J., Gnyp, M. L. and Bareth, G. 2015. Combining UAV-based plant height from crop surface models, visible, and near infrared vegetation indices for biomass monitoring in barley. *International Journal of Applied Earth Observation and Geoinformation* 39, 79–87.

Berger, K., Verrelst, J., Féret, J.-B., Hank, T., Wocher, M., Mauser, W. and Camps-Valls, G. 2020. Retrieval of aboveground crop nitrogen content with a hybrid machine learning method. *International Journal of Applied Earth Observation and Geoinformation* 92, 102174.

Blackburn, G. A. 2007. Hyperspectral remote sensing of plant pigments. *Journal of Experimental Botany* 58(4), 855–867.

Boegh, E., Soegaard, H., Broge, N., Hasager, C. B., Jensen, N. O., Schelde, K. and Thomsen, A. 2002. Airborne multispectral data for quantifying leaf area index, nitrogen concentration, and photosynthetic efficiency in agriculture. *Remote Sensing of Environment* 81(2–3), 179–193.

Borra-Serrano, I., De Swaef, T., Quataert, P., Aper, J., Saleem, A., Saeys, W., Somers, B., Roldán-Ruiz, I. and Lootens, P. 2020. Closing the phenotyping gap: high resolution UAV time series for soybean growth analysis provides objective data from field trials. *Remote Sensing* 12(10), 1644.

Brown, T. B., Cheng, R., Sirault, X. R., Rungrat, T., Murray, K. D., Trtilek, M., Furbank, R. T., Badger, M., Pogson, B. J. and Borevitz, J. O. 2014. TraitCapture: genomic and environment modelling of plant phenomic data. *Current Opinion in Plant Biology* 18, 73–79.

Buchaillot, M. L., Gracia-Romero, A., Vergara-Diaz, O., Zaman-Allah, M. A., Tarekegne, A., Cairns, J. E., Prasanna, B. M., Araus, J. L. and Kefauver, S. C. 2019. Evaluating maize genotype performance under low nitrogen conditions using RGB UAV phenotyping techniques. *Sensors* 19(8), 1815.

Burkart, A., Hecht, V. L., Kraska, T. and Rascher, U. 2018. Phenological analysis of unmanned aerial vehicle based time series of barley imagery with high temporal resolution. *Precision Agriculture* 19(1), 134–146.

Bustos-Korts, D., Boer, M. P., Malosetti, M., Chapman, S., Chenu, K., Zheng, B. and Van Eeuwijk, F. A. 2019. Combining crop growth modeling and statistical genetic modeling to evaluate phenotyping strategies. *Frontiers in Plant Science* 10, 1491.

Cabrera-Bosquet, L., Molero, G., Stellacci, A., Bort, J., Nogués, S. and Araus, J. 2011. NDVI as a potential tool for predicting biomass, plant nitrogen content and growth in wheat genotypes subjected to different water and nitrogen conditions. *Cereal Research Communications* 39(1), 147–159.

Cairns, J. E., Sanchez, C., Vargas, M., Ordonez, R. and Araus, J. L. 2012. Dissecting maize productivity: ideotypes associated with grain yield under drought stress and well-watered conditions. *Journal of Integrative Plant Biology* 54(12), 1007–1020.

Camino, C., Gonzalez-Dugo, V., Hernandez, P., Sillero, J. C. and Zarco-Tejada, P. J. 2018a. Improved nitrogen retrievals with airborne-derived fluorescence and plant traits quantified from VNIR-SWIR hyperspectral imagery in the context of precision agriculture. *International Journal of Applied Earth Observation and Geoinformation* 70, 105–117.

Camino, C., Zarco-Tejada, P. J. and Gonzalez-Dugo, V. 2018b. Effects of heterogeneity within tree crowns on airborne-quantified SIF and the CWSI as indicators of water stress in the context of precision agriculture. *Remote Sensing* 10(4), 604.

Cammarano, D., Fitzgerald, G. J., Casa, R. and Basso, B. 2014. Assessing the robustness of vegetation indices to estimate wheat N in Mediterranean environments. *Remote Sensing* 6(4), 2827-2844.

Campbell, M., Walia, H. and Morota, M. 2018. Utilizing random regression models for genomic prediction of a longitudinal trait derived from high-throughput phenotyping. *The Plant Journal* 2, 1-11.

Casadesús, J., Kaya, Y., Bort, J., Nachit, M. M., Araus, J. L., Amor, S., Ferrazzano, G., Maalouf, F., Maccaferri, M., Martos, V., Ouabbou, H. and Villegas, D. 2007. Using vegetation indices derived from conventional digital cameras as selection criteria for wheat breeding in water-limited environments. *Annals of Applied Biology* 150(2), 227-236.

Chandra, A. L., Desai, S. V., Balasubramanian, V. N., Ninomiya, S. and Guo, W. 2020. Active learning with point supervision for cost-effective panicle detection in cereal crops. *Plant Methods* 16, 34.

Che, Y., Wang, Q., Xie, Z., Zhou, L., Li, S., Hui, F., Wang, X., Li, B. and Ma, Y. 2020. Estimation of maize plant height and leaf area index dynamics using an unmanned aerial vehicle with oblique and nadir photography. *Annals of Botany* 126(4), 765-773.

Chenu, K., Chapman, S. C., Tardieu, F., McLean, G., Welcker, C. and Hammer, G. L. 2009. Simulating the yield impacts of organ-level quantitative trait loci associated with drought response in maize: a "gene-to-phenotype" modeling approach. *Genetics* 183(4), 1507-1523.

Christopher, J. T., Veyradier, M., Borrell, A. K., Harvey, G., Fletcher, S. and Chenu, K. 2014. Phenotyping novel stay-green traits to capture genetic variation in senescence dynamics. *Functional Plant Biology* 41(11), 1035-1048.

Clarke, T. R., Moran, M. S., Barnes, E. M., Pinter, P. J. and Qi, J. 2001. Planar domain indices: a method for measuring a quality of a single component in two-component pixels. IGARSS 2001. Scanning the Present and Resolving the Future. Proceedings. IEEE 2001 International Geoscience and Remote Sensing Symposium (Cat. No.01CH37217), 9-13 July 2001. 1279-1281.

Comar, A., Burger, P., De Solan, B., Baret, F. D. R., Daumard, F. and Hanocq, J. O. 2012. A semi-automatic system for high throughput phenotyping wheat cultivars in-field conditions: description and first results. *Functional Plant Biology* 39(11), 914-924.

Cooper, M., Messina, C. D., Podlich, D., Totir, L. R., Baumgarten, A., Hausmann, N. J., Wright, D. and Graham, G. 2014. Predicting the future of plant breeding: complementing empirical evaluation with genetic prediction. *Crop and Pasture Science* 65(4), 311-336.

Cooper, M., Powell, O., Voss-Fels, K. P., Messina, C. D., Gho, C., Podlich, D. W., Technow, F., Chapman, S. C., Beveridge, C. A., Ortiz-Barrientos, D. and Hammer, G. L. 2021. Modelling selection response in plant-breeding programs using crop models as mechanistic gene-to-phenotype (CGM-G2P) multi-trait link functions. *In Silico Plants* 3(1), diaa016.

Corti, M., Cavalli, D., Cabassi, G., Marino Gallina, P. and Bechini, L. 2018. Does remote and proximal optical sensing successfully estimate maize variables? A review. *European Journal of Agronomy* 99, 37-50.

Costa-Neto, G., Fritsche-Neto, R. and Crossa, J. 2021a. Nonlinear kernels, dominance, and envirotyping data increase the accuracy of genome-based prediction in multi-environment trials. *Heredity* 126(1), 92–106.

Costa-Neto, G., Galli, G., Carvalho, H. F., Crossa, J. and Fritsche-Neto, R. 2021b. EnvRtype: a software to interplay enviromics and quantitative genomics in agriculture. *G3* 11(4), jkab040.

Costa-Neto, G. M. F., Júnior, O. P. M., Heinemann, A. B., De Castro, A. P. and Duarte, J. B. 2020. A novel GIS-based tool to reveal spatial trends in reaction norm: upland rice case study. *Euphytica* 216, 1–16.

Crain, J., Mondal, S., Rutkoski, J., Singh, R. P. and Poland, J. 2018. Combining high-throughput phenotyping and genomic information to increase prediction and selection accuracy in wheat breeding. *Plant Genome* 11(1), 1–14.

Crossa, J., Fritsche-Neto, R., Montesinos-Lopez, O. A., Costa-Neto, G., Dreisigacker, S., Montesinos-Lopez, A. and Bentley, A. R. 2021. The modern plant breeding triangle: optimizing the use of genomics, phenomics, and enviromics data. *Frontiers in Plant Science* 12, 651480.

Crossa, J., Perez-Rodriguez, P., Cuevas, J., Montesinos-Lopez, O., Jarquin, D., De Los Campos, G., Burgueno, J., Gonzalez-Camacho, J. M., Perez-Elizalde, S., Beyene, Y., Dreisigacker, S., Singh, R., Zhang, X. C., Gowda, M., Roorkiwal, M., Rutkoski, J. and Varshney, R. K. 2017. Genomic selection in plant breeding: methods, models, and perspectives. *Trends in Plant Science* 22(11), 961–975.

Cuevas, J., Montesinos-López, O., Juliana, P., Guzmán, C., Pérez-Rodríguez, P., González-Bucio, J., Burgueño, J., Montesinos-López, A. and Crossa, J. 2019. Deep Kernel for genomic and near infrared predictions in multi-environment breeding trials. *G3* 9(9), 2913–2924.

David, E., Madec, S., Sadeghi-Tehran, P., Aasen, H., Zheng, B., Liu, S., Kirchgessner, N., Ishikawa, G., Nagasawa, K., Badhon, M. A., Pozniak, C., De Solan, B., Hund, A., Chapman, S. C., Baret, F., Stavness, I. and Guo, W. 2020. Global Wheat Head Detection (GWHD) dataset: a large and diverse dataset of high resolution RGB labelled images to develop and benchmark wheat head detection methods. *Plant Phenomics* 2020, 3521852.

De Los Campos, G., Pérez-Rodríguez, P., Bogard, M., Gouache, D. and Crossa, J. 2020. A data-driven simulation platform to predict cultivars' performances under uncertain weather conditions. *Nature Communications* 11(1), 4876.

Deery, D., Jimenez-Berni, J., Jones, H., Sirault, X. and Furbank, R. 2014. Proximal remote sensing buggies and potential applications for field-based phenotyping. *Agronomy* 4(3), 349–379.

Dobrowski, S. Z., Pushnik, J. C., Zarco-Tejada, P. J. and Ustin, S. L. 2005. Simple reflectance indices track heat and water stress-induced changes in steady-state chlorophyll fluorescence at the canopy scale. *Remote Sensing of Environment* 97(3), 403–414.

Duan, L., Huang, C., Chen, G., Xiong, L., Liu, Q. and Yang, W. 2015. Determination of rice panicle numbers during heading by multi-angle imaging. *The Crop Journal* 3(3), 211–219.

Duan, T., Zheng, B., Guo, W., Ninomiya, S., Guo, Y. and Chapman, S. C. 2016. Comparison of ground cover estimates from experiment plots in cotton, sorghum and sugarcane based on images and Ortho-mosaics captured by UAV. *Functional Plant Biology* 44(1), 169–183.

Duncan, J. M. A., Dash, J. and Atkinsonz, P. M. 2015. The potential of satellite-observed crop phenology to enhance yield gap assessments in smallholder landscapes. *Frontiers in Environmental Science* 3, 56. doi: 10.3389/fenvs.2015.00056.

Ecarnot, M., Compan, F. and Roumet, P. 2013. Assessing leaf nitrogen content and leaf mass per unit area of wheat in the field throughout plant cycle with a portable spectrometer. *Field Crops Research* 140, 44–50.

El-Hendawy, S. E., Al-Suhaibani, N. A., Elsayed, S., Hassan, W. M., Dewir, Y. H., Refay, Y. and Abdella, K. A. 2019a. Potential of the existing and novel spectral reflectance indices for estimating the leaf water status and grain yield of spring wheat exposed to different irrigation rates. *Agricultural Water Management* 217, 356 -373.

El-Hendawy, S. E., Alotaibi, M., Al-Suhaibani, N., Al-Gaadi, K., Hassan, W., Dewir, Y. H., Emam, M. A. E., Elsayed, S. and Schmidhalter, U. 2019b. Comparative performance of spectral reflectance indices and multivariate modeling for assessing agronomic parameters in advanced spring wheat lines under two contrasting irrigation regimes. *Frontiers in Plant Science* 10, 1537.

Elsayed, S., Barmeier, G. and Schmidhalter, U. 2018. Passive reflectance sensing and digital image analysis allows for assessing the biomass and nitrogen status of wheat in early and late tillering stages. *Frontiers in Plant Science* 9, 1478.

Elsayed, S., Elhoweity, M., Ibrahim, H. H., Dewir, Y. H., Migdadi, H. M. and Schmidhalter, U. 2017. Thermal imaging and passive reflectance sensing to estimate the water status and grain yield of wheat under different irrigation regimes. *Agricultural Water Management* 189, 98–110.

Elsayed, S., Rischbeck, P. and Schmidhalter, U. 2015. Comparing the performance of active and passive reflectance sensors to assess the normalized relative canopy temperature and grain yield of drought-stressed barley cultivars. *Field Crops Research* 177, 148–160.

Erdle, K., Mistele, B. and Schmidhalter, U. 2013. Spectral high-throughput assessments of phenotypic differences in biomass and nitrogen partitioning during grain filling of wheat under high yielding Western European conditions. *Field Crops Research* 141, 16–26.

Evain, S., Flexas, J. and Moya, I. 2004. A new instrument for passive remote sensing: 2. Measurement of leaf and canopy reflectance changes at 531 nm and their relationship with photosynthesis and chlorophyll fluorescence. *Remote Sensing of Environment* 91(2), 175–185.

Fernandez-Gallego, J. A., Kefauver, S. C., Gutierrez, N. A., Nieto-Taladriz, M. T. and Araus, J. L. 2018. Wheat ear counting in-field conditions: high throughput and low-cost approach using RGB images. *Plant Methods* 14, 22.

Fischer, R. A. and Rebetzke, G. J. 2018. Indirect selection for potential yield in early-generation, spaced plantings of wheat and other small-grain cereals: a review. *Crop and Pasture Science* 69(5), 439–459.

Fitzgerald, G., Rodriguez, D. and O'Leary, G. 2010. Measuring and predicting canopy nitrogen nutrition in wheat using a spectral index-the canopy chlorophyll content index (CCCI). *Field Crops Research* 116(3), 318–324.

Frels, K., Guttieri, M., Joyce, B., Leavitt, B. and Baenziger, P. S. 2018. Evaluating canopy spectral reflectance vegetation indices to estimate nitrogen use traits in hard winter wheat. *Field Crops Research* 217, 82–92.

Friedli, M., Kirchgessner, N., Grieder, C., Liebisch, F., Mannale, M. and Walter, A. 2016. Terrestrial 3D laser scanning to track the increase in canopy height of both monocot and dicot crop species under field conditions. *Plant Methods* 12, 9.

Fu, P., Meacham-Hensold, K., Guan, K. and Bernacchi, C. J. 2019. Hyperspectral leaf reflectance as proxy for photosynthetic capacities: an ensemble approach based on multiple machine learning algorithms. *Frontiers in Plant Science* 10, 730.

Fu, P., Meacham-Hensold, K., Guan, K., Wu, J. and Bernacchi, C. 2020. Estimating photosynthetic traits from reflectance spectra: a synthesis of spectral indices, numerical inversion, and partial least square regression. *Plant, Cell and Environment* 43(5), 1241-1258.

Gamon, J. A., Huemmrich, K. F., Wong, C. Y. S., Ensminger, I., Garrity, S., Hollinger, D. Y., Noormets, A. and Peñuelas, J. 2016. A remotely sensed pigment index reveals photosynthetic phenology in evergreen conifers. *Proceedings of the National Academy of Sciences of the United States of America* 113(46), 13087-13092.

Garriga, M., Romero-Bravo, S., Estrada, F., Escobar, A., Matus, I. A., Del Pozo, A., Astudillo, C. A. and Lobos, G. A. 2017. Assessing wheat traits by spectral reflectance: do we really need to focus on predicted trait-values or directly identify the Elite Genotypes Group? *Frontiers in Plant Science* 8, 280.

Ghosal, S., Zheng, B., Chapman, S. C., Potgieter, A. B., Jordan, D. R., Wang, X., Singh, A. K., Singh, A., Hirafuji, M., Ninomiya, S., Ganapathysubramanian, B., Sarkar, S. and Guo, W. 2019. A weakly supervised deep learning framework for sorghum head detection and counting. *Plant Phenomics* 2019, 1525874.

Gitelson, A. and Solovchenko, A. 2018. Non-invasive quantification of foliar pigments: possibilities and limitations of reflectance- and absorbance-based approaches. *Journal of Photochemistry and Photobiology. B, Biology* 178, 537-544.

Gonçalves, M. T. V., Morota, G., Costa, P. M. A., Vidigal, P. M. P., Barbosa, M. H. P. and Peternelli, L. A. 2021. Near-infrared spectroscopy outperforms genomics for predicting sugarcane feedstock quality traits. *PLoS ONE* 16(3), e0236853.

Gonzalez-Dugo, V., Zarco-Tejada, P. J. and Fereres, E. 2014. Applicability and limitations of using the crop water stress index as an indicator of water deficits in citrus orchards. *Agricultural and Forest Meteorology* 198-199, 94-104.

Gracia-Romero, A., Kefauver, S. C., Vergara-Diaz, O., Zaman-Allah, M. A., Prasanna, B. M., Cairns, J. E. and Araus, J. L. 2017. Comparative performance of ground vs. aerially assessed RGB and multispectral indices for early-growth evaluation of maize performance under phosphorus fertilization. *Frontiers in Plant Science* 8, 2004.

Grieder, C., Hund, A. and Walter, A. 2015. Image based phenotyping during winter: a powerful tool to assess wheat genetic variation in growth response to temperature. *Functional Plant Biology* 42(4), 387-396.

Guo, W., Fukatsu, T. and Ninomiya, S. 2015. Automated characterization of flowering dynamics in rice using field-acquired time-series RGB images. *Plant Methods* 11, 7.

Guo, W., Rage, U. K. and Ninomiya, S. 2013. Illumination invariant segmentation of vegetation for time series wheat images based on decision tree model. *Computers and Electronics in Agriculture* 96, 58-66.

Gutierrez, M., Norton, R., Thorp, K. R. and Wang, G. 2012. Association of spectral reflectance indices with plant growth and lint yield in upland cotton. *Crop Science* 52(2), 849-857.

Haboudane, D., Miller, J. R., Pattey, E., Zarco-Tejada, P. J. and Strachan, I. B. 2004. Hyperspectral vegetation indices and novel algorithms for predicting green LAI of crop canopies: modeling and validation in the context of precision agriculture. *Remote Sensing of Environment* 90(3), 337–352.

Hamuda, E., Glavin, M. and Jones, E. 2016. A survey of image processing techniques for plant extraction and segmentation in the field. *Computers and Electronics in Agriculture* 125, 184–199.

Han, L., Yang, G. J., Yang, H., Xu, B., Li, Z. H. and Yang, X. D. 2018. Clustering field-based maize phenotyping of plant-height growth and canopy spectral dynamics using a UAV remote-sensing approach. *Frontiers in Plant Science* 9, 1638.

Hayes, B. J., Panozzo, J., Walker, C. K., Choy, A. L., Kant, S., Wong, D., Tibbits, J., Daetwyler, H. D., Rochfort, S., Hayden, M. J. and Spangenberg, G. C. 2017. Accelerating wheat breeding for end-use quality with multi-trait genomic predictions incorporating near infrared and nuclear magnetic resonance-derived phenotypes. *TAG. Theoretical and Applied Genetics. Theoretische und Angewandte Genetik* 130(12), 2505–2519.

Hocevar, M., Sirok, B., Godesa, T. and Stopar, M. 2014. Flowering estimation in apple orchards by image analysis. *Precision Agriculture* 15(4), 466–478.

Holman, F. H., Riche, A. B., Castle, M., Wooster, M. J. and Hawkesford, M. J. 2019. Radiometric calibration of 'commercial off the shelf' cameras for UAV-based high-resolution temporal crop phenotyping of reflectance and NDVI. *Remote Sensing* 11(14), 1657.

Holman, F. H., Riche, A. B., Michalski, A., Castle, M., Wooster, M. J. and Hawkesford, M. J. 2016. High throughput field phenotyping of wheat plant height and growth rate in field plot trials using UAV based remote sensing. *Remote Sensing* 8(12), 23.

Hosoi, F., Nakai, Y. and Omasa, K. 2013. 3-D voxel-based solid modeling of a broad-leaved tree for accurate volume estimation using portable scanning lidar. *ISPRS Journal of Photogrammetry and Remote Sensing* 82, 41–48.

Hoyos-Villegas, V., Houx, J. H., Singh, S. K. and Fritschi, F. B. 2014. Ground-based digital imaging as a tool to assess soybean growth and yield. *Crop Science* 54(4), 1756–1768.

Jarquín, D., Crossa, J., Lacaze, X., Du Cheyron, P., Daucourt, J., Lorgeou, J., Piraux, F., Guerreiro, L., Pérez, P., Calus, M., Burgueño, J. and de los Campos, G. 2014. A reaction norm model for genomic selection using high-dimensional genomic and environmental data. *TAG. Theoretical and Applied Genetics. Theoretische und Angewandte Genetik* 127(3), 595–607.

Jarquín, D., Howard, R., Xavier, A., Das Choudhury, S. 2018. Increasing predictive ability by modeling interactions between environments, genotype and canopy coverage image data for soybeans. *Agronomy* 8, 51. https://doi.org/10.3390/agronomy8040051.

Jay, S., Rabatel, G., Hadoux, X., Moura, D. and Gorretta, N. 2015. In-field crop row phenotyping from 3D modeling performed using structure from motion. *Computers and Electronics in Agriculture* 110, 70–77.

Jimenez-Berni, J. A., Deery, D. M., Rozas-Larraondo, P., Condon, A. T. G., Rebetzke, G. J., James, R. A., Bovill, W. D., Furbank, R. T. and Sirault, X. R. R. 2018. High throughput determination of plant height, ground cover, and above-ground biomass in wheat with LiDAR. *Frontiers in Plant Science* 9, 237.

Jin, X., Madec, S., Dutartre, D., De Solan, B., Comar, A. and Baret, F. 2019. High-throughput measurements of stem characteristics to estimate ear density and above-ground biomass. *Plant Phenomics* 2019, 4820305.

Jones, H. G., Serraj, R., Loveys, B. R., Xiong, L. Z., Wheaton, A. and Price, A. H. 2009. Thermal infrared imaging of crop canopies for the remote diagnosis and quantification of plant responses to water stress in the field. *Functional Plant Biology* 36(11), 978–989.

Juliana, P., Montesinos-Lopez, O. A., Crossa, J., Mondal, S., Gonzalez Perez, L., Poland, J., Huerta-Espino, J., Crespo-Herrera, L., Govindan, V., Dreisigacker, S., Shrestha, S., Perez-Rodriguez, P., Pinto Espinosa, F. and Singh, R. P. 2019. Integrating genomic-enabled prediction and high-throughput phenotyping in breeding for climate-resilient bread wheat. *TAG. Theoretical and Applied Genetics. Theoretische und Angewandte Genetik* 132(1), 177–194.

Kefauver, S. C., Vicente, R., Vergara-Díaz, O., Fernandez-Gallego, J. A., Kerfal, S., Lopez, A., Melichar, J. P. E., Serret Molins, M. D. and Araus, J. L. 2017. Comparative UAV and field phenotyping to assess yield and nitrogen use efficiency in hybrid and conventional barley. *Frontiers in Plant Science* 8, 1733.

Kelly, J., Kljun, N., Olsson, P.-O., Mihai, L., Liljeblad, B., Weslien, P., Klemedtsson, L. and Eklundh, L. 2019. Challenges and best practices for deriving temperature data from an uncalibrated UAV thermal infrared camera. *Remote Sensing* 11(5), 567. https://doi.org/10.3390/rs11050567.

Kipp, S., Mistele, B. and Schmidhalter, U. 2014. Identification of stay-green and early senescence phenotypes in high-yielding winter wheat, and their relationship to grain yield and grain protein concentration using high-throughput phenotyping techniques. *Functional Plant Biology* 41(3), 227–235.

Krause, M. R., González-Pérez, L., Crossa, J., Pérez-Rodríguez, P., Montesinos-López, O., Singh, R. P., Dreisigacker, S., Poland, J., Rutkoski, J., Sorrells, M., Gore, M. A. and Mondal, S. 2019. Hyperspectral reflectance-derived relationship matrices for genomic prediction of grain yield in wheat. *G3* 9(4), 1231–1247.

Kronenberg, L., Yu, K., Walter, A. and Hund, A. 2017. Monitoring the dynamics of wheat stem elongation: genotypes differ at critical stages. *Euphytica* 213(7), 157.

Kyratzis, A. C., Skarlatos, D. P., Menexes, G. C., Vamvakousis, V. F. and Katsiotis, A. 2017. Assessment of vegetation indices derived by UAV imagery for durum wheat phenotyping under a water limited and heat stressed Mediterranean environment. *Frontiers in Plant Science* 8, 1114.

Lane, H. M., Murray, S. C., MontesinosLópez, O. A., MontesinosLópez, A., Crossa, J., Rooney, D. K., Barrero-Farfan, I. D., De La Fuente, G. N. and Morgan, C. L. S. 2020. Phenomic selection and prediction of maize grain yield from near-infrared reflectance spectroscopy of kernels. *The Plant Phenome Journal* 3(1), e20002.

Li, J., Shi, Y., Veeranampalayam-Sivakumar, A. N. and Schachtman, D. P. 2018. Elucidating sorghum biomass, nitrogen and chlorophyll contents With spectral and morphological traits derived From unmanned aircraft system. *Frontiers in Plant Science* 9, 1406.

Li, Z. and Sillanpää, M. J. 2015. Dynamic quantitative trait locus analysis of plant phenomic data. *Trends in Plant Science* 20(12), 822–833.

Liu, Y., Cen, C., Che, Y., Ke, R., Ma, Y. and Ma, Y. 2020. Detection of maize tassels from UAV RGB imagery with faster R-CNN. *Remote Sensing* 12(2), 338.

Lobos, G., Escobar-Opazo, A., Estrada, F., Romero, S., Garriga, M., Del Pozo, A., Poblete-Echeverría, C., Gonzalez-Talice, J., González-Martinez, L. and Caligari, P. 2019. Spectral reflectance modeling by wavelength selection: studying the scope for blueberry physiological breeding under contrasting water supply and heat conditions. *Remote Sensing* 11(3), 329. https://doi.org/10.3390/rs11030329.

Lopez-Cruz, M., Olson, E., Rovere, G., Crossa, J., Dreisigacker, S., Mondal, S., Singh, R. and De Los Campos, G. 2020. Regularized selection indices for breeding value prediction using hyper-spectral image data. *Scientific Reports* 10, 1.

Lozada, D. N., Godoy, J. V., Ward, B. P. and Carter, A. H. 2020. Genomic prediction and indirect selection for grain yield in US Pacific Northwest winter wheat using spectral reflectance indices from high-throughput phenotyping. *International Journal of Molecular Sciences* 21(1), 165.

Lu, H., Cao, Z., Xiao, Y., Zhuang, B. and Shen, C. 2017. TasselNet: counting maize tassels in the wild via local counts regression network. *Plant Methods* 13, 79.

Lu, J. S., Cheng, D. L., Geng, C. M., Zhang, Z. T., Xiang, Y. Z. and Hu, T. T. 2021. Combining plant height, canopy coverage and vegetation index from UAV-based RGB images to estimate leaf nitrogen concentration of summer maize. *Biosystems Engineering* 202, 42–54.

Lyra, D. H., Virlet, N., Sadeghi-Tehran, P., Hassall, K. L., Wingen, L. U., Orford, S., Griffiths, S., Hawkesford, M. J. and Slavov, G. T. 2020. Functional QTL mapping and genomic prediction of canopy height in wheat measured using a robotic field phenotyping platform. *Journal of Experimental Botany* 71(6), 1885–1898.

Madec, S., Jin, X. L., Lu, H., De Solan, B., Liu, S. Y., Duyme, F., Heritier, E. and Baret, F. 2019. Ear density estimation from high resolution RGB imagery using deep learning technique. *Agricultural and Forest Meteorology* 264, 225–234.

Maes, W. H. and Steppe, K. 2012. Estimating evapotranspiration and drought stress with ground-based thermal remote sensing in agriculture: a review. *Journal of Experimental Botany* 63(13), 4671–4712.

Magney, T. S., Eitel, J. U. H., Huggins, D. R. and Vierling, L. A. 2016. Proximal NDVI derived phenology improves in-season predictions of wheat quantity and quality. *Agricultural and Forest Meteorology* 217, 46–60.

Maimaitijiang, M. and Sidike, P. 2019. Vegetation Index Weighted Canopy Volume Model (CVMVI) for soybean biomass estimation from Unmanned Aerial System-based RGB imagery. *ISPRS Journal of Photogrammetry and Remote Sensing* 151, 27–41.

Malosetti, M., Ribaut, J. M. and Van Eeuwijk, F. A. 2013. The statistical analysis of multi-environment data: modeling genotype-by-environment interaction and its genetic basis. *Frontiers in Physiology* 4, 44.

Mantilla-Perez, M. B. and Salas Fernandez, M. G. 2017. Differential manipulation of leaf angle throughout the canopy: current status and prospects. *Journal of Experimental Botany* 68(21–22), 5699–5717.

Meacham-Hensold, K., Fu, P., Wu, J., Serbin, S., Montes, C. M., Ainsworth, E., Guan, K., Dracup, E., Pederson, T., Driever, S. and Bernacchi, C. 2020. Plot-level rapid screening for photosynthetic parameters using proximal hyperspectral imaging. *Journal of Experimental Botany* 71(7), 2312–2328.

Millet, E. J., Kruijer, W., Coupel-Ledru, A., Prado, S. A., Cabrera-Bosquet, L., Lacube, S., Charcosset, A., Welcker, C., Van Eeuwijk, F. and Tardieu, F. 2019. Genomic prediction of maize yield across European environmental conditions. *Nature Genetics* 51(6), 952–956.

Mir, R. R., Reynolds, M., Pinto, F., Khan, M. A. and Bhat, M. A. 2019. High-throughput phenotyping for crop improvement in the genomics era. *Plant Science* 282, 60–72.

Momen, M., Campbell, M. T., Walia, H. and Morota, G. 2019. Predicting longitudinal traits derived from high-throughput phenomics in contrasting environments using genomic Legendre polynomials and B-splines. *G3* 9(10), 3369-3380.

Montazeaud, G., Karatogma, H., Ozturk, I., Roumet, P., Ecarnot, M., Crossa, J., Ozer, E., Ozdemir, F. and Lopes, M. S. 2016. Predicting wheat maturity and stay-green parameters by modeling spectral reflectance measurements and their contribution to grain yield under rainfed conditions. *Field Crops Research* 196, 191-198.

Montes, J. M., Technow, F., Dhillon, B. S., Mauch, F. and Melchinger, A. E. 2011. High-throughput non-destructive biomass determination during early plant development in maize under field conditions. *Field Crops Research* 121(2), 268-273.

Montesinos-Lopez, O. A., Montesinos-Lopez, A., Crossa, J., De Los Campos, G., Alvarado, G., Suchismita, M., Rutkoski, J., Gonzalez-Perez, L. and Burgueno, J. 2017a. Predicting grain yield using canopy hyperspectral reflectance in wheat breeding data. *Plant Methods* 13, 4.

Montesinos-Lopez, A., Montesinos-Lopez, O. A., Cuevas, J., Mata-Lopez, W. A., Burgueno, J., Mondal, S., Huerta, J., Singh, R., Autrique, E., Gonzalez-Perez, L. and Crossa, J. 2017b. Genomic Bayesian functional regression models with interactions for predicting wheat grain yield using hyper-spectral image data. *Plant Methods* 13, 62.

Montesinos-Lopez, A., Montesinos-Lopez, O. A., De Los Campos, G., Crossa, J., Burgueno, J. and Luna-Vazquez, F. J. 2018. Bayesian functional regression as an alternative statistical analysis of high-throughput phenotyping data of modern agriculture. *Plant Methods* 14, 46.

Moran, M. S., Clarke, T. R., Inoue, Y. and Vidal, A. 1994. Estimating crop water-deficit using the relation between surface-air temperature and spectral vegetation index. *Remote Sensing of Environment* 49(3), 246-263.

Moreira, F. F., Oliveira, H. R., Volenec, J. J., Rainey, K. M. and Brito, L. F. 2020. Integrating high-throughput phenotyping and statistical genomic methods to genetically improve longitudinal traits in crops. *Frontiers in Plant Science* 11, 681.

Morota, G. and Gianola, D. 2014. Kernel-based whole-genome prediction of complex traits: a review. *Frontiers in Genetics* 5, 363.

Morota, G., Jarquin, D., Campbell, M. T. and Iwata, H. 2019. Statistical methods for the quantitative genetic analysis of high-throughput phenotyping data. arXiv preprint arXiv:1904.12341. https://arxiv.org/abs/1904.12341.

Munns, R., James, R. A., Sirault, X. R. R., Furbank, R. T. and Jones, H. G. 2010. New phenotyping methods for screening wheat and barley for beneficial responses to water deficit. *Journal of Experimental Botany* 61(13), 3499-3507.

Pancorbo, J. L., Camino, C., Alonso-Ayuso, M., Raya-Sereno, M. D., Gonzalez-Fernandez, I., Gabriel, J. L., Zarco-Tejada, P. J. and Quemada, M. 2021. Simultaneous assessment of nitrogen and water status in winter wheat using hyperspectral and thermal sensors. *European Journal of Agronomy* 127, 126287.

Pask, A., Pietragalla, J., Mullan, D. and Reynolds, M. P. 2012. *Physiological Breeding II: A Field Guide to Wheat Phenotyping*. Mexico, D.F.: CIMMYT.

Pavuluri, K., Chim, B. K., Griffey, C. A., Reiter, M. S., Balota, M. and Thomason, W. E. 2015. Canopy spectral reflectance can predict grain nitrogen use efficiency in soft red winter wheat. *Precision Agriculture* 16(4), 405-424.

Peguero-Pina, J. J., Morales, F., Flexas, J., Gil-Pelegrín, E. and Moya, I. 2008. Photochemistry, remotely sensed physiological reflectance index and de-epoxidation state of the xanthophyll cycle in Quercus coccifera under intense drought. *Oecologia* 156(1), 1–11.

Pérez-Rodríguez, P., Crossa, J., Bondalapati, K., De Meyer, G., Pita, F. and Campos, Gdl 2015. A pedigree-based reaction norm model for prediction of cotton yield in multienvironment trials. *Crop Science* 55(3), 1143–1151.

Piao, S. L., Liu, Q., Chen, A. P., Janssens, I. A., Fu, Y. S., Dai, J. H., Liu, L. L., Lian, X., Shen, M. G. and Zhu, X. L. 2019. Plant phenology and global climate change: current progresses and challenges. *Global Change Biology* 25(6), 1922–1940.

Porcar-Castell, A., Garcia-Plazaola, J. I., Nichol, C. J., Kolari, P., Olascoaga, B., Kuusinen, N., Fernández-Marín, B., Pulkkinen, M., Juurola, E. and Nikinmaa, E. 2012. Physiology of the seasonal relationship between the photochemical reflectance index and photosynthetic light use efficiency. *Oecologia* 170(2), 313–323.

Porker, K., Coventry, S., Fettell, N., Cozzolino, D. and Eglinton, J. 2020. Using a novel PLS approach for envirotyping of barley phenology and adaptation. *Field Crops Research* 246, 107697.

Prasad, B., Carver, B. F., Stone, M. L., Babar, M. A., Raun, W. R. and Klatt, A. R. 2007. Genetic analysis of indirect selection for winter wheat grain yield using spectral reflectance indices. *Crop Science* 47(4), 1416–1425.

Prey, L., Hu, Y. and Schmidhalter, U. 2020. High-throughput field phenotyping traits of grain yield formation and nitrogen use efficiency: optimizing the selection of vegetation indices and growth stages. *Frontiers in Plant Science* 10, 1672.

Prey, L., Kipp, S., Hu, Y. and Schmidhalter, U. 2019. Nitrogen use efficiency and carbon traits of high-yielding european hybrid vs. line winter wheat cultivars: potentials and limitations. *Frontiers in Plant Science* 9, 1988.

Prey, L. and Schmidhalter, U. 2019. Temporal and spectral optimization of vegetation indices for estimating grain nitrogen uptake and late-seasonal nitrogen traits in wheat. *Sensors* 19(21), 27.

Prey, L. and Schmidhalter, U. 2020. Deep phenotyping of yield-related traits in wheat. *Agronomy* 10(4), 603.

Rascher, U., Nichol, C. J., Small, C. and Hendricks, L. 2007. Monitoring spatio-temporal dynamics of photosynthesis with a portable hyperspectral imaging system. *Photogrammetric Engineering and Remote Sensing* 73(1), 45–56.

Rebetzke, G. J., Jimenez-Berni, J., Fischer, R. A., Deery, D. M. and Smith, D. J. 2019. Review: high-throughput phenotyping to enhance the use of crop genetic resources. *Plant Science* 282, 40–48.

Rebetzke, G. J., Jimenez-Berni, J. A., Bovill, W. D., Deery, D. M. and James, R. A. 2016. High-throughput phenotyping technologies allow accurate selection of stay-green. *Journal of Experimental Botany* 67(17), 4919–4924.

Resende, R. T., Piepho, H.-P., Rosa, G. J., Silva-Junior, O. B., e Silva, F. F., De Resende, M. D. V. and Grattapaglia, D. 2021. Enviromics in breeding: applications and perspectives on envirotypic-assisted selection. *Theoretical and Applied Genetics* 134, 95–112.

Reynolds, M., Foulkes, J., Furbank, R., Griffiths, S., King, J., Murchie, E., Parry, M. and Slafer, G. 2012. Achieving yield gains in wheat. *Plant, Cell and Environment* 35(10), 1799–1823.

Reynolds, M. and Langridge, P. 2016. Physiological breeding. *Current Opinion in Plant Biology* 31, 162–171.

Rincent, R., Charpentier, J. P., Faivre-Rampant, P., Paux, E., Le Gouis, J., Bastien, C. and Segura, V. 2018. Phenomic selection is a low-cost and high-throughput method based on indirect predictions: proof of concept on wheat and poplar. *G3* 8(12), 3961-3972.

Rischbeck, P., Elsayed, S., Mistele, B., Barmeier, G., Heil, K. and Schmidhalter, U. 2016. Data fusion of spectral, thermal and canopy height parameters for improved yield prediction of drought stressed spring barley. *European Journal of Agronomy* 78, 44-59.

Rodriguez, D., Fitzgerald, G. J., Belford, R. and Christensen, L. K. 2006. Detection of nitrogen deficiency in wheat from spectral reflectance indices and basic crop eco-physiological concepts. *Australian Journal of Agricultural Research* 57(7), 781-789.

Runcie, D. E., Qu, J., Cheng, H. and Crawford, L. 2021. MegaLMM: mega-scale linear mixed models for genomic predictions with thousands of traits. *Genome Biology* 22(1), 213.

Rutkoski, J., Poland, J., Mondal, S., Autrique, E., Perez, L. G., Crossa, J., Reynolds, M. and Singh, R. 2016. Canopy temperature and vegetation indices from high-throughput phenotyping improve accuracy of pedigree and genomic selection for grain yield in wheat. *G3* 6(9), 2799-2808.

Sadeghi-Tehran, P., Virlet, N., Sabermanesh, K. and Hawkesford, M. J. 2017a. Multi-feature machine learning model for automatic segmentation of green fractional vegetation cover for high-throughput field phenotyping. Plant Methods, 13(1): 103.

Sadeghi-Tehran, P., Sabermanesh, K., Virlet, N. and Hawkesford, M. J. 2017b. Automated method to determine two critical growth stages of wheat: heading and flowering. *Frontiers in Plant Science* 8, 252.

Sadeghi-Tehran, P., Virlet, N., Ampe, E. M., Reyns, P. and Hawkesford, M. J. 2019. DeepCount: in-field automatic quantification of wheat spikes using simple linear iterative clustering and deep convolutional neural networks. *Frontiers in Plant Science* 10, 1176.

Salas Fernandez, M. G., Bao, Y., Tang, L. and Schnable, P. S. 2017. A high-throughput, field-based phenotyping technology for tall biomass crops. *Plant Physiology* 174(4), 2008-2022.

Santini, F., Kefauver, S. C., Resco De Dios, V., Araus, J. L. and Voltas, J. 2019. Using unmanned aerial vehicle-based multispectral, RGB and thermal imagery for phenotyping of forest genetic trials: a case study in *Pinus halepensis*. *Annals of Applied Biology* 174(2), 262-276.

Schaeffer, L. 2016. Random regression models. http://animalbiosciences.uoguelph.ca/~lrs/BOOKS/rrmbook.pdf.

Schaeffer, L. R. 2004. Application of random regression models in animal breeding. *Livestock Production Science* 86(1-3), 35-45.

Scossa, F., Alseekh, S. and Fernie, A. R. 2021. Integrating multi-omics data for crop improvement. *Journal of Plant Physiology* 257, 153352.

Silva-Perez, V., Molero, G., Serbin, S. P., Condon, A. G., Reynolds, M. P., Furbank, R. T. and Evans, J. R. 2018. Hyperspectral reflectance as a tool to measure biochemical and physiological traits in wheat. *Journal of Experimental Botany* 69(3), 483-496.

Sun, J., Rutkoski, J. E., Poland, J. A., Crossa, J., Jannink, J. L. and Sorrells, M. E. 2017. Multitrait, random regression, or simple repeatability model in high-throughput phenotyping data improve genomic prediction for wheat grain yield. *Plant Genome* 10(2), 1-12.

Tardieu, F., Cabrera-Bosquet, L., Pridmore, T. and Bennett, M. 2017. Plant phenomics, from sensors to knowledge. *Current Biology* 27(15), R770-R783.

Thorp, K. R. and Dierig, D. A. 2011. Color image segmentation approach to monitor flowering in *Lesquerella*. *Industrial Crops and Products* 34(1), 1150-1159.

Thorp, K. R., Gore, M. A., Andrade-Sanchez, P., Carmo-Silva, A. E., Welch, S. M., White, J. W. and French, A. N. 2015. Proximal hyperspectral sensing and data analysis approaches for field-based plant phenomics. *Computers and Electronics in Agriculture* 118, 225-236.

Van Eeuwijk, F. A., Bustos-Korts, D., Millet, E. J., Boer, M. P., Kruijer, W., Thompson, A., Malosetti, M., Iwata, H., Quiroz, R., Kuppe, C., Muller, O., Blazakis, K. N., Yu, K., Tardieu, F. and Chapman, S. C. 2019. Modelling strategies for assessing and increasing the effectiveness of new phenotyping techniques in plant breeding. *Plant Science* 282, 23-39.

Van Eeuwijk, F. A., Bustos-Korts, D. V. and Malosetti, M. 2016. What should students in plant breeding know about the statistical aspects of genotype×environment interactions? *Crop Science* 56(5), 2119-2140.

Varela, S., Pederson, T., Bernacchi, C. J. and Leakey, A. D. B. 2021. Understanding growth dynamics and yield prediction of sorghum using high temporal resolution UAV imagery time series and machine learning. *Remote Sensing* 13(9), 17.

Vargas, J. J. Q., Zhang, C. Y., Smitchger, J. A., McGee, R. J. and Sankaran, S. 2019. Phenotyping of plant biomass and performance traits using remote sensing techniques in pea (*Pisum sativum*, L.). *Sensors* 19, 23.

Velumani, K., Madec, S., De Solan, B., Lopez-Lozano, R., Gillet, J., Labrosse, J., Jezequel, S., Comar, A. and Baret, F. 2020. An automatic method based on daily in situ images and deep learning to date wheat heading stage. *Field Crops Research* 252, 107793.

Vigneau, N., Ecarnot, M., Rabatel, G. and Roumet, P. 2011. Potential of field hyperspectral imaging as a non destructive method to assess leaf nitrogen content in wheat. *Field Crops Research* 122(1), 25-31.

Violle, C., Navas, M. L., Vile, D., Kazakou, E., Fortunel, C., Hummel, I. and Garnier, E. 2007. Let the concept of trait be functional! *Oikos* 116(5), 882-892.

Virlet, N., Lebourgeois, V., Martinez, S., Costes, E., Labbe, S. and Regnard, J. L. 2014. Stress indicators based on airborne thermal imagery for field phenotyping a heterogeneous tree population for response to water constraints. *Journal of Experimental Botany* 65(18), 5429 5442.

Virlet, N., Sabermanesh, K., Sadeghi-Tehran, P. and Hawkesford, M. J. 2016. Field Scanalyzer: an automated robotic field phenotyping platform for detailed crop monitoring. *Functional Plant Biology* 44(1), 143-153.

Wallace, J. G., Rodgers-Melnick, E. and Buckler, E. S. 2018. On the road to breeding 4.0: unraveling the good, the bad, and the boring of crop quantitative genomics. *Annual Review of Genetics* 52, 421-444.

Wang, X., Xuan, H., Evers, B., Shrestha, S., Pless, R. and Poland, J. 2019. High-throughput phenotyping with deep learning gives insight into the genetic architecture of flowering time in wheat. *GigaScience* 8(11), giz120, https://doi.org/10.1093/gigascience/giz120.

Watanabe, K., Guo, W., Arai, K., Takanashi, H., Kajiya-Kanegae, H., Kobayashi, M., Yano, K., Tokunaga, T., Fujiwara, T., Tsutsumi, N. and Iwata, H. 2017. High-throughput phenotyping of sorghum plant height using an unmanned aerial vehicle and its application to genomic prediction modelling. *Frontiers in Plant Science* 8, 421.

Weber, V. S., Araus, J. L., Cairns, J. E., Sanchez, C., Melchinger, A. E. and Orsini, E. 2012. Prediction of grain yield using reflectance spectra of canopy and leaves in maize plants grown under different water regimes. *Field Crops Research* 128, 82-90.

Winterhalter, L., Mistele, B., Jampatong, S. and Schmidhalter, U. 2011. High throughput phenotyping of canopy water mass and canopy temperature in well-watered and drought stressed tropical maize hybrids in the vegetative stage. *European Journal of Agronomy* 35(1), 22–32.

Wyber, R., Malenovský, Z., Ashcroft, M., Osmond, B. and Robinson, S. 2017. Do daily and seasonal trends in leaf solar induced fluorescence reflect changes in photosynthesis, growth or light exposure? *Remote Sensing* 9(6), 604.

Wyber, R., Osmond, B., Ashcroft, M. B., Malenovsky, Z. and Robinson, S. A. 2018. Remote monitoring of dynamic canopy photosynthesis with high time resolution light-induced fluorescence transients. *Tree Physiology* 38(9), 1302–1318.

Xiao, S., Chai, H., Shao, K., Shen, M., Wang, Q., Wang, R., Sui, Y. and Ma, Y. 2020. Image-based dynamic quantification of aboveground structure of sugar beet in field. *Remote Sensing* 12(2), 269.

Xiong, X., Duan, L., Liu, L., Tu, H., Yang, P., Wu, D., Chen, G., Xiong, L., Yang, W. and Liu, Q. 2017. Panicle-SEG: a robust image segmentation method for rice panicles in the field based on deep learning and superpixel optimization. *Plant Methods* 13, 104.

Xu, R., Li, C., Paterson, A. H., Jiang, Y., Sun, S. and Robertson, J. S. 2018. Aerial images and convolutional neural network for cotton bloom detection. *Frontiers in Plant Science* 8, 2235.

Xu, Y. 2016. Envirotyping for deciphering environmental impacts on crop plants. *TAG. Theoretical and Applied Genetics. Theoretische und Angewandte Genetik* 129(4), 653–673.

Yamasaki, E., Altermatt, F., Cavender-Bares, J., Schuman, M. C., Zuppinger-Dingley, D., Garonna, I., Schneider, F. D., Guillen-Escriba, C., Van Moorsel, S. J., Hahl, T., Schmid, B., Schaepman-Strub, G., Schaepman, M. E. and Shimizu, K. K. 2017. Genomics meets remote sensing in global change studies: monitoring and predicting phenology, evolution and biodiversity. *Current Opinion in Environmental Sustainability* 29, 177–186.

Yu, K., Anderegg, J., Mikaberidze, A., Karisto, P., Mascher, F., McDonald, B. A., Walter, A. and Hund, A. 2018. Hyperspectral canopy sensing of wheat Septoria tritici blotch disease. *Frontiers in Plant Science* 9, 1195.

Yue, J. B., Yang, G. J., Li, C. C., Li, Z. H., Wang, Y. J., Feng, H. K. and Xu, B. 2017. Estimation of winter wheat above-ground biomass using unmanned aerial vehicle-based snapshot hyperspectral sensor and crop height improved models. *Remote Sensing* 9(7), 708.

Zhang, Z., Chen, C., Rutkoski, J., Schnable, J., Murray, S., Wang, L., Jin, X., Stich, B., Crossa, J. and Hayes, B. 2021. *Harnessing Agronomics Through Genomics and Phenomics in Plant Breeding: A Review*, 2021030519. doi: 10.20944/preprints202103.0519.v1.

Zhou, C. Q., Ye, H. B., Hu, J., Shi, X. Y., Hua, S., Yue, J. B., Xu, Z. F. and Yang, G. J. 2019. Automated counting of rice panicle by applying deep learning model to images from unmanned aerial vehicle platform. *Sensors* 19(14), 16.

Chapter 10

The role of crop growth models in crop improvement: integrating phenomics, envirotyping and genomic prediction

Jana Kholová, International Crops Research Institute for the Semi-Arid Tropics (ICRISAT), India; Amir Hajjarpoor, UMR DIADE, Université de Montpellier, Institut de Recherche pour le Développement (IRD), France; Vincent Garin, International Crops Research Institute for the Semi-Arid Tropics (ICRISAT), Mali; William Nelson, Gottingen University, Germany; Madina Diacoumba, International Crops Research Institute for the Semi-Arid Tropics (ICRISAT), Mali; Carlos D. Messina, Pioneer Hi-Bred International, USA; Graeme L. Hammer, Queensland Alliance for Agriculture and Food Innovation - The University of Queensland, Australia; Yunbi Xu, Chinese Academy of Agricultural Sciences, China and International Maize and Wheat Improvement Center (CIMMYT), Mexico; Milan O. Urban, International Center for Tropical Agriculture (CIAT), Colombia; and Jan Jarolímek, Czech University of Life Sciences (CZU), Czech Republic

1 Introduction

The improvement of agricultural systems requires a quantitative understanding of individual components in the soil-plant-atmosphere continuum and their interplay. Crop growth models (CGMs) are *in silico* representations (or so-called 'digital twins') of these essential elements of agricultural systems

http://dx.doi.org/10.19103/AS.2022.0102.13

and their inter-dependencies. They integrate information retrieved at the level of the phenotype, the environment and the genotype. CGMs can help design environment-specific crop cultivars and their management. The chapter highlights the interplay of high-throughput phenotyping (HTP) with CGMs. We also describe the limitations of these approaches, particularly the necessity for hypothesis-driven phenomics which is the key to the improvement of crops and cropping systems. This chapter provides a generic overview of this topic, illustrated by examples of successful model applications.

2 Crop growth models to understand gene × environment × management interactions

Crop physiology-related disciplines study and describe live plant processes and functions, interactions within plants and the interactions of plants with their environment (natural environment ('E') and their management ('M')). Provided this understanding is quantitative, it can be translated into a series of algorithms expressing causalities within these relationships. This approach helps us to translate our understanding into a digital reflection of reality - known as a 'digital twin' (reviewed in e.g. Hammer et al., 2019; Kholová et al., 2020). This technology is being used to dissect, understand and characterize the natural or man-made environment (E - natural ecosystems and atmospheric environmental conditions, or M - human interventions in the natural systems, e.g. crop management practices) and their effects on crops (G × E × M interactions) (see e.g. Xu, 2012, 2016; Xu et al., 2012; Cooper et al., 2014). Crop simulation models make it possible to translate static, measurable information relating to G, E and M variables in the cropping system into a dynamic framework representing their cause–effect interactions (G × E × M). These interactions, determining agri-system outcomes (e.g. crop production and soil water depletion), are usually hard to capture and understand experimentally. Crop models are being used to evaluate season- and site-specific outcomes of agricultural interventions and quantify and manage the associated trade-offs.

Modelling and simulation systems are among the very few ways to extrapolate quantitative relations observed in the soil-plant–atmosphere continuum at the required scale and across the necessary time–space boundaries, something which would be difficult to achieve by *in vivo* experimentation (e.g. in Hammer et al., 2020; Kholová et al., 2020). The need for crop models initially arose from the realisation that very limited insight into the dynamics of experimental or cropping systems is gained by simply 'running the trials' (de Wit, 1965; Keating et al., 2003; Jones et al., 2016; Chenu et al., 2017; Kholová et al., 2020). Dynamic crop growth and development models were developed to help explore how to relate genotype, environment and management to yield (de Wit, 1970; Arkin et al., 1976; Vanderlip and Arken, 1977).

Crop modelling and simulation approaches now provide an invaluable tool to quantitatively characterise the effect of the environment experienced by crops. They can assist in the development of adaptation strategies, for example, alternative cultivation methods and optimal crop varieties to mitigate current and future climatic risks. Models can connect crop performance more directly with raw environmental variables (such as weather data) across spatial–temporal scales. Digital agri-system twins also provide us the unique opportunity to assess agri-systems in situations in which it would be hard to replicate via *in vivo* experimentation and where target environments for study are complex or relatively inaccessible (classically the low-input systems of semi-arid tropics; see the importance of crop simulation modelling in crop improvement (CI) in e.g. Kholová et al., 2021). There are now numerous CGMs addressing these challenges. Users need to be aware of their limitations and the fact that the right CGM tool needs to be selected to answer particular questions (reviewed in e.g. Boote et al., 2021; Kholová et al., 2020). Models lacking some mechanistic, non-linear feedback loops might misinterpret the system outcomes in particular situations (e.g. Hay and Porter, 2006).

3 The role of crop simulation modelling in envirotyping

Currently, the concept of 'envirotyping' (i.e. environmental characterization) is recognized as a third 'typing' technology, complementing genotyping and phenotyping (Xu, 2016). There are several approaches to envirotyping discussed in the literature which can be broadly termed 'independent' and 'integrative,' based on how information on environmental characteristics has been used.

'Independent' envirotyping typically considers environmental characteristics 'one factor at the time', for example, climatic characteristics, soils, diseases, pest infestation and even certain management practices. Such defined environmental units, sometimes called mega-environments, have been successfully used by many breeding programs, especially those targeting more stable production agri-systems (e.g. International Wheat and Maize Improvement Centre (CIMMYT) wheat breeding program (Sonder, 2016)).

Problems in this original approach to envirotyping usually arise when trying to characterize more unstable climatic/input systems: typically rain-fed production systems. Examples include:

- the University of Queensland sorghum breeding program (Chapman et al., 2000a,b; Chauhan and Rachaputi, 2014; Chenu et al., 2011, 2013; Lobell et al., 2015; Chauhan et al., 2008, 2013);
- the Corteva Agriscience US-corn breeding program (Cooper et al., 2014, 2021; Hammer et al., 2006; Messina et al., 2020, 2015); and

- the International Crops Research Institute for the Semi-Arid Tropics programs for rain-fed sorghum, chickpea and peanuts (TPE-workshop, 2020; Kholová et al., 2013; Hajjarpoor et al., 2018, 2021).

In these complex systems, separate information on, for example, rainfall, soils and pests might all have very different consequences on crop production depending on their complex interactions during the cropping season. 'Integrative' envirotyping therefore considers agri-system characteristics and their interactions at each point of time (i.e. emergent properties of the system).

We can demonstrate the difference between these approaches with a simple example. 'Independent' envirotyping might classify environments with the same amount of rainfall regardless of its consequences on crops grown on deep or shallow soil. An 'integrative' approach is better able to take into account the consequences of rainfall on the crop, that is, the context of the particular soil where the availability of water to the crop is dependent on soil characteristics. In this simplified example, the 'independent' approach would detect only one type of environment based on the amount of rainfall, while the 'integrative' approach would be able to segregate the environments based on interactions between rainfall and soil based on its effects on the crop (e.g. Chapman et al., 2000a,b,c; Chauhan and Rachaputi, 2014; Chenu et al., 2011, 2013; Lobell et al., 2015; Chauhan et al., 2008, 2013; Kholová et al., 2013; Hajjarpoor et al., 2018, 2021).

The crop breeding community was initially sceptical of the use of simulation models (see Kholová et al., 2020). However, the transition from *in vivo* experimental systems towards their digital twins in agriculture-related disciplines seems inevitable. Similar transitions have already occurred in other sciences like meteorology, astro- and particle-physics or quantitative chemistry. The agricultural sciences are also now progressing slowly but steadily towards greater digitization. There are now several examples of digital tools like CGMs benefitting breeders and agronomists and helping to identify priorities, for example, in breeding programs (Cooper et al., 2014; Hammer et al., 2006; Messina et al., 2020, 2015; Sinclair et al., 2010; 2020; Harrison et al., 2014).

Integrative envirotyping methods were first developed in Australia, where erratic rainfall caused severe losses in agricultural production in the 1990s (Muchow et al., 1996). Integrative envirotyping methods were initially applied to sorghum and wheat systems in Australia (Chapman et al., 2000a,b,c). Chauhan and Rachaputi (2014) also demonstrated the value of integrative envirotyping for mung bean production systems. The authors used Agricultural Production Systems Simulator (APSIM) outputs to identify distinct agro-ecoregions (or target production environments (TPEs)) based on clustering water stress index and seasonal yield variation data at different locations. The approach also demonstrated how subdividing agri-systems based on integrating their characteristics could reduce the production variation caused by G × E effects, which is critical in improving the efficiency of breeding. Model-defined agro-ecoregions (TPEs) are now used to assess G × M interactions to optimize production. A similar TPE approach is also being used with success in wheat, chickpea and maize breeding programs in Australia (Chenu et al., 2011, 2013; Chapman et al., 2000a,b,c; Lobell et al., 2015; Chauhan et al., 2008, 2013). It is also being used in European and US corn and soybean breeding programmes (e.g. Harrison et al., 2014; Sinclair et al., 2010, 2020; Messina et al., 2020; Cooper et al., 2021).

Envirotyping approaches have also been harnessed by scientists and breeders to aid agriculture in developing countries via the Consultative Group for International Agricultural Research (CGIAR). Digital twins of agri-production systems in India revealed system diversity with unprecedented clarity and accuracy (Fig. 1). These results challenged long-held dogmas about the intensity and frequency of drought stress in certain production regions and its

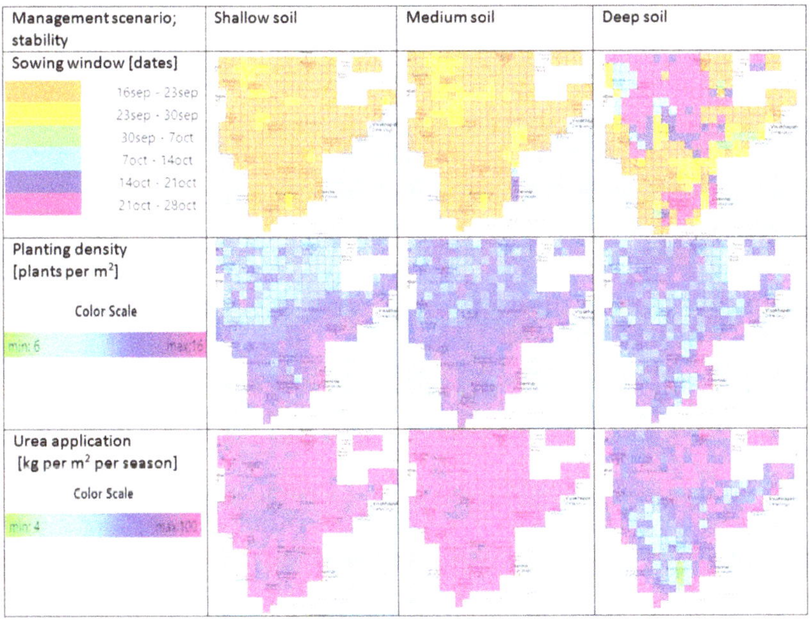

Figure 1 The visualization of the APSIM simulation outputs (summary of 14.7 Tb of data, Jarolímek et al., 2019). Each of the panels shows the target post-rainy sorghum-growing region in the peninsular part of the Indian subcontinent. The colour grids (50 × 50 km) signify the geographical variation in management (M) practices expected to contribute to the improved resilience of post-rainy sorghum production on three types of soils. For each of the tested M, the colour scale of a particular grid indicates the level of M contributing to enhanced system stability. The exact value of M involved in optimal G × M combination for each grid, scenario and soil can be inspected at https://ls40.pef.czu.cz/maps.Source: Redrawn from Ronanki et al., FCR (under review).

effects on post-rainy season sorghum cultivation (Kholová et al., 2013), chickpea (Hajjarpoor et al., 2018) and rainy season peanut cultivation (Hajjarpoor et al., 2021). Using the CGM approach, the abiotic limitations of agri-systems are now well documented, quantified, classified and linked to production losses. In some geographic regions, analyses revealed that drought was not a limiting factor, contrary to common perceptions in the plant breeding community (Kholová et al., 2013; Hajjarpoor et al., 2018, 2021). The TPE approach is now being integrated into CI programs to, for example, optimize the number and distribution of multi-location trials to representatively sample production environments and apply appropriate statistical treatment to the data collected. The TPE analysis has also indicated that there is a need for different breeding and/or agronomic interventions in each TPE to optimize crop production, especially in a more complex environmental context (e.g. Xu, 2016; Kholová et al., 2021; Fig. 1).

It is important to note that the digital twin of cropping systems can also help predict future climate effects on agri-systems dynamics. These predictions might be critical to the success of CI programs to develop climate-resilient cultivars (Saini, 2020; Searchinger et al., 2014; Harrison et al., 2014; Kadiyala et al., 2021; more below). The process of breeding and releasing new cultivars can easily take 10–15 years. This means that the crop breeders have to begin the development of climate-resilient cultivars at least a decade in advance. CGMs are one of the few tools that can help predict crop ideotypes suited to the future climate.

4 The role of crop models in defining phenotyping methods and targets

Breeding programs, particularly in large multinational seed companies, have long benefited from envirotyping approaches. These are slowly but steadily finding their place in the public sector (e.g. CGIAR: www.cgiar.org). The knowledge of how environmental characteristics drive crop agronomic performance in given locations is indispensable for defining phenotyping protocols (e.g. variables such as soil, temperature, light, drought conditions, etc.). For many key traits in breeding (such as yield), it is the phenotyping protocol that will determine how crop genetics are expressed in the crop phenotype. In other words, environmental characteristics of agri-systems have to be reflected in phenotyping protocols as these will determine the relevance of generated phenotypic data to CI. In short, precise phenotyping needs to be driven by precise envirotyping.

Integrative envirotyping can also be used to design crop-management (G × M) interventions for optimized/sustainable crop production in the context of a particular TPE. However, designing the agri-system requires CGM tools with very particular features. While many modelling tools are suitable for optimization of crop management interventions (M) in a particular environmental context (E), only some types of models can reliably capture plant processes at the level of detail required to take account of the G component (i.e. crop phenotype) (see e.g. Boote et al., 2021; Hammer et al., 2020; Kholová et al., 2013; Messina et al., 2019; Fig. 2). Unfortunately, there are numerous studies that have not used appropriate tools and/or the simulation set-ups and thus fail to take proper account of crop biology.

Generic models with no reference to species, or species-specific models with no reference to genotypes, are generally not suitable models for such a task. Suitable models require the capacity to reconstruct the way different crop functions interact with each other (G × G) and with E × M so that system outputs reflect these interactions. Genetic differences need to be represented by cultivar-specific parameters in CGMs that allow them to simulate unique growth and development responses of specific genotypes under a given environment.

Currently, some mechanistic CGMs (e.g. APSIM, simple simulation model (SSM); see the review on mechanistic models by Tardieu et al., 2020) can handle the level of genetic detail required. This makes them powerful tools to identify and prioritize phenotypes based on their effects on agronomic production in a given context (TPE). The knowledge of phenotypic targets (in terms of trait consequences in an M × E context) is necessary to design *in vivo* experiments to explore these traits (reviewed by Tardieu, 2012, 2018; Kholová et al., 2014, 2021; Kaloki et al., 2019; Harrison et al., 2014). The characterization of crop and genotype traits and specific responses to differing conditions are increasingly required for different areas of research (e.g. CI, climate-change impacts, crop genetics). HTP methods will pave the way to deliver this information at scale.

As an example, Kholová et al. (2014) compared the value of particular sorghum traits related to stay-green expression (traits related to canopy structural and functional characters) for post-rainy season cultivation in India. This approach made it possible to establish a screening protocol and prioritize selection criteria in the post-rainy season sorghum product development pipeline. Kaloki et al. (2019) used a similar approach to assess the suitability of chickpea ideotypes in Australia. In this study, the range of traits was assessed manually in *in vivo* field experiments and consequently used to develop a chickpea ideotype *in silico*. The study identified the chickpea ideotype best suited for water-limited environments (i.e. combination of rate of canopy establishment and canopy structure, phenology, chlorophyll content and partitioning of assimilates). A similar study is currently progressing for chickpea systems in India (Fig. 2, Hajjarpoor et al. in prep.).

The study by Harrison et al. (2014) is an excellent example showing how the knowledge of TPE can support our understanding of crop traits needed to adapt to projected future drought stress conditions (including the effects

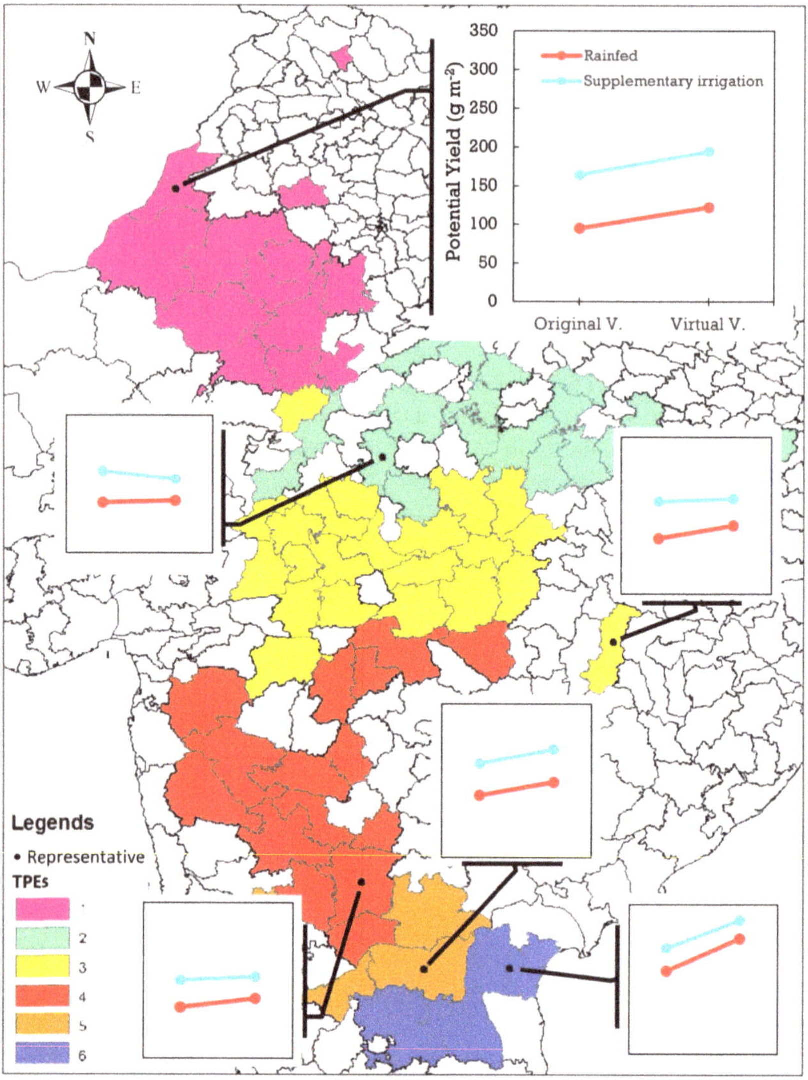

Figure 2 Map showing the portion of the Indian subcontinent where the majority of chickpea is grown (colour-highlighted target population of environments; TPEs). The inserted graphs illustrate the results of initial simulations showing the potential yield of chickpea crop in particular E × M context (i.e. TPE). The chickpea crop (original variety ('Original V') and original variety virtually introgressed with the QTL hotspot ('Virtual V') was evaluated in the context of rain-fed (red line) and supplementary irrigation (blue line) conditions in six locations within the TPEs as defined in Hajjarpoor et al., 2018.Source: adapted from Hajjarpoor et al. (in prep).

of elevated CO_2). They studied the patterns of drought stress conditions that European maize crops may face as well as the frequency with which they occur. Their environmental analysis revealed four dominant drought stress seasonal patterns and the probability of their future occurrence. The authors provided the evidence that the currently targeted ideotype (in this case testing the combination of phenology, synchrony of anthesis and seed set) would meet future climate scenarios although predicted yields might differ, depending on frequencies of particular stress patterns. Since the development and release of new cultivars can easily take a decade, these types of findings help plan the development of relevant climate-ready crop products for the years to come.

Understanding of target phenotypes would be, however, significantly limited without the ability to gather and process relevant data at scale (e.g. in breeding, genetic mapping, studying G × E interactions, etc.). Most scientists still struggle to analyse the large amount of genetic material for many important traits. Since the early 2000s, high-throughput sensor-based phenotyping technologies have become available to plant biologists. These technologies include, for example, ground-bound sensors (in- and outdoors) and air-borne sensors (UAV- and satellite imaging-based methods). These novel HTP tools and technologies, often labelled as 'phenomics,' can assess plant phenotypes with high spatial–temporal and plant-specific resolution. These phenomics approaches now enable us to study 100s–1000s of genotypes by direct measurements of traits and by building response curves to environmental conditions, thus removing a major phenotyping bottleneck (Tardieu et al., 2017). However, it is critical to emphasise again that any phenomics approach (i.e. technology, protocol and data analysis) has to be driven by rigorous hypotheses (e.g. the target phenotype needs to be identified prior to building phenomics tools). This is an absolute 'must-have' particularly where a phenomics approach is targeted at applications such as breeding programs. The reverse approach – common in the early development of phenomics – that is starting with an apparently promising hi-tech solution – has damaged phenomics research and its reputation amongst non-specialists (e.g. Bernardo, 2016).

This was especially the case with phenotyping under controlled conditions (e.g. indoors), where it is particularly difficult to link observations generated in such artificial environments to real conditions in natural environments. Indoor phenotyping has, however, contributed to essential methodological development to advance the development of phenomics. We are now witnessing the transition of phenotyping from indoor 'plant-to-sensor' systems to outdoor 'sensor-to-plant' systems. However, even outdoor 'sensor-to-plant' facilities require rigorous hypotheses and testing for information relevant to CI. Much of the phenomics community can now point to the practical usefulness of phenomics data, and some teams have been able to validate their use to assist CI programs (Sivashakti et al., 2018; Tharanya et al., 2018; Trapp et al., 2016;

Tardieu et al., 2018; Ranjan et al., 2018, 2019; Sankaran et al., 2019; Zhang et al., 2021; Selvaraj et al., 2020; Naito et al., 2017).

5 Crop models of the future: how can they gain from the current developments in phenotyping?

CGMs now provide the quantitative approach necessary to define phenotyping targets. Phenomic technologies also have the potential to assess these phenotypes at scale. But can results obtained by phenomics platforms potentially enhance the use of CGMs?

According to Boote et al. (2021), most crop models only allow the definition of genetic variability between cultivars using very coarse cultivar/genotype-specific parameters (e.g. days to flowering and physiological maturity). Very few models can describe genotype specificity at the level of, for example, biochemical processes, crop functional–structural characteristics, genotype response curves whilst, at the same time, remaining realistically connected to the agronomic performance of a crop (see the debate on CGM complexity, scale, relevance e.g. in Hammer et al., 2019; Tardieu et al., 2020; Kholová et al., 2020). Traditionally, the determination of those genotype-specific parameters and responses requires long-term and intensive phenotyping work (Adam et al., 2018; Bustos-Korts et al., 2019a). This is still considered a serious bottleneck.

With the maturation of CGM tools and phenomics tools capable of measuring precisely those plant features needed to define genotype-specific coefficients in CGMs, roles can change and phenomics might become the driver for modelling. This approach has been pioneered by researchers from institutions such as l'Institut national de recherche pour l'agriculture, l'alimentation et l'environnement and Wageningen University and Research (e.g. Tardieu, 2003; Tardieu et al., 2005, 2020; Chenu et al., 2009; Bustos-Korts et al., 2019a). For example, Tardieu et al. (2005) have developed a model of leaf elongation rate as a function of meristem temperature, vapor pressure deficit (VPD) and soil water potential. Differences between genotypes can then be expressed through differences in response curves for these parameters. Figure 3b shows the example of plant transpiration response to VPD which can be directly translated into the crop coefficients for some models such as APSIM and SSM (see also Kronenberg et al., 2021; Roth et al., 2020).

With modern data-processing pipelines, it is now possible to automate optimisation of model coefficients from genotypic response curves at the scale required, for example, for CI. It might not be possible to optimize all the critical coefficients required by the model through a single phenotyping approach. However, in some breeding populations, for example, generic population characteristics have already been defined and fixed by selection (e.g. phenology, growth habit, harvest index). The generic coefficients of

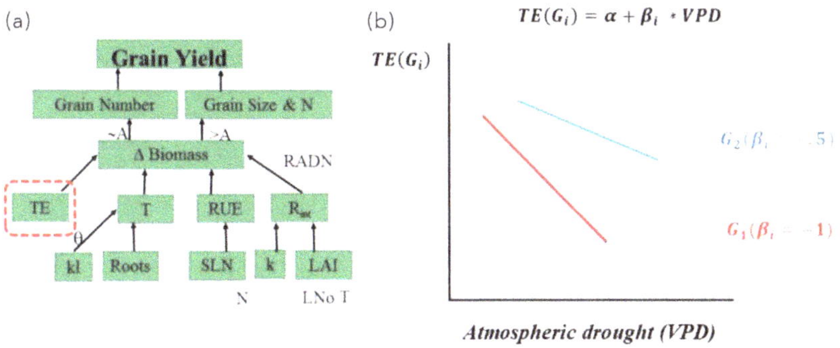

Figure 3 (a) A crop growth model (CGM) illustration (re-drawn from Hammer et al., 2020) and (b) CGM module expressing plant transpiration efficiency (TE) as a function of the atmospheric drought with genotype-specific slopes. Source: re-drawn from Kholová et al. (2014).

such a population can be thus defined *a priori*, while the genotype-specific response curves (e.g. canopy growth rate, transpiration response to VPD) can be automatically integrated into this generic population-specific background for many genotypes using phenomics approaches. In this way, the added agronomic/economic value of every single genotype (group of genotypes) could be predicted for the target E × M context to facilitate further selection. A few research groups are currently testing the feasibility of such an approach to support CI programs.

6 Integrating statistical genetic models and crop growth models (SGM-CGM)

The automated generation of genotype-specific response curves could, logically, be taken one step further. Tardieu (2003) has proposed linking genotype-specific physiological responses (which correspond to CGM coefficients) to DNA polymorphisms (e.g. quantitative trait loci (QTL)). This approach was initially tested by Tardieu et al. (2005) and Chenu et al. (2009). For example, Tardieu et al. (2005) developed an expression of leaf elongation rate as a function of meristem temperature, VPD and soil water potential. They used a maize RIL population to express the genotype response curve in QTL models. The combination of QTL (which determines the genotype-specific physiological response curve and thus the CGM coefficient) and crop models then make it possible to predict the behaviour of existing or novel genotypic variability in a specific G × M context. The integration of genetic variation into crop models is therefore an important area of current research recently labelled as 'genotype-to-phenotype' modelling (also G2P; f(G) = >P; e.g. Cooper et al., 2021).

The use of statistical genetic models (SGM) allows the linkage of genetic information to plant physiological functions. This means CGMs are further augmented with biological information. SGM-CGM integration is, therefore, an elegant way to connect the genetics of simple crop characteristics (e.g. canopy traits) with higher-order agronomic traits (e.g. grain yield which is an emergent property of SGM-CGM). Since SGM-CGM expresses the non-linear transfer function from crop genetics and traits to crop agronomic production in a given E × M context, it is able to capture and understand these complex relationships (Bustos-Korts et al., 2019b) much better than approaches using machine learning (ML) methods (e.g. Montesinos-López et al., 2018). This is because mechanistic CGM enables us to incorporate existing quantitative knowledge of simple crop traits and their impact on crop production.

It is true that ML or genomic prediction can build connections between geno type, crop trait and crop production. However, the artificially built linkages in ML typically do not incorporate any understanding of underlying biological processes, limiting the value of this approach. Although ML models can offer good prediction ability, they provide limited insight into the mechanisms of processes (e.g. Millet et al., 2019). ML algorithms also require a large amount of training data, and models might not work beyond the tested conditions. ML could, however, help to overcome knowledge gaps - that is, where the traits mechanisms are not fully understood (see example below).

As a result, mechanistic SGM-CGM has considerable advantages over ML-based methods, especially in predicting the agronomic performance of particular genotypes (or QTL effects) under untested or predicted future E × M scenarios. One of the main advantages of SGM-CGM frameworks is that these can also be used to predict the performance of unknown future crops (and crop genetics) in unknown future environments (Fig. 4). This is being increasingly used for breeding because it allows *in silico* exploration of untested trait combinations (Luquet et al., 2016; Kholová et al., 2014; Ronanki et al., 2022, under review).

Different solutions to integrate genetic information in SGM-CGM have been explored. Among them, the use of whole genome information (genomic prediction) to translate genetic information into quantitative phenotype seems to be the most promising (Technow et al., 2015, 2019; Cooper et al., 2021). The use of whole genome information to define a G2P transfer function for SGM-CGM set-ups might allow a more accurate prediction of highly complex and quantitative traits compared to discrete models using QTLs. For example, Technow et al. (2015) used Bayesian computation to directly parametrise CGM maize coefficients based on the known linkage of genetic variation to the expression of physiological traits (total leaf number, area of the largest leaf, radiation use efficiency and thermal unit to physiological maturity).

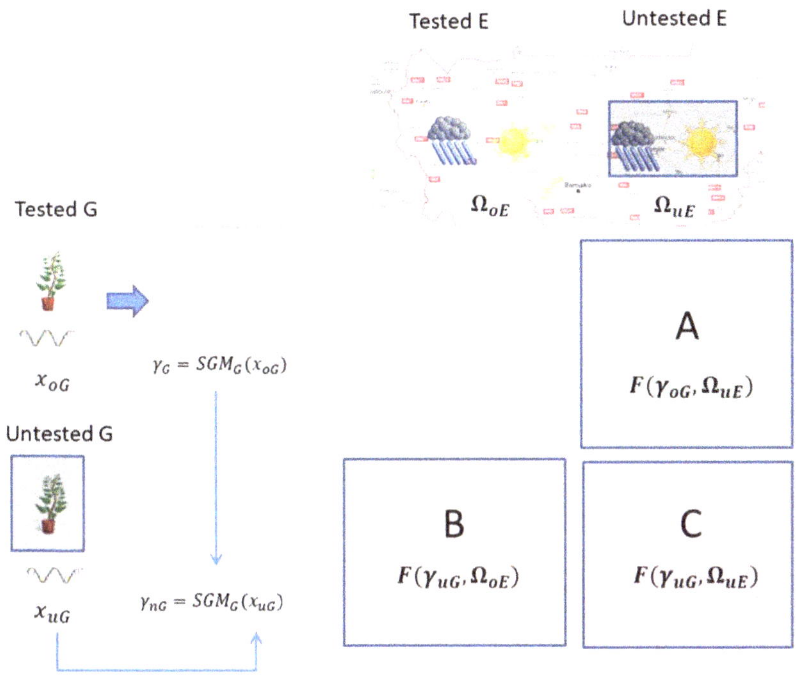

Figure 4 Illustration of main principles underlying an SGM-CGM set-up. These set-ups can be used to predict complex agronomic traits (typically grain yield) by linking the genotype (G) information to the particular crop trait (e.g. canopy size) or function (e.g. canopy transpiration response to VPD) (i.e. phenotype; P) and to reveal its consequence on complex agronomic traits (e.g. crop production) in the context of environmental parameters (E; natural or man-made). Such an SGM-CGM set-up will create a linkage between G and P using a statistical genetic model SGM (QTL or genomic prediction) and combine SGM predictions of P with CGM to create non-linear transfer function between P and E. Such an SGM-CGM set-up can predict the behaviour of observed genotypes (oG) in an untested environment (uE), unobserved genotypes (uG) in an observed environment (oE) and unobserved genotypes (uG) in an untested environment (uE).

ML approaches have also been explored to integrate genetic information in SGM-CGM set-ups (Larue et al., 2020). The ability to estimate simultaneously a large number of parameters with a high degree of accuracy, together with the assumption-free properties of ML, makes this an attractive way to enable the expansion of SGM-CGM approaches. Despite the promise of SGM-CGM approaches, there are challenges limiting their deployment. For example, most of them have been evaluated in simulation conditions (Chenu et al., 2009; Technow et al., 2015; Bustos-Korts et al., 2019a), and only a few G2P pipelines have been validated using real data (Cooper et al., 2016; Uptmoor et al., 2017). Considering the speed of technology development, the current limitations of ML approaches might soon be overcome. In the future, it is quite possible that *validated* SGM-CGM frameworks might take over the research areas which currently require significant input from phenomics (e.g. CI). Even if that happens, the quantitative physiological understanding of the crops supported by phenomics will remain essential to enhance the basic understanding of plant functions and thus define the critical transfer functions for SGM–CGM frameworks.

7 Where to look for further information

The readers interested to further pursue the topics described in this chapter might refer to the following review articles. These reviews synthetize the key research milestones that have shaped our current understanding in these research areas. We can expect these teams would drive the future research and innovations related to the crop simulation modelling, phenomics and genomic predictions, especially their practical applications (for example in crops breeding). Key studies are:

- Cooper, M. et al. (2021). Modelling selection response in plant-breeding programs using crop models as mechanistic gene-to-phenotype (CGM-G2P) multi-trait link functions, *Silico Plants* 3(1). doi: 10.1093/insilicoplants/diaa016.
- Tardieu, F. (2012). Any trait or trait-related allele can confer drought tolerance: Just design the right drought scenario, *Journal of Experimental Botany* 63(1): 25–31. doi: 10.1093/jxb/err269.
- Hammer, G., Cooper, M., Tardieu, F., Welch, S., Walsh, B., van Eeuwijk, F., Chapman, S. and Podlich, D. (2006). Models for navigating biological complexity in breeding improved crop plants, *Trends in Plant Science* 11, 587–593.
- Tardieu, F., Granato, I. S. C., Van Oosterom, E. J., Parent, B. and Hammer, G. L. (2020). Are crop and detailed physiological models equally 'mechanistic' for predicting the genetic variability of whole-plant behaviour? The nexus between mechanisms and adaptive strategies, *Silico Plants* 2(1), diaa011. doi: 10.1093/insilicoplants/diaa011.
- Kholová, J., Urban, M. O., Cock, J., Arcos, J., Arnaud, E., Aytekin, D., Azevedo, V., Barnes, A. P., Ceccarelli, S., et al. (2021). In pursuit of a better world: Crop improvement and the CGIAR, *Journal of Experimental Botany* 72(14), 5158–5179. doi: 10.1093/jxb/erab226.
- Technow, F., Messina, C. D., Totir, L. R. and Cooper, M. (2015). Integrating crop growth models with whole genome prediction through approximate Bayesian computation, *PLoS ONE* 10(6), e0130855.

8 References

Adam, M., Dzotsi, K. A., Hoogenboom, G., Traore, P. C. S., Porte, C. H., Rattunde, H. F. W., Nebie, B., Leiser, W. L., Weltzien, E. and Jones, J. W. (2018). Modelling varietal differences in response to phosphorus in West African sorghum, *European Journal of Agronomy* 100, 35–43. doi: 10.1016/j.eja.2018.04.001.

Arkin, G. F., Vanderlip, R. L. and Ritchie, J. T. (1976). A dynamic grain sorghum growth model, *Transactions of the A.S.A.E. American Society of Agricultural Engineers* 19, 622–626.

Bernardo, R. (2016). Bandwagons I, too, have known, *TAG. Theoretical and Applied Genetics. Theoretische und Angewandte Genetik* 129(12), 2323–2332. doi: 10.1007/s00122-016-2772-5.

Boote, K. J., Jones, J. W. and Hoogenboom, G. (2021). Incorporating realistic trait physiology into crop growth models to support genetic improvement, *In Silico Plants* 3(1), diab002.

Bustos-Korts, D., Boer, M. P., Malosetti, M., Chapman, S., Chenu, K., Zheng, B. and Van Eeuwijk, F. A. (2019a). Combining crop growth modeling and statistical genetic modeling to evaluate phenotyping strategies, *Frontiers in Plant Science* 10, 1491.

Bustos-Korts, D., Malosetti, M., Chenu, K., Chapman, S., Boer, M. P., Zheng, B. and Van Eeuwijk, F. A. (2019b). From QTLs to adaptation landscapes: Using genotype-to-phenotype models to characterize G×E over time, *Frontiers in Plant Science* 10, 1540.

Chapman, S. C., Cooper, M., Hammer, G. L. and Butler, D. G. (2000a). Genotype by environment interactions affecting grain sorghum. II. Frequencies of different seasonal patterns of drought stress are related to location effects on hybrid yields, *Australian Journal of Agricultural Research* 51(2), 209–222. doi: 10.1071/AR99021.

Chapman, S. C., Hammer, G. L., Butler, D. G. and Cooper, M. (2000b). Genotype by environment interactions affecting grain sorghum. III. Temporal sequences and spatial patterns in the target population of environments, *Australian Journal of Agricultural Research* 51(2), 223–234. doi: 10.1071/AR99022.

Chauhan, Y., Wright, G., Rachaputi, N. and McCosker, K. (2008). Identifying chickpea homoclimes using the APSIM chickpea model, *Australian Journal of Agricultural Research* 59(3), 260–269. doi: 10.1071/AR07380.

Chauhan, Y. S. and Rachaputi, R. C. N. (2014). Defining agro-ecological regions for field crops in variable target production environments: A case study on mungbean in the northern grains region of Australia, *Agricultural and Forest Meteorology* 194, 207–217.

Chauhan, Y. S., Solomon, K. F. and Rodriguez, D. (2013). Characterization of north-eastern Australian environments using APSIM for increasing rainfed maize production, *Field Crops Research* 144, 245–255.

Chenu, K., Chapman, S. C., Tardieu, F., McLean, G., Welcker, C. and Hammer, G. L. (2009). Simulating the yield impacts of organ-level quantitative trait loci associated With drought response in maize: A "gene-to-phenotype" modeling approach, *Genetics* 183(4), 1507–1523, doi: 10.1534/genetics.109.105429.

Chenu, K., Cooper, M., Hammer, G. L., Mathews, K. L., Dreccer, M. F. and Chapman, S. C. (2011). Environment characterisation as an aid to wheat improvement-Interpreting genotype-environment interaction by modelling water-deficit patterns in north-eastern Australia, *Journal of Experimental Botany* 62(6), 1743–1755.

Chenu, K., Deihimfard, R. and Chapman, S. C. (2013). Large-scale characterization of drought pattern: A continent-wide modelling approach applied to the Australian wheatbelt - spatial and temporal trends, *New Phytologist* 198(3), 801–820. doi: 10.1111/nph.12192.

Chenu, K., Porter, J. R., Martre, P., Basso, B., Chapman, S. C., Ewert, F., Bindi, M. and Asseng, S. (2017). Contribution of crop models to adaptation in wheat, *Trends in Plant Science* 22(6), 472–490.

Cooper, M., Messina, C. D., Podlich, D., Totir, L. R., Baumgarten, A., Hausmann, N. J., Wright, D. and Graham, G. (2014). Predicting the future of plant breeding: Complementing empirical evaluation with genetic prediction, *Crop and Pasture Science* 65(4), 311–336.

Cooper, M., Powell, O., Voss-Fels, K. P., Messina, C. D., Gho, C., Podlich, D. W., Technow, F., Chapman, S. C., Beveridge, C. A., Ortiz-Barrientos, D. and Hammer, G. L. (2021). Modelling selection response in plant-breeding programs using crop models as mechanistic gene-to-phenotype (CGM-G2P) multi-trait link functions, *In Silico Plants* 3(1), diaa016.

Cooper, M., Technow, F., Messina, C., Gho, C. and Totir, L. R. (2016). Use of crop growth models with whole genome prediction: Application to a maize multienvironment trial, *Crop Science* 56(5), 2141–2156.

de Wit, C. T. (1965). Photosynthesis rate of leaf canopies under various conditions. Agricultural research report no. 663. Pudoc, Wageningen, The Netherlands, 33–44.

de Wit, C. T. (1970). Dynamic concepts in biology. In: *Prediction and Measurement of Photosynthetic Activity*, Setlik, I. (Ed.), Pudoc, Wageningen, 17–23.

Hajjarpoor, A., Kholová, J., Pasupuleti, J., Soltani, A., Burridge, J., Degala, S. B., Gattu, S., Murali, T. V., Garin, V., Radhakrishnan, T. and Vadez, V. (2021). Environmental characterization and yield gap analysis to tackle genotype-by-environment-by-management interactions and map region-specific agronomic and breeding targets in groundnut, *Field Crops Research* 267, 108160. doi: 10.1016/j.fcr.2021.108160.

Hajjarpoor, A., Vadez, V., Soltani, A., Gaur, P. M., Whitbread, A. M., Suresh Babu, D., Gumma, M. K., Diancoumba, M. and Kholova, J. (2018). Characterization of the main chickpea cropping systems in India using a yield gap analysis approach, *Field Crops Research* 223, 93–104.

Hammer, G., Cooper, M., Tardieu, F., Welch, S., Walsh, B., van Eeuwijk, F., Chapman, S. and Podlich, D. (2006). Models for navigating biological complexity in breeding improved crop plants, *Trends in Plant Science* 11(12), 587–593.

Hammer, G., McLean, G., Doherty, A., van Oosterom, E. and Chapman, S. (2019). Sorghum crop modeling and its utility in agronomy and breeding. In: *Sorghum: State of the Art and Future Perspectives*, Ciampitti, I. and Prasad, V. (Eds), American Society of Agronomy and Crop Science Society of America, Madison, WI, 1–25.

Hammer, G. L., McLean, G., van Oosterom, E., Chapman, S., Zheng, B., Wu, A., Doherty, A. and Jordan., D. (2020). Designing crops for adaptation to the drought and high temperature risks anticipated in future climates, *Crop Science* 60(2), 605–621. doi: 10.1002/csc2.20110.

Harrison, M. T., Tardieu, F., Dong, Z., Messina, C. D. and Hammer, G. L. (2014). Characterizing drought stress and trait influence on maize yield under current and future conditions, *Global Change Biology* 20(3), 867–878. doi: 10.1111/gcb.12381.

Hay, R. K. M. and Porter, J. R. (2006). *The Physiology of Crop Yield* (2nd edn.). Blackwell Publishing, Oxford, UK. ISBN-13:978-14051-0859-1.

Jarolímek, J., Pavlík, J., Kholová, J. and Ronanki, S. (2019). Data pre-processing for agricultural simulations, *Agris On-Line Papers in Economics and Informatics* 11(1), 49–53. doi: 10.7160/aol.2019.110105.

Jones, J. W., Antle, J. M., Basso, B. O., Boote, K. J., Conant, R. T., Foster, I., Godfray, H. C. J., Herrero, M., Howitt, R. E., Janssen, S., Keating, B. A., Munoz-Carpena, R., Porter, C. H., Rosenzweig, C. and Wheeler, T. R. (2016). Brief history of agricultural systems modelling, *Agricultural Systems* 155, 240–254.

Kadiyala, M. D. M., Nedumaran, S., Padmanabhan, J., Gumma, M. K., Gummadi, S., Srigiri, S. R., Robertson, R. and Whitbread, A. (2021). Modeling the potential impacts of climate change and adaptation strategies on groundnut production in India, *Science of the Total Environment* 776, 145996. doi: 10.1016/j.scitotenv.2021.145996.

Kaloki, P., Luo, Q., Trethowan, R. and Tan, D. K. Y. (2019). Can the development of drought tolerant ideotype sustain Australian chickpea yield?, *International Journal of Biometeorology* 63(3), 393–403. doi: 10.1007/s00484-019-01672-7.

Keating, B. A., Carberry, P. S., Hammer, G. L., Probert, M. E., Robertson, M. J., Holzworth, D., Huth, N. I., Hargreaves, J. N. G., Meinke, H., Hochman, Z., McLean, G., Verburg, K., Snow, V., Dimes, J. P., Silburn, M., Wang, E., Brown, S., Bristow, K. L., Asseng, S., Chapman, S., McCown, R. L., Freebairn, D. M. and Smith, C. J. (2003). An overview of APSIM, a model designed for farming systems simulation, *European Journal of Agronomy* 18(3–4), 267–288.

Kholová, J., Adam, M., Diancoumba, M., Hammer, G., Hajjarpoor, A., Chenu, K. and Jarolimek, J. (2020). Sorghum: General crop-modelling tools guiding principles and use of crop models in support of crop improvement programs in developing countries. In: *Sorghum in the 21st Century: Food - Fodder - Feed - Fuel for a Rapidly Changing World*, Tonapi, V. A. et al. (Eds). Springer Nature, Singapore. ISBN: 978-981-15-8248-6, 482063_1_En (Chapter 8).

Kholová, J., McLean, G., Vadez, V., Craufurd, P. and Hammer, G. L. (2013). Drought stress characterization of post-rainy season (rabi) sorghum in India, *Field Crops Research* 141, 38–46.

Kholová, J., Murugesan, T., Kaliamoorthy, S., Malayee, S., Baddam, R., Hammer, G. L., McLean, G., Deshpande, S., Hash, C. T., Craufurd, P. Q. and Vadez, V. (2014). Modelling the effect of plant water use traits on yield and stay-green expression in sorghum, *Functional Plant Biology* 41(11), 1019–1034.

Kholová, J., Urban, M. O., Cock, J., Arcos, J., Arnaud, E., Aytekin, D., Azevedo, V., Barnes, A. P., Ceccarelli, S., Chavarriaga, P., Cobb, J. N., Connor, D., Cooper, M., Craufurd, P., Debouck, D., Fungo, R., Grando, S., Hammer, G. L., Jara, C. E., Messina, C., Mosquera, G., Nchanji, E., Ng, E. H., Prager, S., Sankaran, S., Selvaraj, M., Tardieu, F., Thornton, P., Valdes-Gutierrez, S. P., van Etten, J., Wenzl, P. and Xu, Y. (2021). In pursuit of a better world: Crop improvement and the CGIAR, *Journal of Experimental Botany* 72(14), 5158–5179, doi: 10.1093/jxb/erab226.

Kronenberg, L., Yates, S., Boer, M. P., Kirchgessner, N., Walter, A. and Hund, A. (2021). Temperature response of wheat affects final height and the timing of stem elongation under field conditions, *Journal of Experimental Botany* 72(2), 700–717, doi: 10.1093/jxb/eraa471.

Larue, F., Beurier, G., Rouan, L., Pot, D., Rami, J. F. and Luquet, D. (2020). Autoencoding genetic markers to predict the value of ecophysiological model parameters-proof of concept using a sorghum diversity panel, *Multidisciplinary Digital Publishing Institute Proceedings* 36(1), 70.

Lobell, D. B., Thau, D., Seifert, C., Engle, E. and Little, B. (2015). A scalable satellite-based crop yield mapper, *Remote Sensing of Environment* 164, 324–333.

Luquet, D., Rebolledo, C., Rouan, L., Soulie, J. C. and Dingkuhn, M. (2016). Heuristic exploration of theoretical margins for improving adaptation of rice through crop-model assisted phenotyping. In: *Crop Systems Biology*. Springer, 105-127.

Messina, C. D., Cooper, M., Reynolds, M. and Hammer, G. L. (2020). Crop science: A foundation for advancing predictive agriculture, *Crop Science* 60(2), 544-546.

Messina, C. D., Hammer, G. L., McLean, G., Cooper, M., van Oosterom, E. J., Tardieu, F., Chapman, S. C., Doherty, A. and Gho, C. (2019). On the dynamic determinants of reproductive failure under drought in maize, *In Silico Plants* 1(1), diz003, doi: 10.1093/insilicoplants/diz003.

Messina, C. D., Sinclair, T. R., Hammer, G. L., Curan, D., Thompson, J., Oler, Z., Gho, C. and Cooper, M. (2015). Limited-transpiration trait may increase maize drought tolerance in the US corn belt, *Agronomy Journal* 107(6), 1978-1986.

Millet, E. J., Kruijer, W., Coupel-Ledru, A., Alvarez Prado, S., Cabrera-Bosquet, L., Lacube, S., Charcosset, A., Welcker, C., van Eeuwijk, F. and Tardieu, F. (2019). Genomic prediction of maize yield across European environmental conditions, *Nature Genetics* 51(6), 952-956.

Montesinos-López, A., Montesinos-López, O. A., Gianola, D., Crossa, J. and Hernández-Suárez, C. M. (2018). Multi-environment genomic prediction of plant traits using deep learners With dense architecture, *G3 (Bethesda, Md.)* 8(12), 3813-3828. doi: 10.1534/g3.118.200740.

Muchow, R. C., Cooper, M. and Hammer, G. L. (1996). *'Characterizing environmental challenges using models', Plant Adaptation and Crop Improvement* (vol. 1996). CAB International, Wallingford, 349-364.

Naito, H., Ogawa, S., Valencia, M. O., Mohri, H., Urano, Y., Hosoi, F., Shimizu, Y., Chavez, A. L., Ishitani, M., Selvaraj, M. G. and Omasa, K. (2017). Estimating rice yield related traits and quantitative trait loci analysis under different nitrogen treatments using a simple tower-based field phenotyping system with modified single-lens reflex cameras, *ISPRS Journal of Photogrammetry and Remote Sensing* 125, 50-62. doi: /j.isprsjprs.2017.01.010.

Ranjan, R., Chandel, A. K., Khot, L. R., Bahlol, H. Y., Zhou, J., Boydston, R. A. and Miklas, P. N. (2019). Irrigated Pinto bean crop stress and yield assessment using ground based low altitude remote sensing technology, *Information Processing in Agriculture*. 6(4), 502-514. doi: 10.1016/j.inpa.2019.01.005.

Ronanki, S., Pavlík, J., Masner, J., Jarolímek, J., Stočes, M., Subhash, D., Talwar, H. S., Tonapi, V. A., Srikanth, M., Baddam, R. and Kholová, J. (2022). An APSIM-powered framework for post-rainy sorghum-system design in India, *Field Crops Research* 277, 108422.

Roth, L., Camenzind, M., Aasen, H., Kronenberg, L., Barendregt, C., Camp, K. H., Walter, A., Kirchgessner, N. and Hund, A. (2020). Repeated multiview imaging for estimating seedling tiller counts of wheat genotypes using drones, *Plant Phenomics* 2020, Article 3729715, doi: 10.34133/2020/3729715.

Saini, S. (2020). Multi-disciplinary approaches to crop improvement for faster climate change adaptation, *CCAFS Www Pages*. https://ccafs.cgiar.org/research-highlight/multi-disciplinary-approaches-crop-improvement-faster-climate-change-adaptation#.X1bv79N7m4G (accessed 12 October 2021).

Sankaran, S., Quirós, J. J. and Miklas, P. N. (2019). Unmanned aerial system and satellite-based high resolution imagery for high-throughput phenotyping in dry bean, *Computers and Electronics in Agriculture* 165(October): 104965. doi: 10.1016/j.compag.2019.104965.

Sankaran, S., Zhou, J., Khot, L. R., Trapp, J. J., Mndolwa, E. and Miklas, P. N. (2018). High-throughput field phenotyping in dry bean using small unmanned aerial vehicle based multispectral imagery, *Computers and Electronics in Agriculture* 151(August), 84–92. doi: 10.1016/j.compag.2018.05.034.

Searchinger, T., Hanson, C. and Lacape, J. M. (2014). Crop breeding: Renewing the global commitment, World Resources Institute Working Paper. https://www.wri.org/publication/crop-breeding-renewing-global-commitment (accessed 12 October 2021).

Selvaraj, M. G., Valderama, M., Guzman, D., Valencia, M., Ruiz, H. and Acharjee, A. (2020). Machine learning for high-throughput field phenotyping and image processing provides insight into the association of above and below-ground traits in cassava (Manihot esculenta Crantz), *Plant Methods* 16(1), 87. doi: 10.1186/s13007-020-00625-1.

Sinclair, T. R., Messina, C. D., Beatty, A. and Samples, M. (2010). Assessment across the United States of the benefits of altered soybean drought traits, *Agronomy Journal* 102(2), 475–482. doi: 10.2134/agronj2009.0195.

Sinclair, T. R., Soltani, A., Marrou, H., Ghanem, M. and Vadez, V. (2020). Geospatial assessment for crop genetic and management improvements, *Crop Science* 60(2), 700– 708. doi: 10.1002/csc2.20106.

Sivashakti, K., Thudi, M., Tharanya, M., Kale, S. M., Kholová, J., Halime, M. H., Jaganathan, D., Baddam, R., Thirunalasundari, T., Gaur, P. M., Varshney, R. K. and Vadez, V. (2018). Plant vigour QTLs co-map with an earlier reported QTL hotspot for drought tolerance while water saving QTLs map in other regions of the chickpea genome, *BMC Plant Biology* 18(1), 29. doi: 10.1186/s12870-018-1245-1.

Sonder, K. (2016). Global map of wheat mega-environments. *CIMMYT Research Data & Software Repository Network* 4, https://hdl.handle.net/11529/10625.

Tardieu, F. (2003). Virtual plants: Modelling as a tool for the genomics of tolerance to water deficit, *Trends in Plant Science* 8(1), 9–14.

Tardieu, F. (2012). Any trait or trait-related allele can confer drought tolerance: Just design the right drought scenario, *Journal of Experimental Botany* 63(1), 25–31. doi: 10.1093/jxb/err269.

Tardieu, F., Cabrera-Bosquet, L., Pridmore, T. and Bennett, M. (2017). Plant phenomics, From sensors to knowledge, *Current Biology* 27(15), R770–R783. doi: 10.1016/J.cub.2017.05.055.

Tardieu, F., Granato, I. S. C., Van Oosterom, E. J., Parent, B. and Hammer, G. L. (2020). 'Are crop and detailed physiological models equally 'mechanistic' for predicting the genetic variability of whole-plant behaviour? The nexus between mechanisms and adaptive strategies', in silico Plants, *In Silico Plants* 2(1), diaa011. doi: 10.1093/insilicoplants/diaa011.

Tardieu, F., Reymond, M., Muller, B., Granier, C., Simonneau, T., Sadok, W. and Welcker, C. (2005). Linking physiological and genetic analyses of the control of leaf growth under changing environmental conditions, *Australian Journal of Agricultural Research* 56(9):937946.

Tardieu, F., Simonneau, T. and Muller, B. (2018). The physiological basis of drought tolerance in crop plants: A scenario-dependent probabilistic approach, *Annual Review of Plant Biology* 69, 733–759.

Technow, F., Messina, C. D., Totir, L. R. and Cooper, M. (2015). Integrating crop growth models with whole genome prediction through approximate Bayesian computation, *PLoS ONE* 10(6), e0130855.

Tharanya, M., Kholová, J., Sivasakthi, K., Seghal, D., Hash, C. T., Raj, B., Srivastava, R. K., Baddam, R., Thirunalasundari, T., Yadav, R. and Vadez, V. (2018). Quantitative trait loci (QTLs) for water use and crop production traits co-locate with major QTL for tolerance to water deficit in a fine-mapping population of pearl millet (*Pennisetum glaucum* LR Br.), *TAG. Theoretical and Applied Genetics. Theoretische und Angewandte Genetik* 131(7), 1509–1529.

TPE-webinar: Target Population of Environments (TPE) and beyond: Helping make better crop improvement practice. TPE-workshop. https://bigdata.cgiar.org/event/webinar-target-population-of-environments-tpe-beyond-helping-make-better-crop-improvement-practice/.

Trapp, J. J., Urrea, C. A., Zhou, J., Khot, L. R., Sankaran, S. and Miklas, P. N. (2016). Selective phenotyping traits related to multiple stress and drought response in dry bean, *Crop Science* 56(4), 1460–1472.

Uptmoor, R., Pillen, K. and Matschegewski, C. (2017). Combining genome-wide prediction and a phenology model to simulate heading date in spring barley, *Field Crops Research* 202, 84–93.

Vanderlip, R. L. and Arken, G. F. (1977). Simulating accumulation and distribution of dry matter in grain sorghum 1, *Agronomy Journal* 69(6), 917–923.

White, J. W., Alagarswamy, G., Ottman, M. J., Porter, C. H., Singh, U. and Hoogenboom, G. (2015). An overview of CERES-sorghum as implemented in the cropping System Model version 4.5, *Agronomy Journal* 107(6), 1987–2002.

Xu, Y. (2016). Envirotyping for deciphering environmental impacts on crop plants, *TAG. Theoretical and Applied Genetics. Theoretische und Angewandte Genetik* 129(4), 653–673.

Xu, Y., Lu, Y., Xie, C., Gao, S., Wan, J. and Prasanna, B. M. (2012). Whole-genome strategies for marker-assisted plant breeding, *Molecular Breeding* 29(4), 833–854.

Zhang, C., McGee, R. J., Vandemark, G. J. and Sankaran, S. (2021). Crop performance evaluation of chickpea and dry pea breeding lines Across seasons and locations using phenomics data, *Frontiers in Plant Science* 12, 640259. doi: 10.3389/fpls.2021.640259.

Part 5

Case studies

Chapter 11

Using phenotyping techniques to analyse crop functionality and photosynthesis

Eva Rosenqvist, University of Copenhagen, Denmark

1 Introduction

Global food production has shown increased harvest yields during most of the twentieth century. However, at around the year 2000 the increase in wheat levelled off in the most high-yielding regions of the world (Rosenqvist et al., 2019). With the current increase in global human population, there is a need for higher yield especially as the area of available arable land is decreasing. Climate change is predicted to result in more extreme weather events with a resulting need for more stress-resilient as well as higher-yielding crops. Breeding for stress resilience in the face of climate change will inevitably involve understanding climate change effects on photosynthesis. Photosynthesis is firstly controlled by the light level as energy input and secondly by the CO_2 concentration in the leaf as substrate, the latter controlled via stomatal regulation.

The aim of phenotyping is to speed up the selection of promising candidates for breeding. Primary breeding targets have traditionally been harvest yield

http://dx.doi.org/10.19103/AS.2022.0102.15

and disease resistance. Phenotyping in the future needs to incorporate stress resilience to develop crops that can cope with climate change. However, as recently as 2015 quantification of the photosynthetic activity of a crop was not mentioned as a target for new technologies in phenotyping (Araus et al., 2015). When phenotyping the photosynthetic process in genotypes, one should keep in mind that photosynthesis responds both directly and indirectly to most abiotic parameters. Therefore, the measuring protocol must be carefully designed to ensure that it is differences in the actual stress response that are detected between genotypes and not just responses to inconsistencies in measuring conditions.

2 Understanding photosynthesis and its relationship to crop growth and stress response

2.1 Photosynthesis mechanisms and improvement

The potential for improving the photosynthesis capacity in crops to improve yield has been extensively discussed in the literature (e.g. Long et al., 2006; Parry et al., 2011; Wilhelm and Selmar, 2011; Reynolds et al., 2012; Ort et al., 2015; Furbank et al., 2020). The obvious initial target of interest is the ribulose-1,5-bisphosphate (RuBP) carboxylase/oxygenase (Rubisco) enzyme. Oxygenation by Rubisco in photorespiration can act to protect the photosynthetic machinery in times when energy input exceeds utilisation for CO_2 assimilation (Niyogi, 2000); for example, when photosynthesis is limited by stomatal conductance (g_s). However, it is also responsible for lowering the quantum efficiency of photosynthesis in C_3 plants in general (Zhu et al., 2010). Another rate-limiting step is the sedoheptulose bisphosphatase (SBPase) enzyme in the Calvin-Benson cycle. In transgenic wheat with over-expression of SBPase, photosynthesis increased and, in one of the transgenic lines, this was translated into increased biomass and number of grains (Driever et al., 2017).

In the electron transport chain, the two photosystems are spatially separated with photosystem II (PSII) in the appressed grana stacks and photosystem I (PSI) in the non-appressed stroma lamellae. The two photosystems are connected by the cytochrome b/f complex, via mobile electron carriers. Electron transport through cytochrome b/f is a rate-limiting step (Price et al., 1998). Photosynthesis can also be improved at the whole plant level by improving light interception by the canopy structure with more erect upper leaves (Long et al., 2006; Zhu et al., 2010), which has been shown to be the case in modern rice cultivars (Burgess et al., 2017). This means that photosynthesis is affected by several levels of a plant's response to the surrounding environment. It is crucial to understand photosynthetic response in detail in order to investigate ways of improving it via breeding.

2.2 Photosynthesis and growth

Even though photosynthesis is the basis for both energy harvest and carbon gain in plants, there is no direct correlation between net assimilation rate (NAR, photosynthesis per leaf area) and relative growth rate (RGR) in different species. Relative growth rate does correlate to both the leaf area ratio (LAR, leaf area per total shoot biomass) and specific leaf area (SLA, leaf area per leaf mass) (Poorter and Remkes, 1990). This suggests that RGR is more dependent on biomass allocation patterns to photosynthesising tissue within the plant than on photosynthesis per se.

This means that phenotyping different genotypes, for example, maximum carboxylation capacity of Rubisco ($V_{c,max}$) or maximum electron transport capacity (J_{max}) per leaf area at light saturation will not necessarily prove very useful for predicting or increasing the harvest yield. The problem is that measuring whole plant photosynthesis (Walker et al., 2018) is technically more complicated due to the gradients in gas concentrations, light and temperature in large cuvettes used in experiments. It also prevents us from calculating, for example, stomatal conductance (g_s) and intercellular CO_2 concentration (C_i) on the whole plant level. Analyses have therefore often been done at the leaf level - with the challenge of how to scale them up to the whole plant level (e.g. de Pury and Farquhar, 1997).

The picture is less clear when comparing different cultivars of the same species. The natural variation in several photosynthetic parameters has been phenotyped in 64 wheat genotypes grown in the field (Driever et al., 2014). Photosynthesis parameters varied in the range 22-53% but different parts of the photosynthetic machinery were strongly co-regulated, showing a strong correlation between the maximum rate of electron transport and the maximum carboxylation rate of Rubisco. This prevents single metabolic bottlenecks and agrees with the correlation found between the two processes for 109 species (Wullschleger, 1993). However, grain yield varied by up to 93% between the 64 wheat genotypes, and no correlation was found between photosynthesis and grain yield between the cultivars (Driever et al., 2014). In the subsequent growing season the same cultivars were used in a follow-up experiment. This time the flag leaf photosynthetic rate at a photosynthetic photon flux density (PPFD) of 1000 $\mu mol\ m^{-2}\ s^{-1}$ and the flag leaf longevity correlated to both harvest index and grain yield, both when measured pre-anthesis and post-anthesis (Carmo-Silva et al., 2017).

This shows that whether a correlation between photosynthesis and growth is to be found between cultivars of the same species may depend on whether abiotic resources are sufficiently available. It has been argued that in many ecosystems, growth rates are not driven by the photosynthetic CO_2 uptake but rather by meristematic activity determining the demand for carbon which consequently controls photosynthesis (Körner, 2015). Though intensive agricultural and horticultural production systems are fertilised and sometimes irrigated, even in these more managed conditions there will be variations in both nutrient and water availability in most fields. The lack of correlation between photosynthesis and growth when nutrients are limited is illustrated in Fig. 1.

2.3 Stress effects on photosynthesis

Within genotypes, light harvest and photosynthesis do correlate with biomass production (Monteith, 1977). Canopy photosynthesis can be improved via increased canopy light absorption by increasing the fraction of diffuse light in a *Chrysanthemum* canopy; this was found to increase growth by 9% with unchanged irradiance measured using a horizontal quantum sensor (Markvart et al., 2010). This explains the positive effect seen in plants with more erect upper leaves which allow better light penetration into the canopy (Long et al., 2006; Zhu et al., 2010). If, however, photosynthesis is negatively affected by abiotic stress in a genotype, it will decrease the growth and yield of the crop.

Most of the photosynthesis parameters discussed later in this chapter are different 'capacities' that are measured at light saturation. Modelling indicates

Figure 1 The effect of nutrient availability on photosynthesis and growth. Wheat plants were grown in fertilised peat substrate and watered with full nutrient solution (electric conductivity (EC) 2.0 mS cm^{-1}, blue circles) or tap water (EC 0.8 mS cm^{-1} without any added nitrogen, red triangles), respectively. The photosynthetic parameters light-saturated net photosynthesis at ambient CO_2 concentration (symbols with black borders), carboxylation capacity ($V_{c,max}$), capacity for electron transport (J_{max}) and triose-phosphate utilisation (TPU) all decreased by 23.6–27.3% by the low nutrient availability. Data represent mean values ± SD when larger than the symbol, n = 3 (redrawn from Prag, 2017, with permission). The photograph shows a reduction in the number of spikes by 75–80% (for biomass, even more) from a wheat plant grown with full nutrient solution (left) compared to one grown with tap water (right) at ambient CO_2 concentration. This illustrates how growth can be sink limited rather than source limited if resources like nutrients are not enough available to support growth. This has implications for how nutrients are applied in experiments with, for example, elevated CO_2 concentration, which is expected to increase growth and thus increase the demand for nutrients. (Photographer: Eva Rosenqvist).

that the grain yield of rice can be improved by exploring natural variation in photosynthesis, in particular under light-limiting conditions (Gu et al., 2014). It has been suggested that improving photosynthesis under sub-/super-optimal conditions will be more beneficial than focusing on the photosynthetic process itself (Beadle and Long, 1985). In the context of climate change, increasing temperature affects plants at different organisational levels, ranging from biochemical processes, stomatal regulation and source/sink balance, to reduced time for grain filling. They all interact with atmospheric CO_2 concentration (C_a) which is predicted to increase with climate change (Moore et al., 2021).

The resilience of wheat to climate extremes decreased in most European countries from 2002 to 2009 (Kahiluoto et al., 2018). Considering the predicted and observed increase in extreme weather events in Europe (Christidis et al., 2014), phenotyping for increased stress resilience is an obvious candidate for pre-breeding for future crop development. This means a precise understanding of how to perform and interpret analyses of photosynthesis in plants is of crucial importance.

3 Phenotyping photosynthesis in varying environmental conditions

High-throughput phenotyping requires non-invasive methods to measure mainly morphological traits, particularly using spectral reflectance parameters. Many high-throughput phenotyping infrastructures in greenhouses and climate chambers can simulate drought and nutrient stress and quantify effects by gravimetric analysis and imaging. The primary high-throughput method related directly to photosynthesis is modulated chlorophyll fluorescence, while hyperspectral imaging and thermal imaging are indirectly related to photosynthesis (Li et al., 2014). However, a more mechanistic understanding of stress responses related to photosynthesis requires a low-throughput deep physiological phenotyping approach to establish biomarkers for specific stress responses.

Photosynthesis can be measured with several techniques, among which gas exchange and chlorophyll fluorescence are the most used. Chlorophyll fluorescence can be measured non-invasively and can there be used in high-throughput phenotyping. However, the gas exchange requires that the leaf is inserted into an air-tight cuvette, making it a comparatively slow method. All measurements related to photosynthesis can be supplemented by *in situ* measurements of g_s by porometry or thermal imaging, and by pigment composition using several optical devices in the field/greenhouse/climate chamber. Those methods are non-invasive with measurement taking from a few seconds (optical methods) to sometimes minutes (porometers, dependent on model and climate conditions).

3.1 Climate effects on photosynthesis

Photosynthesis is in general measured at steady state conditions. Most parameters that can be calculated from gas exchange are different maximum capacities at light saturation. However, completely stable environmental conditions often used in controlled environments are unnatural. On sunny days with a clear sky, the change in light and temperature is slow and gradual but scattered clouds can induce changes in irradiance by 1200 μmol m^{-2} s^{-1} from 1 min to the next on a horizontal surface. This is accompanied by corresponding fluctuations in leaf temperature of several degrees (unpublished data). Similar patterns are found in clear sunlight under canopies (Zhu et al., 2004) with leaf movement due to wind increasing fluctuations.

Even though the light response curve follows the equation of a non-rectangular hyperbola with a saturated plateau at high light (Fig. 2a), light harvest is linearly dependent on irradiance. In the linear phase of maximum quantum yield of the light response, all absorbed light is used for photosynthesis. The moment the curve starts to bend, the difference between the linear light absorption and saturating light response curve represents excess light that cannot be processed by photosynthesis (Wilhelm and Selmar, 2011). The plant protects itself against this excess light by the transfer of a sub-population of the light-harvesting complexes (LCH) of PSII from energy funnelling to heat-dissipating antenna through the development of non-photochemical quenching (*NPQ*) of fluorescence (Ruban, 2016). This process is controlled by the pH of the thylakoid lumen and thus by the ΔpH that develops across the thylakoid membrane in light (Li et al., 2000; Niyogi et al., 2005). The formation of a heat-dissipating antenna involves both the pH-controlled de-epoxidation of violaxanthin to zeaxanthin, the protonation of the PsbS protein of PSII and a conformational change of LHCII (Niyogi et al., 2005; Ruban, 2016).

The development and relaxation of *NPQ* take time, so in fluctuating light, there might be a mismatch between the instantaneous light level and *NPQ*. It has been estimated that this mismatch can result in 13–32% loss of daily canopy carbon gain (Zhu et al., 2004). Transgenic *Nicotiana tabacum* has been used as proof of the concept that plants with an accelerated xanthophyll cycle and increased PsbS protein increase both photosynthesis and biomass production by about 15% in fluctuating light compared to the wild type (Kromdijk et al., 2016). This should be kept in mind when evaluating phenotyping data of photosynthesis obtained at steady state and often near light saturation.

Last but not least, g_s can limit CO_2 assimilation. Stomatal conductance is generally regulated to ensure CO_2 supply into the leaf while, at the same time, not losing too much water through transpiration. Some environmental signals like high light, low CO_2 concentration and low leaf-to-air vapour pressure deficit (*VPD*) will open stomata, while low light, high *VPD* and high CO_2

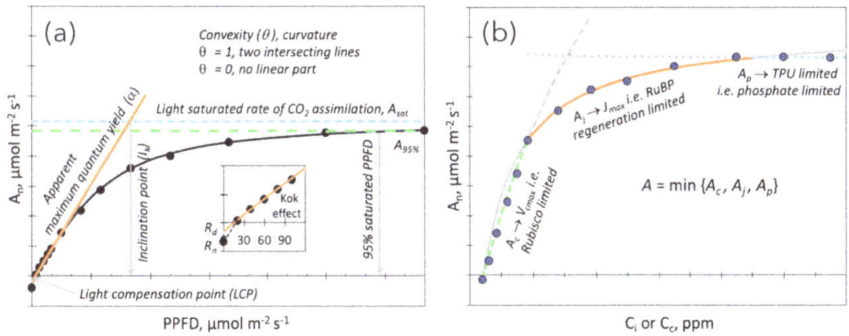

Figure 2 (a) The photosynthetic light response curve and its nomenclature (Ögren, 1993; Long and Hällgren, 1993) where the curve is a non-rectangular hyperbola with the equation:

$$y = \frac{(\alpha\, x + A_{sat})\sqrt{(\alpha\, x + A_{sat})^2 - 4\theta\alpha\, x A_{sat}}}{2\theta} - R_d$$

where x is PPFD, y is A_n i.e. the net CO_2 assimilation rate, α is the apparent maximum quantum yield of CO_2 assimilation (based on incident light, orange), A_{sat} is the light-saturated rate of A_n (blue), θ is the convexity, that is, the curvature of the curve and R_d is the day respiration. The inserted enlargement shows the apparent maximum quantum yield, which is often only linear below a PPFD of 100 µmol m^{-2} s^{-1} and often shows a bend below the light compensation point (*LCP*, where $A_n = R$) towards R_n (night respiration rate) called the Kok effect. For the true maximum quantum yield (Φ) the line must be based on the absorbed instead of incident light. The light response curve of leaves grown in full sun does not always reach saturation at full sunlight and therefore 95% saturation ($A_{95\%}$, green, and corresponding PPFD) is sometimes used. (b) The photosynthetic CO_2 response measured at light saturation is plotted as A_n versus C_i (intercellular CO_2 concentration) or C_c (chloroplast CO_2 concentration). It is composed of three sectional curves (extended by grey lines), where the lowest of the functions ($A_n = \min\{A_c, A_j, A_p\}$) determines the CO_2 assimilation at that C_i (reviewed by von Caemmerer, 2000). To calculate C_c the mesophyll conductance (g_m) must be calculated. The lowest part (green dashed curve) represents the CO_2 limited part where the maximum carboxylation capacity of Rubisco ($A_c \rightarrow V_{c,max}$) is quantified. The intermediate part (orange solid curve) quantifies the maximum rate of electron transport ($A_j \rightarrow J_{max}$), where the produced ATP limits the regeneration of the substrate RuBP. If the upper part of the curve (blue dotted curve) shows a slight decline, the export and utilisation of the triose-P produced in the Calvin-Benson cycle is limited ($A_p \rightarrow TPU$), which limits the counter transport of P_i back into the chloroplast. For equations of the curves, see selected references in Table 2.

concentration lower g_s, and darkness closes stomata. When plants are exposed to simultaneous stomata opening and closing signals, different species respond differently, making it impossible to predict whether stomata will open or close (Merilo et al., 2014). When the leaf is exposed to increasing or decreasing light, the actual net CO_2 assimilation rate (A_n) will depend on light-controlled enzyme activation, development/relaxation of *NPQ* and stomatal regulation. Of those processes, g_s is often the slowest (Slattery et al., 2018).

Leaf photosynthesis can be seen to be limited at different organisational levels; from g_s at leaf level to mesophyll conductance (g_m) at cellular level to photochemical and biochemical limitations in the chloroplast. In angiosperms, these three limitations show balanced co-limitation (Gago et al., 2019). However, in the natural world, light can be the most fluctuating of all climate parameters. On days with scattered clouds, the outdoor light level can fluctuate by up to ± 1200 µmol m^{-2} s^{-1} from 1 min to the next (unpublished data) in a similar way as found in canopies on clear days (Zhu et al., 2004). This means that the regulation of photosynthesis by formation and relaxation of heat dissipation through *NPQ* in the antennae of PSII, activation level of the enzymes of the Calvin-Benson cycle, as well as stomatal regulation, is highly dynamic (Kaiser et al., 2019).

4 Using gas exchange to analyse photosynthesis

Analysing gas exchange is based on enclosing an attached leaf in a sealed cuvette, in which the concentration of CO_2 and water vapour is precisely controlled. Air is pumped through the cuvette at a constant flow rate and the gas composition is measured before and after the air moves through the cuvette. Air temperature is controlled, and leaf temperature is measured in the cuvette. Photosynthetic photon flux density is provided by natural daylight or by lamps. Light-emitting diodes (LED) are used as a light source in modern gas exchange systems. A common combination is some per cent blue in red light, sometimes supplemented with white LEDs. The precise rate of photosynthesis can only be compared between plants when the same spectral composition of light is used in both cases. Blue and red photons mostly drive photosynthesis in the surface cell layers because of their high absorption in chlorophyll, while green photons penetrate deeper into the leaf and contribute to photosynthesis in the lower layers of the leaf (Brodersen and Vogelmann, 2010). This means different models of gas exchange equipment with different lamps cannot be mixed in the same experiment.

The kind of photosynthesis measurement needed depends on the working hypothesis and the time for measuring each plant. Most information can be extracted if full response curves are measured but this is time consuming and can generally only be done on a limited number of genotypes and treatments. For full response curves, photosynthesis can be plotted against CO_2 concentration (A_n/C_i) or light (A_n/Q), where A_n is net CO_2 assimilation rate, C_i is the intercellular CO_2 concentration and Q represents quanta of light, that is, short for PPFD.

4.1 Point measurements

If a large number of plants is included in the experiment, point measurements can be taken for one predetermined set of climate conditions. In comparing genotypes and their response to any given stress, most stress responses show the strongest effect at saturating light (e.g. Sharma et al., 2015). When photosynthesis rates in different genotypes and experimental treatments are compared, it is recommended to supply PPFD in a controlled way from a lamp since photosynthesis is extremely light dependent and natural light is below light saturation for most of the day

In all gas exchange analyses, whether response curves or point measurements are used, it is important to keep the leaf in the cuvette long enough to allow the measured rate of photosynthesis to reach a steady state. This means that A_n can be related to the climate parameters in the cuvette: PPFD, CO_2 concentration, leaf temperature and air humidity. The steady-state rate of A_n is determined by a combination of the chosen climate parameters (particularly PPFD), on one hand, and g_s and activation state of Rubisco, on the other, with the latter two fitting the given level of PPFD. How long time it takes to reach steady state photosynthesis depends on species (Ögren and Sundin, 1996), on which combination of climate parameters are used (Kaiser et al., 2017), and on how different the measuring conditions are from the climatic conditions in the growth environment. If the plants are grown in a greenhouse or the field, steady state photosynthesis at high light is often reached faster on sunny than on overcast days. On sunny days, the leaf will have higher g_s and Rubisco already activated when it is inserted into the cuvette, as compared to dull days. On the other hand, if sunny weather is combined with high air temperature and low relative air humidity, the leaf-to-air *VPD* will be high and this can induce local water deficit symptoms in the leaves that lower g_s and consequently A_n even in well-watered plants (e.g. Muraoka et al., 2000). The g_s issue is discussed further in Fig. 3.

4.2 Light response curve of photosynthesis (A/Q)

The A_n/Q curve (Fig. 2a) is measured at the CO_2 concentration at which the plants have been grown. From this curve, the light-saturated rate of A_n at ambient CO_2 (A_{sat}), the apparent (based on incident light) maximum quantum yield of photosynthesis (α), the light compensation point (LCP), the inclination point (I_k) between light-limited and light-saturated photosynthesis and the light saturation point ($PPFD_{sat}$) can be determined. The insert in Fig. 2a shows an enlargement of the lower part of the curve, in which the day respiration (R_d) is the linear extension of the curve to darkness, while the night respiration (R_n) is a measured value. The bend between the *LCP* and R_n is called the Kok effect and is, among other things, caused by the downregulation of mitochondrial respiration in light (Yin et al., 2020).

If gas exchange is included in phenotyping experiments, pilot studies are essential. By growing control plants and measuring light response curves

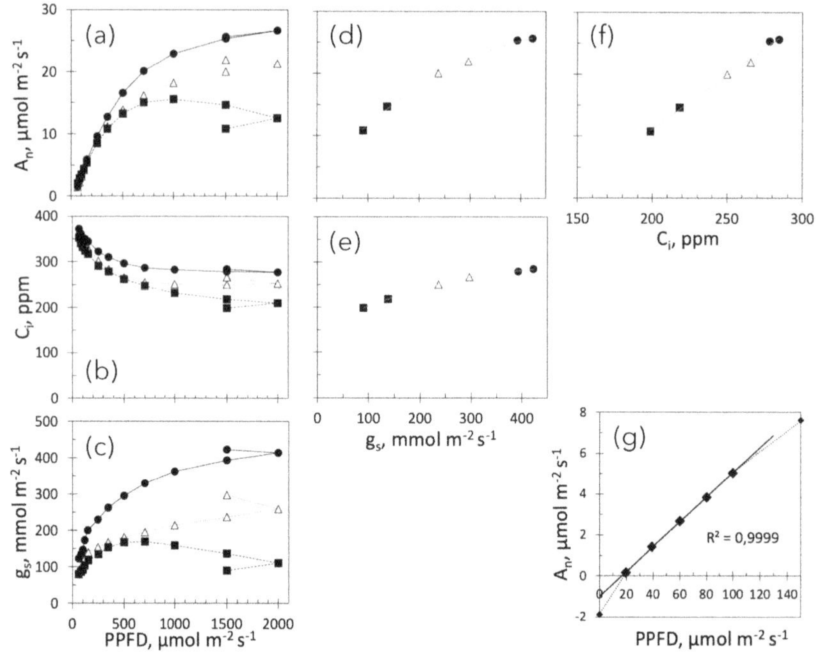

Figure 3 Examples of three light response curves based on raw data of a cutting propagated clone of *Chrysanthemum morifolium* 'Coral Charm' grown under optimum greenhouse conditions, where the data points have been connected in the order they were measured, starting at a PPFD of 1500 µmol m^{-2} s^{-1} for initial steady state. The light response of (a) net CO_2 assimilation rate (A_n), (b) intercellular CO_2 concentration (C_i) and (c) stomatal conductance (g_s) are illustrated. Furthermore, the dependences of (d) A_n and (e) C_i on g_s and (f) A_n on C_i at a PPFD of 1500 µmol m^{-2} s^{-1} are shown. (g) An enlargement of the apparent (based on incident light) maximum quantum yield (from another plant) shows that the linear part often is restricted to a PPFD ≤ 100 µmol m^{-2} s^{-1} and that the mitochondrial dark respiration cannot be included in the linear curve fit because of the Kok effect, which is partly caused by the downregulation of mitochondrial respiration in light (Yin et al., 2020).

to establish the response times when changing the climate parameters, it is possible to establish which measuring protocol fits best for the species or genotype. There are several protocols for light response curves that can be used (Table 1), depending on the aim of the measurement and the previous light conditions. Because of the shape of the curve, the increase in light should be smaller in the lower part of the curve, increasing towards light saturation to give the best mathematical curve fit.

In general, leaves are slower to reach steady state when light is increased in steps since this involves both enzyme activation and stomatal opening (Ögren and Sundin, 1996). If the interaction between A_n and g_s is being studied, it is best to start at low light, starting just after light is introduced when g_s is still close

Table 1 Measuring protocols (A–D) with suggested PPFD for light response (A_n/Q) and (E) atmospheric CO_2 concentration for CO_2 response (A_n/C_i) curves. Dark respiration is often measured as the last point because it triggers stomatal closure. An exception is if chlorophyll fluorescence is measured simultaneously for full quenching analysis, including *NPQ*. The dark-adapted F_o and F_m should be measured first, after 30 min dark adaptation, since the measurement of the light response curve itself may induce photoinhibition, which affects the quantification of *NPQ*. The red PPFD is the starting point where the initial steady state A_n is established. The grey levels of PPFD are often omitted to save time. (A) High light-grown plants in the field and greenhouse with full light activation at the beginning of the measurement (fast). Dark respiration is measured last. (B) Measurements where the relationship between steady state A_n and g_s is investigated (slow). (C) Starting at intermediate light (500–1000 μmol m^{-2} s^{-1}, italic in the table) to prevent photoinhibition when waiting for steady state, utilising the high g_s and speed of decreasing the light level by taking the high light first, before decreasing along the light response. (D) Starting at intermediate light for steady state, starting by decreasing the light to prevent photoinhibition, with a new waiting time at the original PPFD before ending at high light. The grey light levels are often omitted if the maximum quantum yield of CO_2 assimilation is estimated from the curve fit of the non-rectangular hyperbola of the light reaction. Protocol C is the fastest if there is a risk that photoinhibition is induced by waiting for steady state at the highest light level, while protocol D is preferred in shade-adapted plants, starting at 500 μmol m^{-2} s^{-1}. (E) The CO_2 response curve is measured at saturating PPFD, starting at the growth CO_2 concentration to ensure the correct activation state of Rubisco. By starting by decreasing the CO_2 concentration the g_s will increase, returning to ambient CO_2 before measuring the high CO_2 concentrations where g_s may decrease. If the plants are grown at elevated CO_2, the CO_2 levels may have to be changed.

Light Response (A_n/Q)							CO_2 Response (A_n/C_i)		
Growth CO_2 conc.	Protocols, measuring order						*Saturating PPFD*		
PPFD, μmol m^{-2} s^{-1}	A	B	C		D		Reference CO_2, ppm	E	
2000	1	14	4			14	1500		12
1500	2	13	3			13	1200		11
1000	3	12	*2*			*12*	900		10
700	4	11	*1*	5	*1*	*11*	600		9
500	5	10		6	*2*		400	1	8
350	6	9		7	3		300	2	
250	7	8		8	4		250	3	
150	8	7		9	5		200	4	
100	9	6		10	6		150	5	
80	10	5		11	7		100	6	
60	11	4		12	8		50	7	
40	12	3		13	9				
20	13	2		14	10				
0	14	1		15	15				

to zero. If the focus is the potential photosynthesis at each light level without stomatal limitation, it is better to start at high light conditions. At the first light level, the leaf needs enough time to reach steady state with fully open stomata and this can take from 5 min to an hour, depending on species and which enzyme activation state and g_s the leaf had when placed in the cuvette. When g_s is high from the beginning, decreasing light will lead to an immediate decrease in A_n and 1–3 min on each light level is often sufficient. This protocol is suitable for plants grown in the field or in non-shaded greenhouses during the natural growing season due to the natural highlight acclimation of photosynthesis.

For plants grown in lower light levels in climate chambers, or in greenhouses with supplementary light from autumn to spring at Northern latitudes, the situation is somewhat different. Low and intermediate light-acclimated leaves are more susceptible to photoinhibition than high light-acclimated leaves (Ögren, 1988) and even sun-acclimated leaves can be > 10% photoinhibited after just 30 min exposure to a PPFD of 1800 μmol m^{-2} s^{-1} (Ögren and Rosenqvist, 1992). This means that waiting for the first steady-state level at a PPFD of 1500–2000 μmol m^{-2} s^{-1} has the potential of photoinhibiting the leaf.

Photoinhibition will lower the maximum quantum yield of photosynthesis, which will lower the whole light response curve, especially at low and intermediate light levels (Ögren and Sjöström, 1990). To minimise this risk a third protocol can be adopted where the first light level is chosen in the bending part of the curve in the range 500–1000 μmol m^{-2} s^{-1}, dependent on the light acclimation of the plant, related to the PPFD of the growth environment. After reaching steady state, the light can be increased in steps up to light saturation, back to the first light level. One then needs to wait for the same value as before, after which light can be decreased to finalise the curve. If plants have been grown under low-to-moderate light conditions, and therefore are more susceptible to photoinhibition, the lower part of the curve under the bend can be measured before the light-saturated part. This will give a double point at the first PPFD, which works as a reference point, ensuring that the lower and upper parts of the curve have comparable g_s and enzyme activation of Rubisco (see discussion in Fig. 3).

4.3 CO_2 response of photosynthesis (A_n/C_i)

The response of photosynthesis to CO_2 cannot be described by one single curve but is composed of three equations (Fig. 2b), each describing a different rate-limiting process for different ranges of C_i. When C_i is used instead of the C_a in the cuvette, any limitation by g_s is eliminated. At light saturation, biochemical models of C_3 photosynthesis can be used to quantify several metabolic limitations to photosynthesis (Farquhar et al., 1980; von Caemmerer and Farquhar, 1981, further reviewed by von Caemmerer, 2000 and 2013). At low C_i, Rubisco is far from saturated and A_n is determined by the maximum carboxylation capacity of Rubisco ($V_{c,max}$). When C_i increases, Rubisco will become saturated and the concentration of the RuBP will become limiting, which in turn is limited by the capacity for electron transport (J_{max}) that produces the ATP that is necessary to regenerate the RuBP by the Calvin-Benson cycle.

When the curve reaches saturation, a decrease in A_n is sometimes seen. To maintain the highest A_n when both light and CO_2 are saturating, it is necessary to export triose-P produced in the Calvin-Benson cycle to the cytoplasm. This is where the carbohydrate metabolism takes place, and the P-group is removed. The export is a counter transport of triose-P out of the chloroplast and inorganic P_i into the chloroplast. If the leaf is sink limited by the carbohydrate metabolism in a way that down-regulates the export of triose-P from the chloroplast, the light reaction will run out of inorganic P_i for further production of ATP (McClain and Sharkey, 2019). This sometimes happens at high CO_2 concentration and thus A_n will decrease at the highest C_i concentrations in the A_n/C_i curve, due to limitation by triose-P utilisation (*TPU*). If no decrease is seen in the upper part of the A_n/C_i curve, the leaf is not limited by TPU. Thus, measurements of full A_n/C_i curves provide information about the three major limitations to the photosynthetic capacity: electron transport, J_{max}, Rubisco carboxylation, $V_{c,max}$ and *TPU*. For references for calculations, see Table 2.

The A_n/C_i response is measured at light saturation and steady state should be established at the growth CO_2 concentration. Rubisco does not just use CO_2 as a substrate, but a separate CO_2 molecule is needed for the activation of the enzyme, together with Mg^{2+} and Rubisco Activase (Bracher et al., 2017). At the growth CO_2 concentration, Rubisco will be activated to the same state as in the growth environment. From there, the CO_2 concentration is progressively decreased, returned to the growth CO_2 concentration, and subsequently increased progressively to saturation level (Table 1).

The need for CO_2 for activation of Rubisco also means that the measurements of the lowest CO_2 concentrations should be done as quickly as possible, in order not to lose activation. In most gas exchange equipment, the sensor cells for measuring CO_2 and water vapour in the air as it enters and leaves the cuvette need to be adjusted in relation to each other (e.g. called zero point in Walz, matching in LI-COR and differential balance in PP Systems), so that zero ΔCO_2 and ΔH_2O, that is, the difference in gas concentration before and after the cuvette is registered as zero. This should be done when changing CO_2 concentration and, even though it takes some time, it is a necessary safeguard to ensure comparable measurements.

An alternative measuring protocol for A_n/C_i curves has been developed, in which the gas exchange is measured continuously during the transient change of reference CO_2 concentration from the highest to the lowest values (Stinziano et al., 2017). This protocol requires approximately 5 min, but the accuracy of

Table 2 Selected references on photosynthesis models, methods and calculations of gas exchange data as entrance points to the literature

Topic	References
Original Farquhar-von Caemmerer-Berry photosynthesis model	Farquhar et al. (1980), von Caemmerer and Farquhar (1981)
Some later reviews of the models	von Caemmerer (2013, 2020), von Caemmerer et al. (2009)
Light response curves and θ	Ögren (1993)
Measurements of Γ^*	Moualeu-Ngangue et al. (2017)
Statistics of the FvCB model	Dubios et al. (2007)
Principles of gas exchange and practical measurements with some tips and tricks	Long and Hällgren (1993); Long et al. (1996); Long and Bernacchi (2003); Flexas et al. (2007); Haworth et al. (2018), Walker et al. (2018)
Mesophyll conductance, g_m	Harley et al. (1992), Flexas et al. (2008), Hanson et al. (2016), Gago et al. (2020)
The Kok effect of the light response curve	Sharp et al. (1984), Yin et al. (2020)
Rapid A/C_i curves, with comments on the method	Stinziano et al. (2017), Taylor and Long (2019), Stinziano et al. (2019)
Excel tools for fitting gas exchange data	Sharkey et al. (2007), Bellasio et al. (2016a,b)
SAS programme for fitting gas exchange data	Dubios et al. (2007)
R packages for fitting gas exchange data	Duursma (2015)

the protocol has been debated (Taylor and Long, 2019, replied by Stinziano et al., 2019). Which type of measurement is appropriate for a given experiment depends on whether precise quantification of $V_{c,max}$, J_{max} and *TPU* are needed, or whether the aim is to establish significant differences between genotypes or treatments. In any case, it is important to keep in mind that 5 min per curve does not mean that 12 curves can be measured per hour. The leaf will need time to obtain both a stable g_s and A_n at a steady-state rate at the selected PPFD before the measurement of the response curve can start.

When working with plants with only slightly open stomata, C_i sometimes seems surprisingly high. For example, in darkness, the C_i should only be slightly above C_a, caused by the contribution from dark respiration. However, at low g_s, a C_i of, for example, 500 ppm can be seen when measured at 400 ppm CO_2 in the air. This is caused by a division by a number close to zero in the calculation of C_i (see the manual of the used gas exchange equipment) and is an artefact. It is therefore not possible to make an accurate calculation of C_i when g_s is too low, for example, during severe drought stress when g_s decrease to < 100 mmol m^{-2} s^{-1}, though this number is not exact. Tools for calculation of $V_{c,max}$, J_{max} and *TPU* from gas exchange data are available for several software packages (Table 2).

4.4 Stomatal effects on gas exchange

Up to now the discussion about steady state measurements has been focused on A_n. The A_n/C_i curve is not sensitive towards a moderate decrease in g_s, because A_n is related to C_i, which takes g_s into account in the calculation (Long and Hällgren, 1993). When g_s decreases, a narrower range of C_i will be represented in the curve.

To obtain reproducible single-point measurements or light response curves, one must keep an eye on g_s. During drought stress, the decrease in A_n will be caused by a gradual stomatal closure and this process will also be affected by the CO_2 concentration (Yan et al., 2017). In control plants and stresses that do not primarily involve stomatal regulation, it is important to have comparable g_s during measurements and be aware of the complexity of stomata regulation by numerous opposing environmental cues (Merilo et al., 2014).

To illustrate the effect of different g_s on A_n in gas exchange measurements, three curves have been measured at steady state A_n (Fig. 3). This measuring protocol started at high light to establish steady state photosynthesis with high g_s. All data points were taken when A_n was considered as steady state by the default setting in the GFS-3000 (Heinz Walz, Effeltrich, Germany), which is a change of ≤ 0.2 μmol m^{-2} s^{-1} per minute over 2 min. For plants with high g_s or fast stomatal response, the default setting for automatic evaluation of steady state A_n works well. However, when measurements are accompanied by slow stomatal regulation, an evaluation of the change of g_s has to be included in the determination of steady state A_n. After the initial steady-state PPFD of 1500 μmol m^{-2} s^{-1}, the light level was increased to 2000 μmol m^{-2} s^{-1} and then decreased in steps to darkness.

The closed circles in Fig. 3a–c represent a successful measurement with g_s around 400 mmol m^{-2} s^{-1} at high PPFD. This leaf had excess g_s and, therefore, small changes in g_s at PPFD 1500 μmol m^{-2} s^{-1} did not affect A_n. Under these conditions with initial high g_s, a light response curve of high quality can be obtained in around half an hour. What g_s can be considered as 'high' depends on A_n and species (Körner et al., 1979). The open triangles were measured on a plant with g_s starting around 300 mmol m^{-2} s^{-1} and then rapidly decreasing. The lower initial level of g_s was caused by overcast conditions in the greenhouse and sometimes stomata close when the leaf is placed in the cuvette even if the measuring conditions are optimal. The closed squares had an initial g_s of around 100 mmol m^{-2} s^{-1}, which gradually increased in the cuvette. In the two latter examples, A_n was at steady state at the time of logging the result though, because of the low g_s, the curves do not reflect the full potential of the plant. The lowest two curves would have required considerably longer time for stomatal opening (up to an hour) and initial steady state.

These curves illustrate how g_s affects A_n. The direct relationships are exemplified by the double data points at 1500 µmol m^{-2} s^{-1} PPFD (Fig. 3d), where A_n will reach a constant level when g_s has increased to a level that is sufficient to supply CO_2 to accommodate the potential assimilation rate of CO_2 at that light level. When A_n is limited by g_s it is seen in different levels of C_i (Fig. 3e and f), which is the response seen during drought treatments. For measurements of well-watered control plants grown under optimum conditions, however, the difference in g_s makes A_n incomparable because of stomatal limitation in some measurements. It should be pointed out, though, that, if g_s is sufficiently high, it may vary without affecting A_n. The point is *not* to wait for *steady state* g_s, but to wait for a g_s that is *high enough* not to limit A_n in unstressed leaves at the given light level.

The g_s is co-regulated with photosynthesis, either as a consequence (most conditions) or as a cause (e.g., stomatal closure as response to drought) of the observed rate of photosynthesis. Limiting factors in photosynthesis have been investigated in 13 cultivars of cassava. At steady state, biochemical processes and mesophyll conductance limited photosynthesis by just above 40% each. The stomata limitation was only 16%, while g_s was the dominating limitation during shade to sun transition (De Souza et al., 2020). The rate of light-saturated A_n varied by 18% between the cultivars at steady state, while the variation was three times higher under non-steady state conditions because the stomatal opening was twice as fast in some cultivars compared to others. Information about transient photosynthesis in fluctuating light (Kaiser et al., 2019) is not commonly reported in agricultural crops. This, however, may be an overlooked breeding target for improved photosynthesis (De Souza et al., 2020). In *Arabidopsis* it has even been shown that *ost1* and *PATROL1-ox* mutants with increased g_s have identical growth at constant light but increased biomass production when grown in fluctuating light, compared to the wild type (Kimura et al., 2020).

There is also a diurnal component in stomatal regulation. When measuring gas exchange, lower g_s and A_n are often observed at the end of the day. This phenomenon has been well demonstrated in fluctuating light in banana, including differences between cultivars (Eyland et al., 2021). This must be considered when planning the number of plants that can be measured by gas exchange per day as it seems to be independent of the day length of treatments.

Gas exchange is the 'final target' for phenotyping of photosynthesis since it represents the carbon gain of the crop under different environmental conditions. It can be used to identify which part of the photosynthetic machinery – from electron transport and Rubisco capacity to limitations by the carbohydrate metabolism in the plant - is affected. It is clear, however, that gas exchange analysis will never be a method that can be used in high-throughput phenotyping, for which chlorophyll fluorescence is more appropriate (Flood et al., 2011).

4.5 Measurements in fluctuating light

The aim of most measurements of photosynthesis is to achieve steady state values to get reproducible data. However, since constant light is virtually non-existent in nature, such measurements do not necessarily give a correct picture of natural photosynthetic performance (Lawson et al., 2012). Measuring under fluctuating light requires appropriate adjustment of measuring protocols. The data must be logged at short intervals of a few seconds (Ögren and Sundin, 1996) or minutes (Kaiser et al., 2017) to measure the process. However, there may be a discrepancy between the measured light level and the gas exchange parameters. In a LI-COR the infrared gas analyser (IRGA) is located in the cuvette head while, in Walz and PP Systems for example, it is located in the air conditioning unit. The change in light level is measured instantaneously in the cuvette, while there will be a time lag for measuring A_n and g_s because of the time needed for the air to move from the cuvette to the IRGA, depending on tube length and flow rate. The time lag can be determined by establishing a steady state measurement with an empty cuvette, then briefly opening the cuvette and blowing air into it (≤ 1 s) and then measuring the time until a peak is shown in the CO_2 level. If the time lag is, for example, 15 s, the data can be logged every 1, 3, 5 or 15 s. The data measured in the cuvette (PPFD, leaf temperature and fluorescence data, if it is measured) is shifted 15 s in the data file to create a match between this data and the gas exchange data. It should be noted that the saturating pulses for fluorescence measurements should not be applied more frequently than, for example, every 60 s or more, so as not to affect gas exchange or induce photoinhibition.

5 Using porometry and thermal imaging of g_s and hyperspectral techniques

In addition to calculations from gas exchange measurements, g_s can be measured manually with porometry in low-to-medium throughput experiments. A combined porometer and fluorometer has recently been introduced on the market (LI-160, LI-COR, Lincoln, NE, USA). This simultaneously measures g_s and the fluorescence parameters operation efficiency of PSII (F_q'/F_m') and rate of electron transport (*ETR*) (see Section 6) together with PPFD and leaf temperature, which may prove valuable for various studies on stress. Stomatal conductance will affect leaf cooling through transpiration, so thermal imaging can be used to reveal cultivar differences in stomatal response to stresses (Costa et al., 2013; Jones, 1999).

Thermal imaging has been used to phenotype differences in water use efficiency in ten strawberry cultivars (Grant et al., 2012) and five grapevine cultivars (Costa et al., 2012). It is important to remember, though, that the leaf temperature is determined by the radiation balance of the leaf in combination with the transpiration rate (*E*) from the leaf (Jones, 2004). Within a thermal image, the difference in leaf temperature between leaves reflects differences in g_s. However, g_s is related to *E* and the *VPD* according to Ohm's law ($g_s = E/VPD$). The same g_s can be caused by different *E* if *VPD* is different between measurements, resulting in different leaf temperatures. This means weather conditions during measurement will have a strong impact on the relationship between *E* and g_s. Air temperature, incoming shortwave radiation, wind speed and *VPD* will affect the relationship between g_s and the leaf cooling (Maes and Steppe, 2012; Jones, 2004). Therefore, the leaf temperature measured by thermal imaging under different climate conditions directly reflects *E* and not g_s.

5.1 Hyperspectral techniques

Hyperspectral reflectance in the visible and near-infrared part of the spectrum has been used to indirectly estimate early responses to drought and various leaf nutrient concentrations (Mishra et al., 2020). Hyperspectral imaging is also emerging as a promising non-contact method to estimate $V_{c,max}$ and J_{max} in the field. Using 11 tobacco cultivars, spectral indices based on approximately ten spectral bands were used with partial least square regression to predict $V_{c,max}$ and J_{max} measured by gas exchange with $R^2 = 0.84$ and $R^2 = 0.80$, respectively (Fu et al., 2020). Images at plot level have been correlated to several photosynthetic parameters like $V_{c,max}$, electron transport rate at high light, light-saturated rate of photosynthesis and the chlorophyll a/b ratio with r^2 in the range of 0.54–0.79 (Meacham-Hensold et al., 2020). This is a level of accuracy that could allow this non-contact method to be used for repeated measurements in trial plots in the field to rank cultivars with respect to photosynthetic performance in dark and light conditions.

6 Using chlorophyll fluorescence

The last technique to be discussed here is the chlorophyll fluorescence, which can be used as 'conventional' pulse amplitude modulated (PAM) fluorometry or as fluorescence imaging. Excited chlorophyll has three major competing de-excitation pathways: photochemistry, heat dissipation and chlorophyll fluorescence (Fig. 4a). To be able to calculate parameters related to photochemistry or heat dissipation from a fluorescence signal, we must establish

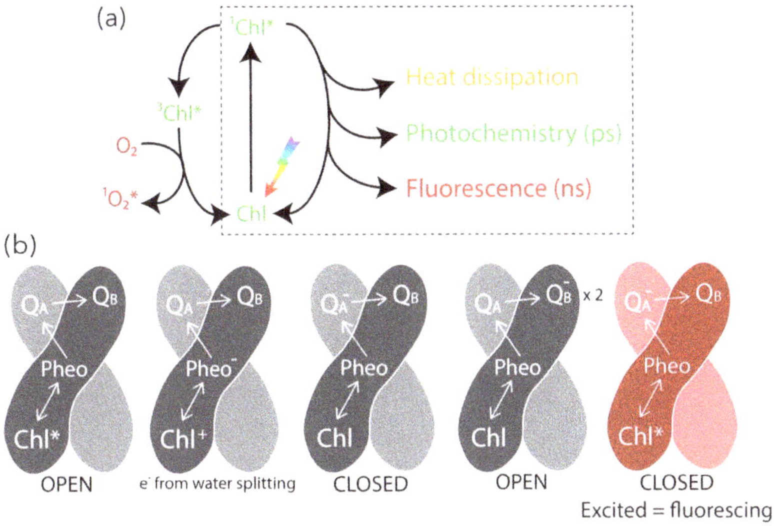

Figure 4 (a) When excited (*) the chlorophyll has three major de-excitation pathways. The excited electron is called an exciton and the energy can be transferred by resonance in the antenna chlorophyll matrix until it reaches the reaction centre. The energy can be dissipated as *heat*, lead to *photochemistry* or be emitted as a red photon, that is, *fluorescence*. Since photochemistry is a faster process (ps) than fluorescence (ns) the de-excitation of chlorophyll will be dominated by photochemistry. If subjected to a high excitation pressure by high light the lifetime of the excited singlet state $^1Chl^*$ can be extended and converted to a triplet state $^3Chl^*$ by inter-system crossing, which can react with oxygen and create singlet oxygen, $^1O_2^*$, which is highly reactive (Harbinson and Rosenqvist, 2003). (b) The PSII reaction centre contains a special pair of chlorophyll *a* (Chl) anchored to the D1 protein (dark grey) followed by the electron acceptors pheophytin (Pheo), Q_A (on the D2 protein, light grey) and Q_B. When an excited electron is emitted from Chl the initial charge separation to Pheo is reversible, while the charge stabilization when the electron is transferred two Q_A is irreversible. Q_A is a one-electron transporter while Q_B receives two electrons, takes up two H^+ from the stroma side of the thylakoid membrane and becomes a mobile part of the PQ pool that transfers the electrons to the cytochrome b/f complex. *When Q_A is oxidised PSII is open* and photochemistry can take place, which means that *no fluorescence* is emitted. *When Q_A is reduced to Q_A^- PSII is closed* and no new photochemistry can take place at that moment, which means that dark red *fluorescence is emitted* if a new exciton reaches the reaction centre Chl.

the value of one in order to calculate the other. This is done by creating open PSII by dark adaptation or closed PSII by applying a saturating light pulse (Fig. 4b). In darkness, all PSII will be open: the moment the excitation energy reaches the reaction centre, photochemistry will take place. If the leaf is exposed to a saturating pulse of light, a 'traffic jam' of electrons will build up in the electron transport chain and all PSII will close, creating a peak of fluorescence (Fig. 5).

Since light absorption in the leaf is linearly related to light level, the leaf will be exposed to excess light as soon as the light response curve starts to bend. With increasing light, the xanthophyll cycle will convert violaxanthin to

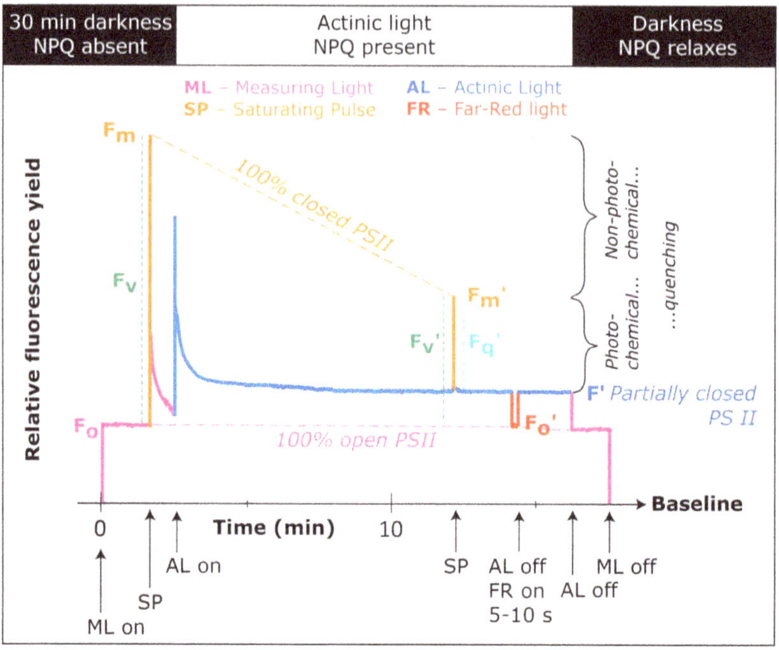

Figure 5 The nomenclature of modulated chlorophyll *a* fluorescence. In leaves that have been dark-adapted for 30 min, heat dissipation through *NPQ* is fully relaxed. When a modulated measuring light (ML) of ≤0.1 μmol m^{-2} s^{-1} is applied, which is weak enough *not* to induce electron transport, a weak fluorescence yield of fully open PSII (oxidised electron acceptor Q_A), F_o, is obtained. By applying a 0.5–1.0 s saturating light pulse (5000–10 000 μmol m^{-2} s^{-1}) PSII will fully close (Q_A^-) giving F_m, and the maximum quantum efficiency of PSII can be calculated as $F_v/F_m = (F_m - F_o)/F_m$. When applying actinic light that induces electron transport a temporary increase of the fluorescence signal (F') is seen when PSII partly close until steady state is reached (sometimes called F_s). After applying a saturating pulse to fully close PSII, the operating efficiency of PSII can be calculated as $F_q'/F_m' = (F_m' - F')/F_m'$ (aka $\Delta F/F_m'$ and Φ_{PSII}), along with non-photochemical quenching $NPQ = (F_m/F_m') - 1$. The electron transport rate $ETR = F_q'/F_m' \times PPFD \times 0.84 \times 0.5$, where 0.84 is the mean value for light absorption in leaves (preferably replaced by a measured value) and 0.5 is the approximate fraction of absorbed PPFD that is absorbed by PSII. For some of the fluorescence parameters, the fluorescence from fully open PSII in the light-adapted state (F_o') is needed. It is measured by briefly turning the actinic light off and in darkness applying far-red light that only excites photosystem I, which drains all electrons from PSII. If it is difficult to briefly darken the leaf for far-red light applications, F_o' can be calculated. Equations of other fluorescence parameters are summarised in the reviews listed on p. 306–307.

zeaxanthin. This will connect to a protonated PsbS protein, which will reshuffle the light-harvesting complex of PSII. This converts the antenna from an energy funnelling to a heat-dissipating antenna (Li et al., 2000; Niyogi et al., 2005; Ruban, 2016). The heat dissipation in the antenna prevents fluorescence from being emitted (cf. Fig. 4a). The mechanism is therefore called non-photochemical quenching (*NPQ*) of fluorescence. This lowers the saturated fluorescence peaks, which allows us to quantify *NPQ* when compared to the dark-adapted peak, where no light-regulated heat dissipation takes place (Fig. 5).

Chlorophyll fluorescence is used for the detection of numerous abiotic stresses (Baker and Rosenqvist, 2004). Since the fluorescence signal originates in (PSII), any stress has to affect PSII to be detectable by fluorescence. One of the most used parameters is the maximum quantum efficiency of PSII (F_v/F_m), which reflects the intactness of PSII. It requires 30 min of dark adaptation of the leaf using dark clips, which can be done on detached leaves, but the measurement is otherwise fast and simple. F_v/F_m correlates linearly to the maximum quantum yield of photosynthesis (Ögren, 1988) and was originally mainly used to measure photoinhibition. A decrease in F_v/F_m and the maximum quantum yield of the light response curve will lead to a decrease in photosynthesis at low and intermediate light levels and, if inhibition is severe enough, also at high light, which will decrease the daily carbon gain (Ögren and Sjöström, 1990; Ögren and Rosenqvist, 1992). When the plant is exposed to any additional climate stress like suboptimal temperature, F_v/F_m will decrease further (Rosenqvist et al., 1991). Therefore, in all uses of F_v/F_m for stress detection, there is a strong interaction with the light level (Sharma et al., 2012).

Light has a direct effect on PSII while temperature has both a direct and indirect effect. It affects biochemical activity, both in PSII itself and by effecting enzyme activity in the rest of the cell, which can then regulate the electron transport chain. Numerous mechanisms protect PSII from damage. With increasing light level, the first line of defence against excess light is established as the increasing ΔpH across the thylakoid membrane induces heat dissipation from the antenna as NPQ is developed (Niyogi et al., 2005; Ruban, 2016). If PSII is over-excited despite the development of NPQ, reactive oxygen species (ROS) will be formed. As a second line of defence antioxidants and scavenging enzymes are activated to protect against ROS (Foyer and Shigeoka, 2011). If damage still occurs to the D1 protein of PSII, the third line of defence is the repair cycle that will replace the damaged D1 of PSII (Baena-González and Aro, 2002). Only when the rate of damage exceeds the rate of repair will F_v/F_m decrease. Finally, photorespiration by Rubisco maintains the sink strength of the Calvin-Benson cycle in C_3 plants if the CO_2 supply is limited due to stomata closure (Niyogi, 2000). The latter protective mechanism makes F_v/F_m a poor indicator of early stages of drought and salinity stress (Baker and Rosenqvist, 2004). In this gradual process, the quenching parameters measured in actinic light will be affected before F_v/F_m, which is often not affected until the leaf loses turgor pressure. Due to the buffering effect of photorespiration when CO_2 availability is limited in the leaf, chlorophyll fluorescence is not very sensitive to either the decrease in g_s in the early stages of drought or to C_a. Therefore, it is not necessary to control the gas composition around the leaf when measuring fluorescence on attached leaves.

For these reasons, F_v/F_m has been used to detect genetic differences in heat stress responses in screening experiments of wheat, followed by quantification of the effect on photosynthesis and growth parameters. By designing a treatment and measuring protocol with precise timing of treatments and dark adaptation, it is possible to do manual high-throughput phenotyping of heat tolerance with greenhouses-grown plants and climate chambers for heat stress. The handling time for attaching dark clips and detaching the leaf, and then measuring F_v/F_m after 30 min, is 1 min per sample for two persons. The only thing to keep in mind is the strong interaction of F_v/F_m with the light level. It may require leaves to be fixed during the heat treatment to ensure equal illumination (Sharma et al., 2012).

Camera-based modulated chlorophyll fluorescence is possible in phenotyping and several systems have been developed for *Arabidopsis* (e.g., Flood et al., 2016; Bresson et al., 2015). High-throughput infrastructures have also been equipped with fluorescence cameras, for which measuring protocols have been designed and technically evaluated (e.g., Tschiersch et al., 2017; Abdelhakim et al., 2021b; Paul et al., 2019). The very weak fluorescence emission from the plants and the attenuation of light with distance limit how far away the camera can be placed. For larger plants, a system must have a saturating light pulse that is strong enough to saturate leaves. In high-throughput systems, a dark adaptation tunnel must be placed before the fluorescence measuring station, if a full quenching analysis should be done. Since the fluorescence signal must originate in the same population of chloroplasts in different physiological states, it is important that plants are not moved during measurement, which puts some limitation to what measuring protocols can be applied.

One should keep in mind that, in one sense, there is a difference between precise physiological measurements and phenotyping. The aim of the first is to establish a precise value of specific parameters so differences between treatments are quantified and mechanisms can be compared to the literature. The aim of the second is to establish differences between genotypes. It means that small errors can be accepted if they are *systematic*. One must ensure, though, that the systematic errors are affected neither by genotype nor by stress treatment. The latter may be problematic since strongly photoinhibited leaves, for example, require lower measuring light levels not to induce electron transport than fully operational control leaves (unpublished data).

For a deeper understanding of chlorophyll fluorescence, there are numerous reviews with explanations of the underlying principles of the technique (e.g. Porcar-Castell et al., 2014; Kalaji et al., 2014; Murchie and Lawson, 2013; Baker, 2008; Klughammer and Schreiber, 2008; Baker and Rosenqvist, 2004; Papageorgiou and Govindjee, 2004; Harbinson and

Rosenqvist, 2003; Rosenqvist and van Kooten, 2003; Maxwell and Johnson, 2000).

Because the fluorescence signal is not very sensitive to the initial stages of drought (Baker and Rosenqvist, 2004), it has also been demonstrated that camera-based high-throughput fluorescence measurements can be done on excised leaves (McAusland et al., 2019). Detached leaf sections of wheat were placed in a transparent chamber with controlled gas composition under a FluorCam CF imager (PSI, Drasov Czech Republic) with measurements undertaken for >3 h without showing differences from intact plants. This kind of set-up will also allow measurements on leaves sampled in field trials with a capacity of >500 samples per day. If excised leaves are used for fluorescence measurements, fluorescence imaging can be extended beyond high-throughput phenotyping infrastructures to smaller fluorescence imaging systems. Data processing may be a bottleneck for fluorescence imaging but web-based software is available for automated image processing (Rousseau et al., 2015).

In the case of low-throughput phenotyping, several devices are available for full quenching analysis, for example, Mini-PAM and PAM-2500 (Heinz Walz GmbH, Effeltrich, Germany), OS1p Ultra-Compact Chlorophyll Fluorometer (ADC BioScientific Ltd., Hoddesdon, UK) and FMS 1^+ and 2^+ (Hansatech Instruments Ltd, UK). In addition, measurement of fluorescence induction and F_v/F_m can be done in a medium- to high-throughput set-up (dependent on manpower) with numerous small devices using dark clips, for example, Handy-PEA (Hansatech Instruments Ltd., UK) and FluorPen 110/D (PSI, Drasov, Czech Republic).

It is possible to follow the long-term photosynthetic performance of plants with modulated chlorophyll fluorescence, for example, by using the Monitoring-PAM (Heinz Walz GmbH, Effeltrich, Germany) and PSP32 Monitoring Fluorometer System (ADC BioScientific Ltd., Hoddesdon, UK). Measuring heads can be mounted on individual leaves and the development of photochemical and non-photochemical processes can be followed in naturally fluctuating daylight over days and weeks (Porcar-Castell et al., 2008).

To be able to measure chlorophyll fluorescence at longer distances than can be used in camera-based fluorescence systems, laser-induced fluorescence transients (LIFT) can be used. A laser is used for the excitation of fluorescence and the signal is collected through a telescope (Keller, 2018; Malenovsky et al., 2009; Ananyev et al., 2005; Kolber et al., 2005). In remote sensing, solar-induced fluorescence (SIF) is a passive remote sensing method that does not require any excitation of fluorescence. Instead, the dark Fraunhofer lines in the solar spectrum are used to detect the weak fluorescence signature from vegetation in and close to the Fraunhofer lines in the range in which fluorescence is emitted by chlorophyll (Porcar-Castell et al., 2014; Malenovsky et al., 2009).

7 Photosynthesis and climate change: accounting for heat stress, drought stress and elevated CO_2

Future climatic conditions are characterised not only by gradually increasing temperature and CO_2 concentration but also by increasing occurrence of extreme weather events. These include flooding and drought, as well as cold winters and heat spells, in the growing season, with such events predicted to increase in frequency in Europe by the 2040s (Christidis et al., 2014). Such events are likely to occur simultaneously, for example, drought may be combined with high temperature. This creates a dilemma for the plant since drought tolerance requires gradual stomatal closure to conserve water, while the high temperature will increase g_s to cool leaves in heat-tolerant plants (Urban et al., 2017; Sharma et al., 2015).

7.1 Heat stress

Assessing heat stress in a phenotyping context can have two different purposes: screening for heat tolerance in many genotypes or establishing the causal effect on physiological responses related, for example, to final grain yield. When F_v/F_m is used as a selection criterion for heat tolerance, a stronger stress is needed to distinguish between tolerant and sensitive genotypes (Sharma et al., 2012). The functionality of PSII is protected by numerous mechanisms, from heat dissipation via *NPQ* to photorespiration when CO_2 supply into the leaf is limited (Niyogi, 2000). To establish the potential differences between genotypes based on F_v/F_m, the stress must be severe. F_v/F_m does not have any direct causal relationship with grain yield. However, all these protective mechanisms ensure that, when F_v/F_m is affected, numerous other heat stress-related responses have already occurred in the plant, which will affect grain yield. In the case of wheat, we have used 40°C for 72 h in climate chambers with a PPFD of 250–300 μmol m^{-2} s^{-1} with a 16 h photoperiod to avoid too much contribution from photoinhibition (Sharma et al., 2012). However, these settings must be established in pilot experiments for each screened species.

When heat stress is applied in deep physiological phenotyping on a limited number of genotypes, a more physiologically relevant treatment must be used to establish temperature effects on carbon metabolism and on final grain yield. For example, in the case of wheat, harvest yield will be affected by any increase in temperature to 25-30°C, because of the increasing developmental rate and subsequent shorter grain-filling period (Wardlaw and Moncur, 1995). However, to detect significant effects on photosynthesis, the temperature has to be around 35°C (cf. Sharma et al., 2015 and Mendanha et al., 2018). For physiological experiments, and where the effect of heat on the grain yield is included, the application of heat and subsequent recovery temperature must be applied gradually to avoid artificial heat shock. One way to do this is to move plants into the greenhouse compartment or climate chamber the day before the heat treatment and start with a gradual increase in the night temperature, before the first day of heat stress in light. For recovery, the plants can be subjected to a high night temperature as the last phase of heat stress. When changing from night to day temperature, the rate of change should not be too steep, allowing for at least 1 h for 5°C change.

7.2 Drought stress

High-throughput phenotyping infrastructures in greenhouses have a weighing and watering station where the water status of the pot can be maintained. However, it is impossible to simulate drought in the field precisely in a pot experiment. The plant cannot explore different depths of the soil profile for water and the drying rate will be faster in the limited volume of a pot. While gradual drying could be achieved by decreasing the amount of water added to the pot, reducing drip irrigation will create unnatural water gradients in the pot. When adding small amounts of water daily, the pot environment will resemble a partial root drying treatment, where roots in the deeper layers of the pot will experience drought and the surface layer in the pot will have free access to water after watering. To create more homogenous drying throughout the root system, the best method is to stop watering and follow the gradual drying of the soil. Water availability can be calculated from gravimetric measurements as a fraction of transpirable soil water (*FTSW*):

$$FTSW = \frac{(WT_n - WT_f)}{TTSW}$$

where WT_n is the pot weight at a given time in the drying cycle, WT_f is the pot weight when g_s in the drought treatment has reached 10% of the control and *TTSW* is total transpirable soil water, which is the difference between the fully watered pot weight and WT_f (Wei et al., 2020). The drying rate needs to be determined in pilot experiments, in which pot size is adjusted to plant size to get the desired drying cycle. However, phenotyping for drought-related traits is further complicated by comparing genotypes with different leaf area.

7.3 Elevated CO_2

Most information about the effect of elevated CO_2 (eCO_2) on the yield of agricultural crops comes from free-air CO_2 enrichment (FACE) experiments (Ainsworth and Long, 2004, 2020; Long et al., 2004). In many ways, eCO_2 may mitigate some of the stress effects related to heat and drought. In commercial greenhouse production in NW Europe, elevated CO_2 is used to enhance growth. These CO_2 application systems work very well but, in commercial production, the CO_2 supply is often shut off or limited when the vents are opened to cool the greenhouse. To be able to work experimentally with elevated CO_2 in this kind of greenhouse set-up, active cooling through air conditioning (AC) is necessary to enable a reasonable temperature control during summer months without opening the vents. Since the temperature of the air from the AC is considerably lower than the temperature set point in the greenhouse, the dew point temperature is also low and the air stream from the AC will have a drying effect on the plants before the air is mixed with the greenhouse air. This means heterogeneity in terms of temperature and humidity in the greenhouse needs to be mapped to avoid tip burns due to low humidity.

The combination of chlorophyll fluorescence and gas exchange has been used in a series of phenotyping experiments on wheat and tomato as discussed in the cases below. They range from high-throughput phenotyping for quantification of heat tolerance to low-throughput deep physiological phenotyping.

8 Case studies

8.1 Case 1: High-throughput phenotyping of heat stress in wheat combined with low-throughput deep physiological phenotyping

To separate heat-tolerant from heat-sensitive genotypes, 1274 spring wheat genotypes were heat stressed in a three-tier experiment with increasing stress severity from the first to the last two screenings, ending with 3 days of exposure to 40°C with a 16 h photoperiod and a PPFD of 300 µmol m^{-2} s^{-1} to minimise the photoinhibitory component of the stress response. In each step, the intermediate genotypes were removed, ending with five heat-tolerant and five heat-sensitive genotypes that had consequently been in the highest and lowest group of F_v/F_m in all three experiments, respectively. In the third heat experiment, the genetic determination of the heat response of F_v/F_m had increased from the original 8.5 ± 2.7% for the 1274 genotypes to 27.9 ± 6.8% in the last screening of 71 genotypes (Sharma et al., 2012). In the heat-tolerant genotypes, three different Quantitative Trait Loci (QTL) related to the maintenance of high F_v/F_m during heat stress were identified, explaining 13–35% of the phenotypic variation in F_v/F_m (Sharma et al., 2017). The genotypes originated from Pakistan and Afghanistan.

The ten genotypes finally selected were used to study what effect the difference in F_v/F_m had on A_n (Sharma et al., 2015). At anthesis, the 5+5 genotypes were moved from 20 ± 2°C/15 ± 1°C day/night temperature to be exposed to 1 week of 36 ± 2°C/30 ± 1°C day/night temperature with 16 h photoperiod and 228 ± 88 μmol m^{-2} s^{-1} PPFD. Photosynthesis was measured in controlled conditions and after 1 week of heat stress at the respective air temperature. The genotypes that could maintain high F_v/F_m after three days at 40°C in the screening were also able to maintain the same A_n after 1 week at 36°C compared to the control at 20°C, accompanied by an increased g_s. In the sensitive genotypes, A_n decreased significantly by 23% after 1 week at high temperature; accompanied by a decreased g_s. The consequence was that the heat-tolerant genotypes could enhance their leaf cooling by transpiration, while the sensitive genotypes were more heat stressed because of the poorer leaf cooling. The dry matter accumulation during the week of heat stress did not correlate to F_v/F_m in the control plants but showed a significant correlation in the stress-treated plants (Sharma et al., 2015). These results indicate that F_v/F_m correlates to and can be used as a proxy for several physiological processes during heat stress.

180 breeding lines from five Nordic breeders have subsequently been screened for heat tolerance with the same protocol and tested for the three QTLs found in genotypes from Afghanistan and Pakistan (Sharma et al., 2017). Especially one QTL was present in the Nordic genotypes (unpublished data). It was noted that several Nordic genotypes showed higher F_v/F_m after heat stress than the genotypes selected in the original high-throughput screening (originating in Afghanistan and Pakistan). Several of the Nordic genotypes did not have any of the three QTLs, indicating several other possible mechanisms for heat tolerance.

Since heat stress rarely occurs in isolation, contrasting genotypes from the Nordic screening have been used to study the effect of combined heat and drought stress both with (Abdelhakim et al., 2021a) and without eCO_2 (Abdelhakim et al., 2021b). When investigated at eCO_2 the four genotypes differed strongly. J_{max} measured at light saturation decreased in heat stress and combined heat and drought stress at both CO_2 concentrations in all four genotypes, while the response of *ETR* measured at the PPFD of 300 μmol m^{-2} s^{-1} (that was used in the climate chambers) differed between genotypes. Two genotypes maintained *ETR* in all treatments except under combined stress at eCO_2, where it decreased by > 30%. One genotype decreased *ETR* by 20% in combined stress at aCO_2, while the last one decreased *ETR* by 50% in combined stress in both CO_2 treatments (Abdelhakim et al., 2021a). These differences in stress response at the two light levels strongly support the argument that measurements of $V_{c,max}$ and J_{max} at light saturation alone may not be sufficient for predicting the photosynthetic performance of different genotypes.

8.2 Case 2: Low-throughput phenotyping of tomato for combined drought and heat stress

The concept of using F_v/F_m to screen for heat tolerance has been transferred to tomato for which 67 genotypes (*Solanum lycopersicum*) were screened in a climate chamber (Zhou et al., 2015). The physiological characterisation of the four most contrasting genotypes with respect to F_v/F_m showed the same clear separation between the two groups in A_n, g_s and leaf cooling as in wheat. In the heat-tolerant genotypes, the chloroplasts kept their normal shape, while they became swollen with decomposing starch grains in the sensitive group. After the one-week experiment, F_v/F_m was still maintained or slightly decreased in the two tolerant genotypes, accompanied by increased pigment content (chlorophylls and carotenoids), while these traits clearly decreased in the sensitive group. The pollen tube length was maintained up to 36°C and only decreased by 50–60% at 40°C in the tolerant genotypes, while it decreased already at 36°C in the sensitive genotype (Zhou et al., 2015).

In a subsequent experiment with the same temperature treatments, the deterioration of pollen viability was accompanied by flower wilting and abnormal abscission in the heat-sensitive genotype, which could only maintain the flowering in one out of three trusses (Zhou et al., 2017a). The heat-tolerant genotype could support full flowering in the three trusses, due to the maintained starch concentration and increased concentration of soluble sugars in mature leaves, while the sensitive genotype had lower starch concentration and unaffected concentration of soluble sugars.

In the field, heat stress rarely happens as a single stress event but is accompanied by gradual soil drying. The common greenhouse cultivar 'Arvento' has been compared to one genotype each of *S. lycopersicum* 'LA1994' and *Solanum pimpinellifolium* 'LA2093' (Zhou et al., 2017b) that were evaluated as heat tolerant in the original screening (Zhou et al., 2015). 'Arveno' was more affected by heat stress with 28% reduction in shoot dry weight, whereas 'LA1994' and 'LA2093' maintained their biomass. 'LA1994' and 'LA2093' increased the stomata pore area by 30% and 40% during heat stress, respectively, while it was unaffected in 'Arveno', which corresponds to the difference in stomatal regulation found in wheat (Sharma et al., 2015).

All cultivars had reduced shoot biomass in drought and combined drought and heat stress but 'LA1994' and 'LA2093' showed a stronger response to stress when drought was involved. The leaf sucrose concentration decreased 40–60% in drought and combined stress, while it more than doubled in 'Arvento' in the two treatments. Also, F_q'/F_m' decreased in the heat-tolerant genotypes when drought was involved, while it was not significantly changed in 'Arvento'. There were also differences between 'LA1994' and 'LA2039' in the combined stress, whereas 'LA1994' showed a stronger decrease in F_v/F_m than 'LA2039'.

These results indicate that the drought component has been dominating over the heat stress at the applied stress levels. The fact that combined stress showed less difference between the three genotypes compared to the individual stresses, also indicates that screening for single stress cannot be extrapolated to combined stress responses.

8.3 Case 3: Screening of heat stress in tomato from climate chambers to field trial

The results from a screening of F_v/F_m in tomato in climate chambers have been verified in a field production trial (Poudyal et al., 2019). Twenty-eight potentially high-yielding genotypes were selected in Nepal and screened twice in climate chambers at 40/28°C and 38/28°C day/night temperature for 2-4 days, respectively. In the second screening, photosynthesis and growth were measured on the four genotypes with the highest, and in the four genotypes with the lowest F_v/F_m. When measured in control conditions, there was no general difference in neither A_n nor any growth parameters. After 4 days at 38/28°C, the heat-tolerant genotypes maintained almost the same A_n, while the heat-sensitive genotypes almost halved A_n. The change in g_s resulted in 50% of the leaf cooling in the sensitive compared to the tolerant genotypes. Compared to the control, the heat-tolerant genotypes produced 10% less dry matter, while the heat-sensitive ones had 33% less biomass. At the same time, the pollen viability decreased from approximately 80% in the control to approximately 50% in the heat-tolerant genotypes but only around 20% in the sensitive genotypes.

Of these genotypes, the two best and the two worst-performing genotypes were grown in an irrigated field trial in Nepal. By chance Nepal experienced a three-month-long heat wave resulting in 38.2/26.1°C mean day/night temperature throughout the experimental period and, without irrigation, the plants would have died. The heat-sensitive genotypes had approximately 50% less dry mass in both leaves and fruits, a twice as high heat injury index for visual damage and a fruit set of 25–28%, compared to 78–84% fruit set in the heat-tolerant genotypes (Poudyal et al., 2019). The results show that the use of F_v/F_m as a selection parameter in controlled heat screening of young plants is related to growth and harvest yield when the crop is experiencing single heat stress in the field.

8.4 Case 4: The effect of drought stress and elevated CO_2 on g_s in tomato

To move forward in phenotyping for climate resilience, more research is needed on the effects of combined drought and heat stress as well as eCO_2. Even though it is not possible to fully simulate a drought situation in the field in pot experiments, because of the limited pot volume, it is possible to obtain useful information about the influence of soil water availability on plant performance. Drought responses are focused on stomatal regulation, which is not only regulated by the major climate factors (Merilo et al., 2014) but also by the concentration of abscisic acid (ABA) synthesised in the roots and transported to the leaves, which is also modulated by C_a in tomato (Liu et al., 2019). Elevated CO_2 decreases g_s in general (e.g. Wei et al., 2018; Liu et al., 2019) but, even though tomato plants may have a lower transpiration rate per leaf area at eCO_2, progressive soil drying can still be faster than in aCO_2 because of the larger plant size at eCO_2 (Yan et al., 2017).

How much higher biomass accumulation is found at eCO_2 depends on the nitrogen status of the plant, but high N fertilisation often results in a more extensive enhancement of A_n than of biomass (Wei et al., 2018), which corresponds to the lack of correlation between A_n and biomass accumulation that is often found (Poorter and Remkes, 1990; Driever et al., 2014; Körner, 2015). Because well-watered plants initially have lower g_s at eCO_2, they can maintain unchanged g_s to a lower *FTSW* compared to plants grown at aCO_2 (Liu et al., 2019). The lower g_s mitigates drought stress but enhances heat stress since heat-tolerant plants need to increase g_s to increase leaf cooling. These relationships must be considered when designing experiments with combined drought and heat stress, especially when eCO_2 is included.

9 Conclusions

Breeding for the future will have to include crop stress resilience on a different scale than today. Phenotyping of the photosynthetic capacity per leaf area for prediction of harvest yield has been done with varying success because the growth of plants is more dependent on the leaf area ratio and specific leaf area than the net assimilation rate per leaf area per se. However, for any given genotype with desirable agronomic traits, screening for stress tolerance and finding biological and/or genetic markers for desirable physiological traits will be necessary for stress resilience to be included in breeding programmes. Direct measurements of photosynthesis by gas exchange are not possible in a high-throughput context, but modulated chlorophyll fluorescence can be applied both via camera-based systems with conveyor belts or manually measured.

Stress detection by F_v/F_m requires dark adaptation but the workload of measuring is under one minute per sample for two persons if plants are arranged efficiently. F_v/F_m reflects the degree of intactness of PSII, but the photosystem is protected by heat dissipation through *NPQ*, the D1 protein repair cycle, antioxidants and scavenging enzymes against ROS in the light reaction and photorespiration in the dark reaction. If the plant is not obviously photoinhibited, a decrease in F_v/F_m will have been accompanied by numerous other physiological stress responses in the plant before the stress affects PSII, which indirectly links F_v/F_m to the final harvest yield. Heat stress is well detected by F_v/F_m, while drought is not because of the protective effect of photorespiration. The reduction of F_v/F_m by heat stress has been shown to be a reliable proxy for stress responses on photosynthesis measured as gas exchange and biomass reduction in wheat, and in reduction of biomass as well as harvest yield in tomato after heat stress, both in climate chambers, greenhouses and irrigated field trials.

Most measurements of photosynthesis are done at steady state and many parameters quantify the maximum capacity for the different parts of the photosynthetic machinery at saturating light. There is a serious knowledge gap in photosynthetic performance in non-saturating light and in fluctuating light, where the accuracy of the regulation of stomata and *NPQ* can limit the daily carbon gain. Our knowledge of genotypic differences in these dynamic responses is very poorly understood. So far, phenotyping of photosynthesis for prediction of harvest yield under optimal growth conditions has not been very successful, but when plants are subjected to suboptimal conditions, photosynthesis may become a limiting process with important implications for the harvest yield.

One dilemma in the context of drought and heat stress is the opposing effects that the two stresses have on stomatal regulation, in which drought-tolerant plants reduce g_s to limit the water loss, while heat-tolerant plants can maintain high g_s for transpiration to cool the canopy. Over time, drought might be mitigated in C_3 plants, while heat stress will be accentuated by gradually increasing CO_2 concentration. Which stress that will dominate in the future will depend on the relative strength of the two in any given weather situation. For stress resilience to be implemented in breeding programmes, the possibility of climate stress other than nutrients and drought must be included in high-throughput infrastructures and extended collaboration between academia and breeding companies is needed to achieve the goals (Rosenqvist et al., 2019).

10 Where to look for further information

Since photosynthesis is measured on live plants that respond to the environment - both the growth environment and the measuring conditions - gas exchange and chlorophyll fluorescence can to some extent be considered a handicraft where experience and skill is built up over time. To get started, six papers on the principles of gas exchange, practical advise and pitfalls are listed in Table 2. In Section 6, Using chlorophyll fluorescence, a list of ten relevant reviews on the technique is listed on p. 306–307.

11 References

Abdelhakim, L. O. A., Palma, C. F. F., Zhou, R., Wollenweber, B., Ottosen, C. O. and Rosenqvist, E. (2021a). The effect of individual and combined drought and heat stress under elevated CO_2 on physiological responses in spring wheat genotypes. *Plant Physiology and Biochemistry* 162: 301-314. doi: 10.1016/j.plaphy.2021.02.015.

Abdelhakim, L. O. A., Rosenqvist, E., Wollenweber, B., Spyroglou, I., Ottosen, C.-O. and Pazarová, K. (2021b). Investigating combined drought- and heat stress effects in wheat under controlled conditions by dynamic image-based phenotyping. *Agronomy* 11(2): 254. doi: 10.3390/agronomy11020364.

Ainsworth, E. A. and Long, S. P. (2004). What have we learned from 15 years of free-air CO_2 enrichment (FACE)? A meta-analytic review of the responses of photosynthesis, canopy properties and plant production to rising CO_2. *New Phytologist* 165(2): 351-371. doi: 10.1111/j.1469-8137.2004.01224.x.

Ainsworth, E. A. and Long, S. P. (2020). 30 years of free-air carbon dioxide enrichment (FACE): what have we learned about future crop productivity and its potential for adaptation? *Global Change Biology* 27(1): 27-49. doi: 10.1111/gcb.15375.

Ananyev, G., Kolber, Z. S., Klimov, D., Falkowski, P. G., Berry, J. A., Rascher, U., Martin, R. and Osmond, B. (2005). Remote sensing of heterogeneity in photosynthetic efficiency, electron transport and dissipation of excess light in *Populus deltoids* stands under ambient and elevated CO_2 concentrations, and in a tropical forest canopy, using a new laser-induced fluorescence transient device. *Global Change Biology* 11(8): 1195-1206. doi: 10.1111/j.1365-2486.2005.00988.x.

Araus, J. L., Elazab, A., Vergara, O., Cabrera-Bosquet, L., Serret, M. D., Zaman-Allah, M. and Cairns, J. E. (2015). New technologies for phenotyping. In: Fritsche-Neto, R. and Borém, A. (Eds). *Phenomics, How Next-Generation Phenotyping is Revolutionizing Plant Breeding*. Springer, Heidelberg, pp. 1-14.

Baena-González, E. and Aro, E. M. (2002). Biogenesis, assembly and turnover of photosystem II units. *Philosophical Transactions of the Royal Society of London. Series B, Biological Sciences* 357(1426): 1451-1460. doi: 10.1098/rstb.2002.1141.

Baker, N. R. (2008). Chlorophyll fluorescence: A probe of photosynthesis *in vivo*. *Annual Review of Plant Biology* 59: 89-113.

Baker, N. R. and Rosenqvist, E. (2004). Applications of chlorophyll fluorescence can improve crop production strategies: An examination of future possibilities. *Journal of Experimental Botany* 55(403): 1607-1621.

Beadle, C. L. and Long, S. P. (1985). Photosynthesis - Is it limiting to biomass production? *Biomass* 8(2): 119-168.

Bellasio, C., Beerling, D. J. and Griffiths, H. (2016a). Deriving C_4 photosynthetic parameters from combined gas exchange and chlorophyll fluorescence using an Excel tool: Theory and practice. *Plant, Cell and Environment* 39(6): 1164-1179. doi: 10.1111/pce.12626.

Bellasio, C., Beerling, D. J. and Griffiths, H. (2016b). An Excel tool for deriving key photosynthetic parameters from combined gas exchange and chlorophyll fluorescence: Theory and practice. *Plant, Cell and Environment* 39(6): 1180-1197. doi: 10.1111/pce.12560.

Bracher, A., Whitney, S. M., Hartl, F. U. and Hayer-Hartl, M. (2017). Biogenesis and metabolic maintenance of RuBisCO. *Annual Review of Plant Biology* 68: 29-60.

Bresson, J., Vasseur, F., Dauzat, M., Koch, G., Granier, C. and Vile, D. (2015). Quantifying spatial heterogeneity of chlorophyll fluorescence during plant growth and in response to water stress. *Plant Methods* 11: 23. doi: 10.1186/s13007-015-0067-5.

Brodersen, C. R. and Vogelmann, T. C. (2010). Do changes in light direction affect absorption profiles in leaves? *Functional Plant Biology* 37(5): 403-412. doi: 10.1071/FP09262.

Burgess, A. J., Retkute, R., Herman, T. and Murchie, E. H. (2017). Exploring relationships between canopy architecture, light distribution, and photosynthesis in contrasting rice genotypes using 3D canopy reconstruction. *Frontiers in Plant Science* 8: 734. doi: 10.3389/fpls.2017.00734.

Carmo-Silva, E., Andralojc, P. J., Scales, J. C., Driever, S. M., Mead, A., Lawson, T., Raines, C. A. and Parry, M. A. J. (2017). Phenotyping of field-grown wheat in the UK highlight contribution of light response of photosynthesis and flag leaf longevity to grain yield. *Journal of Experimental Botany* 68(13): 3473-3486. doi: 10.1093/jxb/erx169.

Christidis, N., Jones, G. S. and Stott, P. A. (2014). Dramatically increasing chance of extremely hot summers since the 2003 European heatwave. *Nature Climate Change* 5(1): 46-50. doi: 10.1038/nclimate2468.

Costa, J. M., Grant, O. M. and Chaves, M. M. (2013). Thermography to explore plant-environment interactions. *Journal of Experimental Botany* 64(13): 3937-3949. doi: 10.1093/jxb/ert029.

Costa, J. M., Ortuño, M. F., Lopes, C. M. and Chaves, M. M. (2012). Grapevine varieties exhibiting differences in stomatal response to water deficit. *Functional Plant Biology* 39(3): 179-189. doi: 10.1071/FP11156.

de Pury, D. G. G. and Farquhar, G. D. (1997). Simple scaling of photosynthesis from leaves to canopies without the errors of big-leaf models. *Plant, Cell and Environment* 20(5): 537-557.

De Souza, A. P., Wang, Y., Orr, D. J., Carmo-Silva, E. and Long, S. P. (2020). Photosynthesis across African cassava germplasm is limited by RuBisCO and mesophyll conductance at steady state, but by stomatal conductance in fluctuating light. *New Phytologist* 225(6): 2498-2512. doi: 10.1111/nph.16142.

Driever, S. M., Lawson, T., Andralojc, P. J., Raines, C. A. and Parry, M. A. J. (2014). Natural variation in photosynthetic capacity, growth and yield in 64 field-grown wheat genotypes. *Journal of Experimental Botany* 65(17): 4959-4973. doi: 10.1093/jxb/eru253.

Driever, S. M., Simkin, A. J., Alotaibi, S., Fisk, S. J., Madgwick, P. J., Sparks, C. A., Jones, H. D., Lawson, T., Parry, M. A. J. and Raines, C. A. (2017). Increased SBPase activity improves photosynthesis and grain yield in wheat grown in greenhouse conditions. *Philosophical Transactions Series B* 372, 20160384. doi: 10.1098/rstb.2016.0384.

Dubios, J.-J. B., Fiscus, E. L., Booker, F. L., Flowers, M. D. and Reid, C. D. (2007) Optimizing the statistical estimation of the parameters of the Farquhar-von Caemmerer-Berry model of photosynthesis. *New Phytologist* 176(2): 402-414. doi: 10.1111/j.1469-8137.2007.02182.x.

Duursma, R. A. (2015). Plantecophys - An R package for analysing and modelling leaf gas exchange data. *PLoS ONE* 10(11): e0143346. doi: 10.1371/journal.pone.0143346.

Eyland, D., van Wesemael, J., Lawson, T. and Carpentier, S. (2021). The impact of slow stomatal kinetics on photosynthesis and water use efficiency under fluctuating light. *Plant Physiology* 186(2): 998-1012. doi: 10.1093/plphys/kiab114.

Farquhar, G. D., von Caemmerer, S. and Berry, J. A. (1980). A biochemical model of photosynthetic CO_2 assimilation in leaves of C_3 species. *Planta* 149(1): 78-90.

Flexas, J., Díaz-Espejo, A., Berry, J. A., Cifre, J., Galmés, J., Kaldenhoff, R., Medrano, H. and Ribas-Carbó, M. (2007). Analysis of leakage in IRGA's leaf chambers of open gas exchange systems: Quantification and its effects in photosynthesis parameterization. *Journal of Experimental Botany* 58(6): 1533–1543. doi: 10.1093/jxb/erm027.

Flexas, J., Ribas-Carbó, M., Diaz-Espejo, A., Galmés, J. and Medrano, H. (2008). Mesophyll conductance to CO_2: Current knowledge and future prospects. *Plant, Cell and Environment* 31(5): 602–621. doi: 10.1111/j.1365-3040.2007.01757.x.

Flood, P. J., Harbinson, J. and Aarts, M. G. M. (2011). Natural genetic variation in plant photosynthesis. *Trends in Plant Science* 16(6): 327–335. doi: 10.1016/j.tplants.2011.02.005.

Flood, P. J., Kruijer, W., Schnabel, S. K., van der Schoor, R., Jalink, H., Snel, J. F. H., Harbinson, J. and Aarts, M. G. M. (2016). Phenomics for photosynthesis, growth and reflectance in *Arabidopsis thaliana* reveals circadian and long-term fluctuations in heritability. *Plant Methods* 12: 14. doi: 10.1186/s13007-016-0113-y.

Foyer, C. H. and Shigeoka, S. (2011). Understanding oxidative stress and antioxidant functions to enhance photosynthesis. *Plant Physiology* 155(1): 93–100. doi: 10.1104/pp.110.166181.

Fu, P., Meacham-Hensold, K., Guan, K., Wu, J. and Bernachi, C. (2020). Estimating photosynthetic traits from reflectance spectra: A synthesis of spectral indices, numerical inversion, and partial least square regression. *Plant, Cell and Environment* 43(5): 1241–1258. doi: 10.1111/pce.13718.

Furbank, R. T., Sharwood, R., Estavillo, G. M., Silva-Perez, V. and Condon, A. G. (2020). Photons to food: Genetic improvement of cereal crop photosynthesis. *Journal of Experimental Botany* 71(7): 2226–2238. doi: 10.1093/jxb/eraa077.

Gago, J., Carriqué, M., Nadal, M., Clemente-Moreno, M. J., Coopman, R. E., Fernie, A. R. and Flexas, J. (2019). Photosynthesis optimized across land plant phylogeny. *Trends in Plant Science* 24(10): 947–958. doi: 10.1016/j.tplants.2019.07.002.

Gago, J., Daloso, D. M., Carriquí, M., Nadal, M., Morales, M., Araújo, W. L., Nunes-Nesi, A. and Flexas, J. (2020). Mesophyll conductance: The leaf corridors for photosynthesis. *Biochemical Society Transactions* 48(2): 429–439. doi: 10.1042/BST20190312.

Grant, O. M., Davies, M. J., James, C. M., Johnson, A. W., Leinonen, I. and Simpson, D. W. (2012). Thermal imaging and carbon isotope composition indicate variation amongst strawberry (*Fragaria* x *ananassa*) cultivars in stomatal conductance and water use efficiency. *Environmental and Experimental Botany* 76: 7–15. doi: 10.1016/j.envexpbot.2011.09.013.

Gu, J., Yin, X., Stompf, T. J. and Struik, P. C. (2014). Can exploiting natural genetic variation in leaf photosynthesis contribute to increasing rice productivity? A simulation analysis. *Plant, Cell and Environment* 37(1): 22–34. doi: 10.1111/pce.12173.

Hanson, D. T., Stutz, S. S. and Boyer, J. S. (2016). Why small fluxes matter: The case and approaches for improving measurements of photosynthesis and (photo)respiration. *Journal of Experimental Botany* 67(10): 3027–3039. doi: 10.1093/jxb/erw139.

Harbinson, J. and Rosenqvist, E. (2003). An introduction to chlorophyll fluorescence. In: DeEll, J. R. and Toivonen, P. M. A. (Eds). *Practical Applications of Chlorophyll Fluorescence in Plant Biology*. Kluwer Academic Publishers, Amsterdam, pp. 1–29.

Harley, P. C., Loreto, F., Di Marco, G. and Sharkey, T. D. (1992). Theoretical considerations when estimating mesophyll conductance to CO_2 flux by analysis of the response of photosynthesis to CO_2. *Plant Physiology* 98(4): 1429–1436.

Haworth, M., Marina, G. and Centritto, M. (2018). An introductory guide to gas exchange analysis of photosynthesis and its application to plant phenotyping and precision irrigation to enhance water use efficiency. *Journal of Water and Climate Change* 9(4): 786–808. doi: 10.2166/wcc.2018.152.

Jones, H. G. (1999). Use of thermography for quantitative studies of spatial and temporal variation of stomatal conductance over leaf surfaces. *Plant, Cell and Environment* 22(9): 1043–1055.

Jones, H. G. (2004). Application of thermal imaging and infrared sensing in plant physiology and ecophysiology. *Advances in Botanical Research* 41: 107–163.

Kahiluoto, H., Kaseva, J., Balek, J., Olesen, J. E., Ruiz-Ramos, M., Gobin, A., Kersebaum, K. C., Takác, J., Ruget, F., Ferrise, R. and Bezak, P. (2018). Decline in climate resilience of European wheat. *PNAS* 116: 123–128. doi: 10.1073/pnas.1804387115.

Kaiser, E., Galvis, V. C. and Armbruster, U. (2019). Efficient photosynthesis in dynamic light environments: A chloroplast's perspective. *Biochemical Journal* 476(19): 2725–2741. doi: 10.1042/BCJ20190134.

Kaiser, E., Kromdijk, J., Harbinson, J., Heuvelink, E. and Marcelis, L. F. (2017). Photosynthetic induction and its diffusional, carboxylation and electron transport processes as affected by CO_2 partial pressure, temperature, air humidity and blue irradiance. *Annals of Botany* 119(1): 191–205. doi: 10.1093/aob/mcw226.

Kalaji, H. M., Schansker, G., Ladle, R. J., Goltsev, V., Bosa, K., Allakhverdiev, S. I., Brestic, M., Bussotti, F., Calatayud, A., Dąbrowski, P., Elsheery, N. I., Ferroni, L., Guidi, L., Hogewoning, S. W., Jajoo, A., Misra, A. N., Nebauer, S. G., Pancaldi, S., Penella, C., Poli, D., Pollastrini, M., Romanowska-Duda, Z. B., Rutkowska, B., Serôdio, J., Suresh, K., Szulc, W., Tambussi, E., Yanniccari, M. and Zivcak, M. (2014). Frequently asked questions about *in vivo* chlorophyll fluorescence: Practical issues. *Photosynthesis Research* 122(2): 121–158. doi: 10.1007/s11120-014-0024-6.

Keller, B. (2018). *Analyzing photosynthetic performance in natural fluctuating environment using light-induced fluorescence transient (LIFT) method in high-throughput*. PhD dissertation, Reinischen Friedrich-Wilhelms-Universität, Bonn, p. 118.

Kimura, H., Hashimoto-Sugimoto, M., Iba, K., Terashima, I. and Yamori, W. (2020). Improved stomatal opening enhances photosynthetic rate and biomass production in fluctuating light. *Journal of Experimental Botany* 71(7): 2339–2350. doi: 10.1093/jxb/eraa090.

Klughammer, C. and Schreiber, U. (2008). Complementary PS II quantum yields calculated from simple fluorescence parameters measured by PAM fluorometry and the saturation pulse method. *PAM Applications Notes* 1: 27–35. https://www.walz.com/files/downloads/pan/PAN078007.pdf.

Kolber, Z., Klimov, D., Ananyev, G., Rascher, U., Berry, J. and Osmond, B. (2005). Measuring photosynthetic parameters at a distance: Laser induced fluorescence transient (LIFT) method for remote measurements of photosynthesis in terrestrial vegetation. *Photosynthesis Research* 84(1–3): 121–129.

Körner, C. (2015). Paradigm shift in plant growth control. *Current Opinion in Plant Biology* 25: 107–114. doi: 10.1016/j.pbi.2015.05.003.

Körner, C., Scheel, J. A. and Bauer, H. (1979). Maximum leaf diffusive conductance in vascular plants. *Photosynthetica* 13: 45–82.

Kromdijk, J., Glowacka, K., Leonelli, L., Gabilly, S. T., Iwai, M., Niyogi, K. K. and Long, S. P. (2016). Improving photosynthesis and crop productivity by accelerating recovery from photoprotection. *Science* 354(6314): 857–861. doi: 10.1126/science.aai8878.

Lawson, T., Kramer, D. M. and Raines, C. A. (2012). Improving yield by exploiting mechanisms underlying natural variation of photosynthesis. *Current Opinion in Biotechnology* 23(2): 215-220. doi: 10.1016/j.copbio.2011.12.012.

Li, L., Zhang, Q. and Huang, D. (2014). A review of imaging techniques for plant phenotyping. *Sensors* 14(11): 20078-20111. doi: 10.3390/s141120078.

Li, X. P., Björkman, O., Shih, C., Grossman, A. R., Rosenquist, M., Jansson, S. and Kiyogi, K. K. (2000). A pigment-binding protein essential for regulation of photosynthetic regulation of photosynthetic light harvesting. *Nature* 403(6768): 391-395.

Liu, J., Hu, T., Fang, L., Peng, X. and Liu, F. (2019). CO_2 elevation modulates the response of gas exchange to progressive soil drying in tomato plants. *Agricultural and Forest Meteorology* 268: 181-188. doi: 10.1016/j.agrformet.2019.01.026.

Long, S. P., Ainsworth, E. A., Rogers, A. and Ort, D. R. (2004). Rising atmospheric carbon dioxide: Plants FACE the future. *Annual Review of Plant Biology* 55: 591-628. doi: 10.1146/annurev.arplant.55.031903.141610.

Long, S. P. and Bernacchi, C. J. (2003). Gas exchange measurements, what can they tell us about the underlying limitations to photosynthesis? Procedures and sources of error. *Journal of Experimental Botany* 54(392): 2393-2401. doi: 10.1093/jxb/erg262.

Long, S. P., Farange, P. K. and Garcia, R. L. (1996). Measurement of leaf and canopy photosynthetic CO_2 exchange in the field. *Journal of Experimental Botany* 47(11): 1629-1642.

Long, S. P. and Hällgren, J.-E. (1993). Measurement of CO_2 assimilation by plants in the field and the laboratory. In: Hall, D. O., Scurlock, J. M. O., Bolhár-Nordenkampf, H. R., Leegood, R. C. and Long, S. P. (Eds). *Photosynthesis and Production in a Changing Environment: A Field and Laboratory Manual*. Chapman & Hall, London, pp. 129-167.

Long, S. P., Zhu, X. G., Naidu, S. L. and Ort, D. R. (2006). Can improvement in photosynthesis increase crop yields? *Plant, Cell and Environment* 29(3): 315-330. doi: 10.1111/j.1365-3040.2005.01493.x.

Maes, W. H. and Steppe, K. (2012). Estimating evapotranspiration and drought stress with ground-based thermal remote sensing in agriculture: A review. *Journal of Experimental Botany* 63(13): 4671-4712. doi: 10.1093/jxb/ers165.

Malenovsky, Z., Mishra, K. B., Zemek, F., Rascher, U. and Nedbal, L. (2009). Scientific and technical challenges in remote sensing of plant canopy reflectance and fluorescence. *Journal of Experimental Botany* 60(11): 2987-3004. doi: 10.1093/jxb/erp156.

Markvart, J., Rosenqvist, E., Aaslyng, J. M. and Ottosen, C.-O. (2010). How is canopy photosynthesis and growth of *Chrysanthemums* affected by diffuse and direct light? *European Journal of Horticultural Science* 75: 253-258.

Maxwell, K. and Johnson, G. N. (2000). Chlorophyll fluorescence - A practical guide. *Journal of Experimental Botany* 51(345): 659-668.

McAusland, L., Atkinson, J. A., Lawson, T. and Murchie, E. H. (2019) High throughput procedure utilizing chlorophyll fluorescence imaging to phenotype dynamic photosynthesis and photoprotection in leaves under controlled gaseous conditions. *Plant Methods* 15:109, 1-15. doi: 10.1186/s13007-019-0485-x.

McClain, A. M. and Sharkey, T. D. (2019). Triose phosphate utilization and beyond: From photosynthesis to end product synthesis. *Journal of Experimental Botany* 70(6): 1755-1766. doi: 10.1093/jxb/erz058.

Meacham-Hensold, K., Fu, P., Wu, J., Serbin, S., Montes, C. M., Ainsworth, E., Guan, K., Dracup, E., Pederson, T., Driever, S. and Bernacchi, C. (2020). Plot-level rapid screening for photosynthetic parameters using proximal hyperspectral imaging. *Journal of Experimental Botany* 71(7): 2312–2328. doi: 10.1093/jxb/eraa068.

Mendanha, T., Rosenqvist, E., Hyldgaard, B. N., Doonan, J. H. and Ottosen, C.-O. (2018). Drought priming effects on alleviating the photosynthetic limitations of wheat cultivars (*Triticum aestivum* L.) with contrasting tolerance to abiotic stresses. *Journal of Agronomy and Crop Science* 206: 651–664. doi: 10.1111/jac.12404.

Merilo, E., Jõesaar, I., Brosché, M. and Kollist, H. (2014). To open or to close: Species-specific stomatal responses to simultaneous applied opposing environmental factors. *New Phytologist* 202(2): 499–508. doi: 10.1111/nph.12667.

Mishra, P., Lohumi, S., Kahn, H. A. and Nordon, A. (2020). Close-range hyperspectral imaging of whole plants for digital phenotyping: Recent applications and illumination correction approaches. *Computers and Electronics in Agriculture* 178: 105780. doi: 10.1016/j.compag.2020.105780.

Monteith, J. L. (1977). Climate and the efficiency of crop production in Britain. *Philosophical Transactions of the Royal Society of London. B, Biological Sciences* 281(980): 277–294.

Moore, C. E., Meacham-Hensold, K., Lemonnier, P., Slattery, R. A., Benjamin, C., Bernacchi, C. J., Lawson, T. and Cavanagh, A. P. (2021). The effect of increasing temperature on crop photosynthesis: From enzymes to ecosystems. *Journal of Experimental Botany* 72(8): 2822–2844. doi: 10.1093/jxb/erab090.

Moualeu-Ngangue, D. P., Chen, T. W. and Stutzel, H. (2017). A method to estimate photosynthetic parameters through net assimilation rate-intercellular space CO_2 concentration (A-C_i) curve and chlorophyll fluorescence measurements. *New Phytologist* 213(3): 1543–1554. doi: 10.1111/nph.14260.

Muraoka, H., Tang, Y., Terashima, I., Koizumi, H. and Washitani, I. (2000). Contributions of diffusional limitation, photoinhibition and photorespiration to midday depression of photosynthesis in *Arisaema heterophyllum* in natural high light. *Plant, Cell and Environment* 23(3): 235–250.

Murchie, E. H. and Lawson, T. (2013). Chlorophyll fluorescence analysis: A guide to good practice and understanding some new applications. *Journal of Experimental Botany* 64(13): 3983–3998. doi: 10.1093/jxb/ert208.

Niyogi, K. K. (2000). Safety valves for photosynthesis. *Current Opinion in Plant Biology* 3(6): 455–460.

Niyogi, K. K., Li, X.-P., Rosenberg, V. and Jung, H.-S. (2005). Is PsbS the site of non-photochemical quenching of photosynthesis? *Journal of Experimental Botany* 65: 375–382. doi: 10.1093/jxb/eri056.

Ögren, E. (1988). Photoinhibition of photosynthesis in willow leaves under field conditions. *Planta* 175(2): 229–236.

Ögren, E. (1993). Convexity of the photosynthetic light-response curve in relation to intensity and direction of light during growth. *Plant Physiology* 101(3): 1013–1019.

Ögren, E. and Rosenqvist, E. (1992). On the significance of photoinhibition and its generality among species. *Photosynthesis Research* 33(1): 63–71.

Ögren, E. and Sjöström, M. (1990). Estimation of the effect of photoinhibition on the carbon gain in leaves of a willow canopy. *Planta* 181(4): 560–567.

Ögren, E. and Sundin, U. (1996). Photosynthetic responses to variable light: A comparison of species from contrasting habitats. *Oecologia* 106(1): 18–27.

Ort, D. R., Merchant, S. S., Alric, J., Barkand, A., Blankenship, R. E., Bock, R., Croce, R., Hanson, M. R., Hibberd, J. M., Long, S. P., Moore, T. A., Moroney, J., Niyogi, K. K., Parry, M. A. J., Peralta-Yahya, P. P., Prince, R. C., Redding, K. E., Spalding, M. H., van Wijk, K.

J., Vermaas, W. F. J., von Caemmerer, S., Weber, A. P. M., Yeates, T. O., Yuan, J. S. and Zhu, X. G. (2015). Redesigning photosynthesis to sustainably meet global food and bioenergy demand. *Proceedings of the National Academy of Sciences of the United States of America* 112(28): 8529–8536. doi: 10.1073/pnas.1424031112.

Papageorgiou, G. C. and Govindjee, G. (Ed.). (2004). *Chlorophyll Fluorescence: A Signature of Photosynthesis*. Kluwer Academic Publishers, Dordrecht.

Parry, M. A. J., Reynolds, M., Salvucci, M. E., Raines, C., Andralojc, P. J., Zhu, X. G., Price, G. D., Condon, A. G. and Furbank, R. T. (2011). Raising yield potential of wheat. II. Increasing photosynthetic capacity and efficiency. *Journal of Experimental Botany* 62(2): 453–467. doi: 10.1093/jxb/erq304.

Paul, K., Sorrentino, M., Lucini, L., Rouphael, Y., Cardarelli, M., Bonini, P., Moreno, M. B. M., Raynaud, H., Canaguier, R., Trtílek, M., Panzarová, K. and Colla, G. (2019). A combined phenotypic and metabolomic approach for elucidating the biostimulant action of a plant-derived protein hydrolysate on tomato grown under limited water availability. *Frontiers in Plant Science* 10: 493. doi: 10.3389/fpls.2019.00493.

Poorter, H. and Remkes, C. (1990). Leaf area ratio and net assimilation rate of 24 wild species differing in relative growth rate. *Oecologia* 83(4): 553–559.

Porcar-Castell, A., Pfündel, E., Korhonen, J. F. J. and Juurola, E. (2008). A new monitoring PAM fluorometer (MONI-PAM) to study the short- and long-term acclimation of photosynthesis II in field conditions. *Photosynthesis Research* 96(2): 173–179.

Porcar-Castell, A., Tyystjärvi, E., Atherton, J., van der Tol, C., Flexas, J., Pfündel, E. E., Moreno, J., Frankenberg, C. and Berry, J. A. (2014). Linking chlorophyll a fluorescence to photosynthesis for remote sensing applications: Mechanisms and challenges. *Journal of Experimental Botany* 65(15): 4065–4095. doi: 10.1093/jxb/eru191.

Poudyal, D., Rosenqvist, E. and Ottosen, C. O. (2019). Phenotyping from lab to field – Tomato lines screened for heat stress using F_v/F_m maintain high fruit yield during thermal stress in the field. *Functional Plant Biology* 46(1): 44–55. doi: 10.1071/FP17317.

Prag, A. A. (2017). *Plant growth under CO_2 enrichment – Limitations and uncertainties*. BSc thesis in Natural Resources, University of Copenhagen, p. 45.

Price, G. D., von Caemmerer, S., Evans, J. R., Siebke, K., Anderson, J. M. and Badger, M. R. (1998). Photosynthesis is strongly reduced by antisense suppression of chloroplastic cytochrome *bf* complex in transgenic tobacco. *Australian Journal of Plant Physiology* 25: 445–452.

Reynolds, M., Foulkes, J., Furbank, R., Griffiths, S., King, J., Murchie, E., Parry, M. and Slafer, G. (2012). Achieving yield gains in wheat. *Plant, Cell and Environment* 35(10): 1799–1823. doi: 10.1111/j.1365-3040.2012.02588.x.

Rosenqvist, E., Großkinsky, D. K., Ottosen, C. O. and van de Zedde, R. (2019). The phenotyping dilemma – The challenges of a diversified phenotyping community. *Frontiers in Plant Science* 10: 163. doi: 10.3389/fpls.2019.00163.

Rosenqvist, E. and van Kooten, O. (2003). Chlorophyll fluorescence: A general description and nomenclature. In: DeEll, J. R. and Toivonen, P. M. A.(Eds). *Practical Applications of Chlorophyll Fluorescence in Plant Biology*. Kluwer Academic Publishers, Amsterdam, pp. 31–77.

Rosenqvist, E., Wingsle, G. and Ögren, E. (1991). Photoinhibition of photosynthesis in intact willow leaves in response to moderate changes in light and temperature. *Physiologia Plantarum* 83: 390–396.

Rousseau, C., Hunault, G., Gaillard, S., Bourbeillon, J., Montiel, G., Simier, P., Campion, C., Jacques, M. A., Belin, E. and Boreau, T. (2015). Phenoplant: A web resource for the exploration of large chlorophyll fluorescence image datasets. *Plant Methods* 11(24): 24. doi: 10.1186/s13007-015-0068-4.

Ruban, A. V. (2016). Nonphotochemical chlorophyll fluorescence quenching: Mechanisms and effectiveness in protecting plant from photodamage. *Plant Physiology* 170(4): 1903–1916. doi: 10.1104/pp.15.01935.

Sharkey, T. D., Bernacchi, C. J., Farquhar, G. D. and Singsaas, E. L. (2007). Fitting photosynthetic carbon dioxide response curves for C_3 leaves. *Plant, Cell and Environment* 30(9): 1035–1040. doi: 10.1111/j.1365-3040.2007.01710.x.

Sharma, D. K., Andersen, S. B., Ottosen, C. O. and Rosenqvist, E. (2012). Phenotyping of wheat cultivars for heat tolerance using chlorophyll a fluorescence. *Functional Plant Biology* 39(11): 936–947, doi: 10.1071/FP12100.

Sharma, D. K., Andersen, S. B., Ottosen, C. O. and Rosenqvist, E. (2015). Wheat cultivars selected for high F_v/F_m under heat stress maintain high photosynthesis, total chlorophyll, stomatal conductance, transpiration and dry matter. *Physiologia Plantarum* 153(2): 284–298. doi: 10.1111/ppl.12245.

Sharma, D. K., Torp, A. M., Rosenqvist, E., Ottosen, C. O. and Andersen, S. B. (2017). QTLs and potential candidate genes for heat stress tolerance identified from the mapping populations specifically segregating for F_v/F_m in wheat. *Frontiers in Plant Science* 8: 1668. doi: 10.3389/fpls.2017.01668.

Sharp, R. E., Matthews, M. A. and Boyer, J. S. (1984). Kock effect and the quantum yield of photosynthesis. Light partially inhibits dark respiration. *Plant Physiology* 75(1): 95–101.

Slattery, R. A., Walker, B. J., Weber, A. P. M. and Ort, D. R. (2018) The impact of fluctuating light on crop performance. *Plant Physiology* 176(2): 990–1003. doi: 10.1104/pp.17.01234.

Stinziano, J. R., McDermitt, D. K., Lynch, D. J., Saathoff, A. J. M., Morgan, P. B. and Hanson, D. T. (2019) The rapid A/C i response: A guide to best practices. *New Phytologist* 221(2): 625–627.

Stinziano, J. R., Morgan, P. B., Lynch, D. J., Saathoff, A. J., McDermitt, D. K. and Hanson, D. T. (2017). The rapid A-C_i response: Photosynthesis in the phenomic era. *Plant, Cell and Environment* 40(8): 1256–1262. doi: 10.1111/pce.12911.

Taylor, S. H. and Long, S. P. (2019). Phenotyping photosynthesis on the limit – A critical examination of RACiR. *New Phytologist* 221(2): 621–624.

Tschiersch, H., Junker, A., Meyer, R. C. and Altman, T. (2017). Establishment of integrated protocols for automated high throughput kinetic chlorophyll fluorescence analyses. *Plant Methods* 13(54): 54. doi: 10.1186/s13007-017-0204-4.

Urban, J., Ingwers, M. W., McGuire, M. A. and Teskey, R. O. (2017). Increase in leaf temperature opens stomata and decouples net photosynthesis from stomatal conductance in *Pinus taeda* and *Populus deltoids x nigra*. *Journal of Experimental Botany* 68(7): 1757–1767. doi: 10.1093/jxb/erx052.

von Caemmerer, S. (2000). *Biochemical Models of Leaf Photosynthesis. Techniques in Plant Science No. 2*. CSIRO Publishing, Collingwood.

von Caemmerer, S. (2013). Steady state models of photosynthesis. *Plant, Cell and Environment* 36(9): 1617–1630. doi: 10.1111/pce.12098.

von Caemmerer, S. (2020). Rubisco carboxylase/oxygenase: From the enzyme to the globe: A gas exchange perspective. *Journal of Plant Physiology* 252: 153240, 1-9. doi: 10.1016/j.jplph.2020.153240.

von Caemmerer, S. and Farquhar, G. D. (1981). Some relationships between the biochemistry of photosynthesis and gas exchange of leaves. *Planta* 153(4): 376–387.

von Caemmerer, S., Farquhar, G. and Berry, J. (2009). Biochemical models of photosynthesis. In: *Advances in Photosynthesis and Respiration* (vol. 2). CSIRO Publishing, Collingwood, pp. 209–230.

Walker, B. J., Busch, F. A., Driever, S. M., Kromdijk, J. and Lawson, T. (2018). Survey of tools for measuring *in vivo* photosynthesis. In: Covshoff, S. (Ed.). *Photosynthesis: Methods and Protocols, Methods in Molecular Biology* (vol. 1770). Springer Science + Bussiness Media, Heidelberg, pp. 3–24. doi: 10.1007/978-1-4939-7786-4_1.

Wardlaw, I. F. and Moncur, L. (1995). The response of wheat to high temperature following anthesis. I. The rate and duration of kernel filling. *Functional Plant Biology* 22(3): 391–397.

Wei, Z., Du, T., Li, X., Fang, L. and Liu, F. (2018). Interactive effects of CO_2 concentration elevation and nitrogen fertilization on water and nitrogen use efficiency of tomato grown under reduced irrigation regimes. *Agricultural Water Management* 202: 174–182. doi: 10.1016/j.agwat.2018.02.027.

Wei, Z., Fang, L., Li, X., Liu, J. and Liu, F. (2020). Effects of elevated atmospheric CO_2 on leaf gas exchange response to progressive drought in barley and tomato plants with different endogenous ABA levels. *Plant and Soil* 447(1–2): 431–446. doi: 10.1007/s11104-019-04393-3.

Wilhelm, C. and Selmar, D. (2011). Energy dissipation is an essential mechanism to sustain the viability of plants: The physiological limits of improved photosynthesis. *Journal of Plant Physiology* 168(2): 79–87. doi: 10.1016/j.jplph.2010.07.012.

Wullschleger, S.D. (1993) Biochemical limitations to carbon assimilation in C3 plants - A retrospective analysis of the A/Ci curves from 109 species. *Journal of Experimental Botany* 44: 907–920.

Yan, F., Li, X. and Liu, F. (2017). ABA signalling and stomatal control in tomato plants exposure to progressive soil drying under ambient and elevated atmospheric CO_2 concentration. *Environmental and Experimental Botany* 139: 99–104. doi: 10.1016/j.envexpbot.2017.04.008.

Yin, X., Niu, Y., van der Putten, P. E. L. and Struik, P. C. (2020). The Kok effect revisited. *New Phytologist* 227(6): 1764–1775, doi: 10.1111/nph.16638.

Zhou, R., Kjaer, K. H., Rosenqvist, E., Yu, X., Wu, Z. and Ottosen, C.-O. (2017a). Physiological response to heat stress during seedling and anthesis stage in tomato genotypes differing in heat tolerance. *Journal of Agronomy and Crop Science* 203(1): 68–80. doi: 10.1111/jac.12166.

Zhou, R., Yu, X., Ottosen, C. O., Rosenqvist, E., Zhao, L., Wang, Y., Yu, W., Zhao, T. and Wu, Z. (2017b). Drought stress had a predominant effect over heat stress on three tomato cultivars subjected to combined stress. *BMC Plant Biology* 17(1): 24. doi: 10.1186/s12870-017-0974-x.

Zhou, R., Yu, X., Kjaer, K. H., Rosenqvist, E., Ottosen, C.-O. and Wu, Z. (2015). Screening and validation of tomato genotypes under heat stress using F_v/F_m to reveal the physiological mechanism of heat tolerance. *Environmental and Experimental Botany* 118: 1–11. doi: 10.1016/j.envexpbot.2015.05.006.

Zhu, X. G., Long, S. P. and Ort, D. R. (2010). Improving photosynthetic efficiency for greater yield. *Annual Review of Plant Biology* 61: 235–261. doi: 10.1146/annurev-arplant-042809-112206.

Zhu, X. G., Ort, D. R., Whitmarch, J. and Long, S. P. (2004). The slow reversibility of photosystem II thermal energy dissipation of transfer from high to low light may cause large losses in carbon gain by crop canopies: A theoretical analysis. *Journal of Experimental Botany* 55(400): 1167–1175. doi: 10.1093/jxb/erh141.

Chapter 12

Using phenotyping techniques to predict and model grain yield: translating phenotyping into genetic gain

Thomas Vatter and José L. Araus, University of Barcelona and AGROTECNIO (Center for Research in Agrotechnology), Spain

1 Introduction

Recent models predict the global population to peak in 2064 at around 9.7 billion and then likely decline to about 8.8 billion by 2100 due to a decline in fertility rates, expedited by improvements in access to modern contraception and female education (Vollset et al., 2020). This requires a significant increase in crop production. The percentage increase in crop production, from current levels, needed to sustainably feed the future world population remains a topic of debate. Based on two food-demand projections (Alexandratos and Bruinsma, 2012; Tilman et al., 2011), many papers state the need to increase production by 60–100% by 2050, while a study by Hunter et al. (2017) concludes that an increase of approximately 25–70% above current production levels may be sufficient to meet 2050 crop demand. Even these lower estimates present a major challenge, particularly as extreme global weather events such as drought, heavy precipitation and high temperatures are expected to further increase with severe implications for stable crop yields (Cogato et al., 2019; Powell and Reinhard, 2016; Rummukainen, 2012). Other factors to consider are limitations in the amount of land under cultivation needed to maintain

http://dx.doi.org/10.19103/AS.2022.0102.16

biodiversity levels and ecosystem functioning, as well as the degradation of land under cultivation, particularly under irrigation, which corresponds to the most productive agrosystems (Hossain et al., 2020; Usubiaga-Liaño et al., 2019).

Increasing grain yield has always been the major goal of most breeding programs. Major milestones in plant breeding resulting in significant grain yield increases include (i) the introduction of hybrid breeding exploiting heterosis, (ii) the introduction of high-yielding semi-dwarf varieties in the process of the 'green revolution' (GR) used alongside improved irrigation methods, novel plant protection agents and efficient fertilizers, and (iii) more recently the introduction of DNA markers and their use in marker-assisted selection (MAS) (Bhargava and Srivastava, 2019; Reeves and Cassaday, 2002). While the GR resulted in rapid and strong increases in grain yield of major crops in most regions, great geographic disparities exist due to the methods not being appropriate where population densities are low and/or market infrastructure is poor, as is the case in Africa (Pingali, 2012). While in the beginning high sequencing costs hampered the application of MAS in plant breeding programs, the development of high-throughput genotyping platforms has drastically reduced the cost and time of sequencing and resulted in their widespread application, most commonly applied in the form of genomic selection (GS) (Bhat et al., 2016; Davey et al., 2011; Poland et al., 2012).

The application of GS in plant breeding considerably increased the efficiency of breeding programs, allowing for reduced phenotyping expenditures by applying sequence-based trait prediction (Jannink et al., 2010; Wang et al., 2018b). Precise phenotyping of large breeding populations

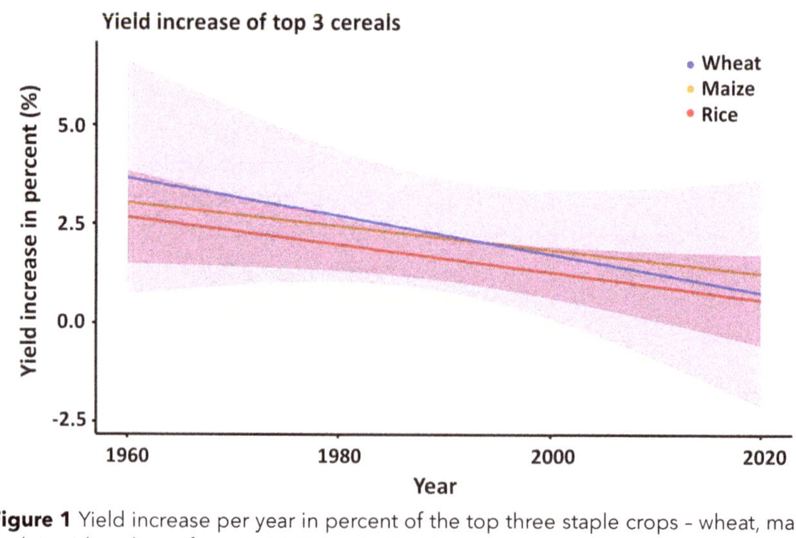

Figure 1 Yield increase per year in percent of the top three staple crops - wheat, maize and rice (data drawn from FAOSTAT FAO, 2020).

under diverse environments is required to harness the full potential of modern breeding techniques and as a result has become the new bottleneck in breeding programs (Araus et al., 2018b; Bhat et al., 2016; Zhao et al., 2019a).

Despite the rapid development and application of techniques in the field of plant breeding, crop yield growth is declining (Xu et al., 2017). Average yield growth (percent per year) for the three major staple crops (wheat, maize and rice) has reduced from 3.19, 2.55 and 2.25 during 1961–1990 to 1.20, 1.71 and 0.97 during 1991-2019, respectively (FAO, 2020). Further declines are predicted (Xu et al., 2017) (Fig. 1).

2 Boosting genetic gain in grain yield by focusing on phenomics

Despite the importance of phenotyping in plant breeding, advances in this field have not kept pace with progress in genotyping (Araus and Cairns, 2014; Furbank and Tester, 2011; Sinclair et al., 2004; Yang et al., 2020). In addition to a lack of research, this can be attributed to the complexity of phenomics, namely the acquisition of high-dimensional phenotypic data on an organism-wide scale, resulting in the inability to apply it to a large number of genotypes (Houle et al., 2010; Sinclair et al., 2004). Development and application of advanced cost- and time-efficient phenotyping techniques on a large scale will have a major impact on the achievable genetic gain in general and grain yield specifically (Fu, 2015; Houle et al., 2010; Lane et al., 2020). This becomes evident when looking at the different components contributing to genetic gain that can be targeted in the breeding process. Genetic gain relates to the increase in performance that is achieved per unit time applying artificial selection (Xu et al., 2017) defined by the equation of Falconer and Mackay (1996) as follows:

$$\Delta G = \frac{i * r * \sigma_A}{y}, \quad (1)$$

where ΔG refers to the response to selection, i to the intensity of selection, r to the selection accuracy, σ_A to the genetic variance and y to the generation interval.

Often, this equation is shown in more condensed form, and due to its importance, it is referred to as the 'breeder's equation' (Lush, 1937):

$$\Delta G = h^2 * S, \quad (2)$$

where ΔG refers to the response to selection, h^2 to the narrow-sense heritability and S to the selection differential.

Looking at the equation above, the dominant role heritability plays becomes evident as it ultimately limits the genetic gain achievable for the trait of interest (Fig. 2).

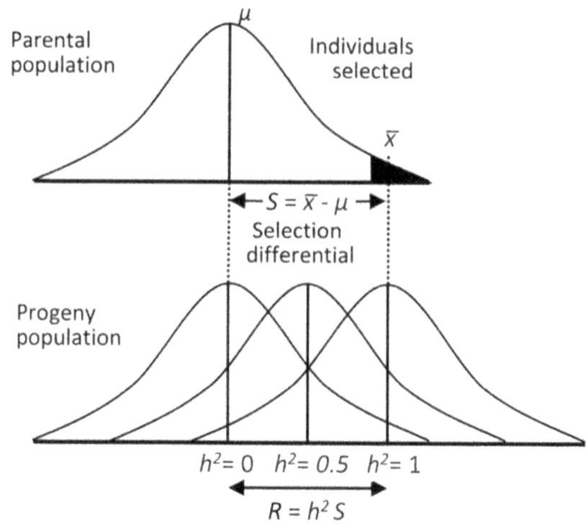

Figure 2 Importance of trait heritability for achievable genetic gain (modified from Xu et al., 2017).

While grain yield is the most important trait in most breeding programs, it is also a highly complex quantitative trait, being the result of the complex interaction between the genetic base and the environment with all growth and development processes that occur throughout the life cycle (Baenziger et al., 2011; Quarrie et al., 2006; Shi et al., 2009). As a result, heritability estimates of grain yield *per se* are often low, especially under drought stress conditions (Michel et al., 2019). However, it has to be considered that while the inherent heritability represents a barrier to achievable genetic gain, heritability in its current state remains an estimate, and a more accurate estimation will result in improved selection and thus obtainable increase in grain yield.

It is evident that the main factor currently hampering genetic gain is limited accuracy in the acquisition of phenotypic data across large breeding populations at reasonable costs (Cobb et al., 2013). As such, the role of high-throughput phenotyping (HTP) techniques that allow rapid and accurate phenotyping of large breeding populations in the field cannot be understated since they allow this gap to be filled (Araus et al., 2018a,b, 2021 (in press)). HTP approaches mostly make use of multispectral, hyperspectral, fluorescence and thermal sensors with simple digital red-green-blue (RGB) cameras being used (Araus et al., 2018b). These sensors are applied in a wide variety of ways, ranging from hand-held devices to satellite-based platforms (Fig. 3).

The application of HTP tools allows the various components of the genetic gain equation to be directly targeted (Araus et al., 2018b). This high-throughput approach increases the size of the breeding population to be phenotyped, which, in turn, allows for higher selection intensity while maintaining sufficient variability. Likewise, the number of locations in which the breeding population

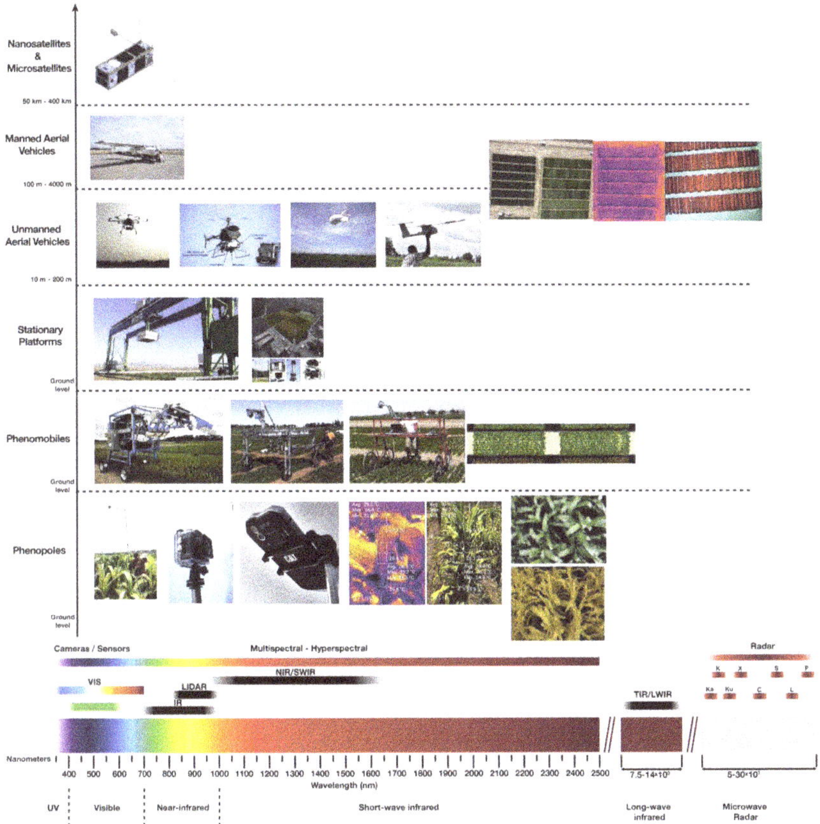

Figure 3 Different categories of potential and actual ground and aerial phenotyping platforms used in phenotyping grain crops, along with the spectral ranges used for different remote-sensing tools, and examples of derived images. RGB cameras (VIS), multispectral and hyperspectral sensors and cameras, light detection and ranging (LiDAR) sensors, thermal sensors and cameras (TIR/LWIR), and the different categories of radars. IR, infrared; LWIR, long-wave infrared; NIR, near-infrared; SWIR, short-wave infrared; TIR, thermal infrared; VIS, visible (modified from Araus et al., 2018b).

is tested can be increased. Moreover, the phenotyping error caused by subjective naked-eye scoring can be reduced, when phenotyping is conducted through an approach based, for example, on automatic assessments through remote sensing. Both of these factors result in higher improved trait heritability estimates and selection accuracy. In addition, HTP can be used to map field variability and has the potential to play an important role in the characterization of sets of environmental variables that can be used to improve prediction models (Araus et al., 2018b). Furthermore, HTP can reveal previously unknown factors influencing yield (Liang et al., 2018). A major potential use of HTP is its ability to predict yield in the early stages of a breeding cycle when limited seed availability prevents the assessment of grain yield in replicated full-sized plots (Krause et al., 2020).

When attempting to increase grain yield, breeders can make use of traits showing correlation with grain yield, to perform indirect selection. A major benefit of this procedure is that the final yield can be estimated before harvest. Among others, prominent yield-related traits are plant height, flowering time, aerial biomass and ear density, as well as physiological traits such as stomatal conductance and functional stay green (SG) traits (Chairi et al., 2018; Chen et al., 2012; Faralli et al., 2019; Fernandez-Gallego et al., 2018; Gautam et al., 2015; Kamal et al., 2019). The benefits of phenotyping for the prediction of grain yield can be seen in analyzing these individual traits and resulting improvements in grain yield. The following sections showcase plant traits that have been related to genetic gains in grain yield using HTP methods. It is beyond the scope of this chapter to give a complete review of recently published research on each trait. Instead, the chapter highlights a few key studies relating a particular trait to yield gain.

2.1 Plant height

Plant height has been shown to be strongly correlated with grain yield (Abdel-Ghani et al., 2016; Baye et al., 2020; Wang et al., 2018a; Zhang et al., 2017). The high heritability of plant height further increases its high value for predicting grain yield (Peiffer et al., 2014; Roy and Shil, 2020; Teich, 1984). The introduction of semi-dwarf varieties of rice and wheat during the GR resulted in unprecedented yield increases by reducing lodging and increasing the number of tillers (Peng et al., 1999). Reduction in plant height was achieved by introgression of genes involved in hormone synthesis, transport and signaling, for example repressing gibberellic acid (GA) biosynthesis and signaling (Peng et al., 1999; Spielmeyer et al., 2002; Wang and Li, 2008).

Plant height in the field is conventionally phenotyped using some form of measuring stick placed in the middle of each plot. While straightforward, this phenotyping approach introduces human bias as the average plot height is obtained from the side of the plot by the eye. This screening method is time consuming and laborious when performed by a single person, to avoid errors introduced by measurements performed by multiple people. This puts an upper limit on the size of the screening population and the number of growth stages plant height is screened for.

Constantly increasing the breeding populations makes the shift from these traditional phenotyping methods toward HTP methods inevitable. In this regard, several studies have focused on the use of both sensors and imaging devices to automate plant height phenotyping in the field. In a study performed by Wang et al. (2018a), five different sensing technologies mounted on a ground vehicle platform and an RGB camera mounted on an unmanned aerial vehicle (UAV) were evaluated for their suitability to replace manual phenotyping of plant height in sorghum. In the study, significant correlations between the sensors ($R^2 \geq 0.90$) and even the simple digital camera-based height measures ($R^2 = 0.73$) and those obtained using the traditional phenotyping approach were observed. Similarly, in a study performed by Holman et al. (2016) on wheat, plant height measures based on 3D digital surface models derived from an RGB camera mounted on a UAV allowed a highly accurate and repeatable collection of plant height data ($R^2 \geq 0.93$ with manual measurements). The study by Wang et al. (2019) represents another suitable example to replace traditional plant height phenotyping with HTP methods. In this study, a digital camera mounted on a UAV was used to obtain a digital elevation model and digital surface models, based on which plant height was obtained (Fig. 4). A high correlation ($R^2 = 0.95$) between UAV-based and manual plant height measurements was observed. In addition, UAV-based data obtained at multiple growth stages allowed for the identification of known and previously unknown plant height-related quantitative trait loci (QTL). As such, the study can be seen as an example of the great potential of combining HTP methods with genomic techniques.

2.2 Flowering time

Flowering time greatly influences the potential grain yield (Cockram et al., 2007; Jung and Müller, 2009). Flowering time is highly heritable, with most

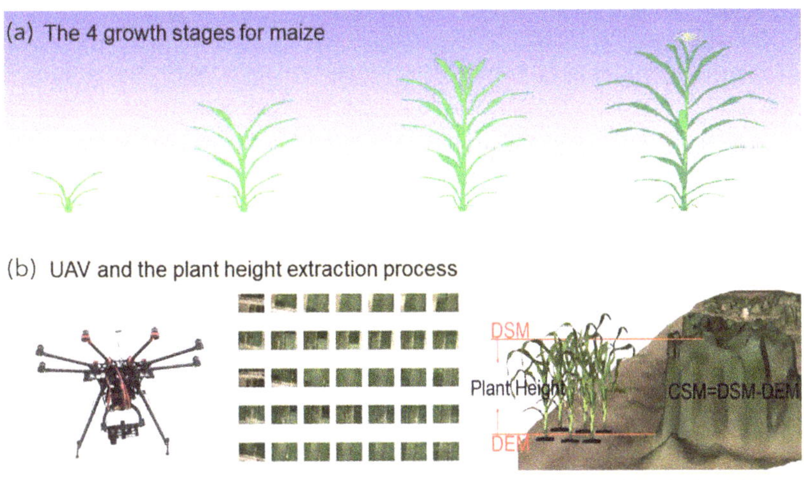

Figure 4 Field high-throughput phenotyping (HTP) for plant height. (a) Digitally designed graphs depicting maize plants at four growth stages. (b) Unmanned aerial vehicle (UAV) equipment and plant height extraction process. DSM, digital surface model, DEM, digital elevation model, CSM, crop surface model (modified from Wang et al., 2019).

genes regulating flowering time being part of the photoperiod (day length) and vernalization (exposure to cold temperatures) pathway (Cockram et al., 2007; Jung and Müller, 2009; Wiegmann et al., 2019). Modifying this trait can result in increased grain yield by adapting it to the specific environmental conditions in which the crop is grown (Jung and Müller, 2009). Generally, early flowering in cereals is beneficial as it extends the grain-filling phase (Jung and Müller, 2009). Furthermore, in growing regions such as the Mediterranean in which abiotic stress is increasingly predominant during the reproductive phase, shifting flowering time to a timepoint before these stresses generally occur (via targeted plant breeding) can contribute to increased grain yield (Hirabayashi et al., 2015; Wiegmann et al., 2019). For example, advancements in phenology have been associated with the breeding success of wheat and other small grain cereals under Mediterranean conditions (Araus et al., 2002; Loss and Siddique, 1994).

Flowering time in field trials is generally defined as the date at which spikes (ears) have emerged from 50% of the tillers (Zadoks et al., 1974). Screening is conventionally carried out by eye on a daily or bi-daily basis. This process is labor intensive and time consuming. With the high, constantly increasing number of genotypes included in breeding trials, screening is often performed by multiple people simultaneously, introducing a significant amount of subjectivity and thus error into the obtained data. Screening flowering time by use of HTP methods reduces the time required for screening as well as error, given the minimal human intervention in the screening process (Sadeghi-Tehran et al., 2017).

The applicability of HTP for screening flowering time was shown in a study by Sadeghi-Tehran et al. (2017) and a study by Wu et al. (2019). In the study by Wu et al. (2019), an array of digital single-lens reflex cameras mounted on a field vehicle was used to obtain image data of the crop canopy of wheat throughout the growing season. In addition, field plots were visually scored for percent heading, serving as a proxy for flowering time at the time of image acquisition used for labeling the images. The labeled images were then used as an input for a convolutional neural network (CNN) to predict flowering time using a training-validation-testing approach. This model allowed for high prediction accuracy (>98%) of flowering time against the values from visual scoring. Similarly, Sadeghi-Tehran et al. (2017) used RGB images of the crop canopy together with manual measures of flowering as an input for a support vector machine model that allowed for 85.45% accuracy for flowering detection.

2.3 Aboveground biomass

Aboveground biomass, comprising all plant parts above ground, has been shown to be an important trait influencing grain yield in a wide array of studies

(Boukerrou and Rasmusson, 1990; Ghassemi-Golezani and Tajbakhsh, 2012; Paul et al., 2019; Sharma, 1993; Tilly et al., 2015; White and Wilson, 2006). The rapid development of a dense plant canopy, referred to as early vigor, reduces soil evaporation, increases radiation interception and suppresses weed growth, all factors positively influencing plant growth (Casadesús and Villegas, 2014; Zhao et al., 2019b). Selecting for greater early vigor, translated to greater biomass production, can contribute to increased grain yield under certain growing conditions (Acuña et al., 2002; Zhao et al., 2019b).

Traditionally, phenotyping for biomass is based on destructive sampling, with data obtained based on oven-dried samples (Golzarian et al., 2011). As well as being laborious and time consuming, a big drawback of this method is that it does not allow for repeated measures across growth stages (Rahaman et al., 2017). By contrast, HTP methods allow for non-destructive screening for biomass at differing growth stages. A common approach applied is the use of RGB images or spectral data. For example, Fu et al. (2014) used a high spectral resolution spectrometer to obtain canopy spectral measurements in wheat. Partial least squares regression using these hyperspectral measurements as an input allowed highly accurate estimations of wheat biomass (R^2 = 0.84, root mean squared error = 0.177 kg/m^2). Similarly, Casadesús and Villegas (2014) used a hand-held conventional digital camera to obtain canopy images used to derive vegetation indices (VI) based on specific wavelengths. The derived VI showed a high correlation with the conventional destructive measurements, when pictures were taken pre-anthesis (R^2 > 0.78). The RGB color parameter u^*, rating the average color of the image between green and red, performed best. Even when using a low-cost approach, a good prediction of biomass can be achieved.

Another non-destructive approach to indirectly phenotype biomass via canopy cover is the use of a ceptometer. The ceptometer measures photosynthetically active radiation (PAR) at the top of the canopy and under the canopy, allowing the proportion of PAR intercepted by the canopy to be estimated and by this to infer canopy cover (Pandey et al., 2016).

2.4 Ear density

Ear density, defined as the number of ears per square meter (ears/m^2) is, due to its strong link with grain yield, one of the most important traits to be evaluated in cereal breeding trials (Laidig et al., 2017; Pietragalla, 2012). Moreover, while the number of grains per unit ground area is usually better correlated with grain yield than ear density, it cannot be phenotyped before harvest (Ferrante et al., 2017; Slafer et al., 2014). In addition, the importance of ear density as a trait is not only because this trait is a major agronomical yield component but also due to the physiological characteristics of the ear. It has been shown that ear photosynthesis has a considerable impact on final grain yield in C_3 cereals, especially under drought conditions (Tambussi et al., 2007). While ear density is related to tiller number per plant, increased tillering is no guarantee of increased ear density (Hiltbrunner and Liedgens, 2008; Madec et al., 2019).

Although there is no clear protocol, traditionally ear density is phenotyped by manually counting ears in a defined area of the plot and extrapolating this data to the whole plot. This phenotyping approach is time consuming and tiring, as well as being dependent on crop uniformity, severely limiting its use as a phenotyping trait when breeding for grain yield (Fernandez-Gallego et al., 2018).

The applicability of HTP for the accurate and rapid screening of ear density in large trials has been shown in several studies. Fernandez-Gallego et al. (2018) developed an automatic ear-counting algorithm capable of estimating ear density under field conditions based on zenithal RGB images of the crop canopy. Several filters applied to the RGB images separate ears from leaves and soil, allowing accurate automatic counting. Slightly adapted, the developed algorithm has been subsequently successfully applied to perform automatic ear counting based on RGB images derived from a UAV-mounted camera (Fernandez-Gallego et al., 2020). In a further study, a similar algorithm based on the temperature differential between the ears and the rest of the canopy was applied to thermal images to obtain accurate ear counts (Fernandez-Gallego et al., 2019) (Fig. 5).

Similar to this approach, Madec et al. (2019) used high-resolution RGB images of the plant canopy as an input for a CNN model. A good agreement of estimates derived from the model with *in situ* ear density was observed, demonstrating that ear density derived from high-resolution RGB imagery

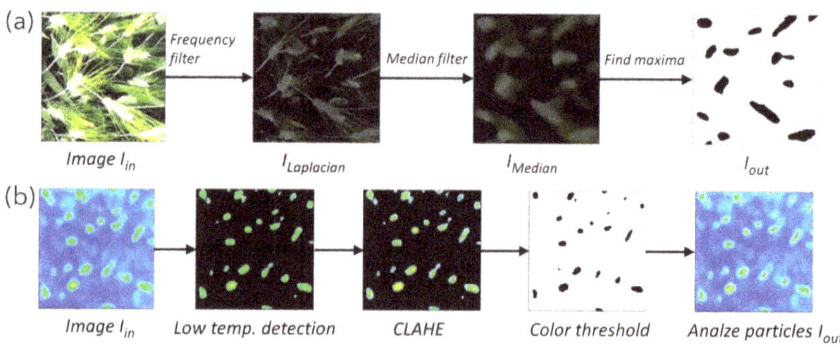

Figure 5 (a) RGB image processing pipeline based on open-source software developed in ImageJ, (1) Laplacian frequency filter, (2) Median filter, (3) Find Maxima segmentation; (b) Thermal image processing pipeline developed in ImageJ open-source software, (1) low-temperature detection, (2) contrast-limited adaptive histogram equalization (CLAHE), (3) color threshold, (4) analyze particles command (Fernandez-Gallego et al., 2018, 2019; Schneider et al., 2012).

could replace the traditional counting method (Madec et al., 2019). A slightly different approach was taken in a study by Zhou et al. (2018). High-definition RGB images and multispectral images were taken of the crop canopy and fused to improve color identification. A segmentation algorithm was then used to obtain ear counts that were shown to be highly accurate when compared with manual counting results (Zhou et al., 2018). However, an implicit limitation of all these approaches is that they only account for the ears directly visible, while obviating the ears underneath other parts of the canopy. Nevertheless, the automatic counting used was shown to correlate better against grain yield than values from manual counting, where all the ears are accounted for (Fernandez-Gallego et al., 2020). This is because the contribution to grain yield of ears from secondary and tertiary tillers is often minor. Irrespective of this, Velumani et al. (2017) showed that the problem of obviating ears underneath other parts of the canopy could be solved by using 3D point clouds from light detection and ranging (LiDAR).

3 Stomatal conductance

Stomatal conductance can be defined as the process of the exchange of carbon dioxide for photosynthesis and water vapor via transpiration through pores called stomata, located on the aerial parts of plants (Murray et al., 2019). Variation in stomatal conductance influences photosynthetic rate and water use and as such the key determinants of crop yield and productivity (Faralli et al., 2019). High stomatal conductance was shown to be positively associated with grain yield under irrigated or non-water-limited conditions (Fischer et al., 1998; Roche, 2015). On the other hand, it has been reported that a moderately high stomatal conductance, slowing water use and increasing transpiration efficiency, maintains high photosynthesis under drought conditions, positively impacting yield and yield stability (Caine et al., 2019; Dunn et al., 2019; El Sabagh et al., 2017; Li et al., 2017). Given a strong genetic basis, exploiting the natural variation in stomatal conductance in breeding programs has significant potential for increasing grain yield (Faralli et al., 2019; Fischer et al., 1998; Rebetzke et al., 2003; Wang et al., 2015).

Traditional phenotyping of stomatal conductance by the use of a porometer or an infrared gas analyzer is a challenging and laborious process (Rebetzke et al., 2012). Furthermore, it is very time consuming as measurements of multiple leaves per genotype are required to clearly distinguish genotypic differences from variation caused by changing environmental factors during data acquisition (Rebetzke et al., 2001). As such, the traditional approach is not suited to large breeding trials.

Thermal imagery of the crop canopy has been shown to be a sensitive method for studying stomatal conductance, while being applicable to HTP platforms (Jones, 2004; Jones et al., 2009; Ludovisi et al., 2017). It is based on the fact that plant surfaces are cooled by transpiration, allowing for a link between canopy temperature and transpiration rates and thereby stomatal conductance (Deery et al., 2019). Fusion of thermal images with RGB images was shown to further increase the informative value of thermal images as RGB images can be used to remove non-canopy areas from the thermal image, thus increasing its association strength with stomatal conductance (Zhang et al., 2019).

Studies applying HTP based on thermal imagery demonstrate that this can be applied to estimate crop yield. Gracia Romero et al. (2019) used a thermal camera mounted on a UAV for yield prediction in wheat. A negative correlation of canopy temperature with grain yield was observed with the best measurements obtained at heading in water-limited trials (R^2 = 0.64), allowing for the identification of genotypes with higher resistance to drought and heat. The study revealed that aerial measurements of canopy temperature outperformed ground measurements. Aerial measurements allowed screening of all plots simultaneously, circumventing bias introduced by temperature fluctuation between measurements. Similarly, Crain et al. (2017) used a thermal camera mounted on a ground-based HTP platform to evaluate its applicability for yield prediction. The study found a robust negative correlation between canopy temperature and yield ($R^2 > 0.64$), although not as strong as those observed between plant height and days to heading. Furthermore, the authors compared the genetic gain achievable at the same selection intensity with thermal image-based selection with that directly selecting for yield. Indirect selection for yield based on canopy temperature at the same selection intensity did not result in a strong gain. However, as highlighted by the authors, the potential to screen a much larger number of plots using the HTP method would allow for a higher selection intensity. Considering this potential increase in selection intensity, indirect selection resulted in significantly higher gains than selection directly on yield (Crain et al., 2017).

4 Functional stay green

Functional SG is a trait that enables plants to maintain their green leaves and photosynthesis capacity for a longer time after anthesis (Kamal et al., 2019). As a consequence, the grain-filling phase is extended, generally allowing for a higher yield than non-SG plants, especially under drought and heat stress conditions (Thomas and Ougham, 2014). Selecting for functional SG in breeding trials can be used to increase stress tolerance, and therefore yield, in non-optimal growth conditions, which is especially important considering future climate change scenarios (Asseng et al., 2011; Kamal et al., 2019; Trachsel et al., 2016).

Screening for the onset of senescence in the flag leaf is widely used to differentiate between SG and early senescence phenotypes (Chapman et al., 2020). Flag leaf senescence is commonly assessed by the eye using a percentage scale or more objectively by leaf chlorophyll content with a hand-held chlorophyll meter (Chapman et al., 2020). More recently, spectral indices linked to chlorophyll content, such as the normalized difference vegetative index (NDVI) or the senescence reflectance index, are applied to determine senescence and through this SG (Anderegg et al., 2020; Lopes and Reynolds, 2012). HTP comes into play here as spectral or RGB cameras mounted on HTP platforms can be used to derive these and other indices linked to SG (Rebetzke et al., 2016).

Deery et al. (2014) used a spectral camera and a LiDAR sensor mounted on a ground-based HTP platform to reliably quantify differences in NDVI between wheat genotypes. The addition of a LiDAR sensor allowed them to create 3D canopy images, making it possible to specifically select flag leaves. Condorelli et al. (2018) used spectral cameras mounted on a UAV and a tractor-based platform to derive NDVI measurements of 248 elite durum wheat lines grown in the field. Results showed high repeatability of NDVI measurements with aerial-derived measurements outperforming ground-based measurements. Obtained NDVI data combined with genomic data allowed for the detection of several NDVI QTL. Some of these overlapped with grain yield QTL. Therefore, the study shows the high potential of HTP of NDVI for identifying suitable candidates for MAS aiming to improve yield under stress conditions.

The great potential of NDVI for indirect selection for grain yield is further supported by a study by Krause et al. (2020). Results showed a significantly higher heritability of NDVI values obtained using a UAV than that observed for yield directly in unreplicated early-stage breeding trials. This provided higher selection accuracy and allowed for higher selection intensity at the early stages of a breeding program, thereby promoting increases in the rate of genetic gain.

5 Case study

One of the main focuses of our research group at the University of Barcelona is the use of remote sensing methods for grain yield prediction in cereals. While applying cutting-edge remote sensing techniques, our research focuses on developing low-cost HTP approaches requiring only RGB cameras. For that purpose, adequate protocols are designed to improve the samples as well as the development of tailor-made software for data extraction. In this regard, we performed a study to evaluate differences between advanced sensors and simple digital cameras, and data obtained from UAV and ground as regards precisely assessing grain yield in wheat. The case study below refers to the work published by Gracia Romero et al. (2019).

We focused on durum wheat (*Triticum turgidum* L. ssp *durum* (Desf) Husn.), widely cultivated in the Mediterranean basin. In order to obtain robust results, replicated trials were conducted at three diverse locations in Spain. In addition, at each location, genotypes were assessed under three diverse growing conditions. Genotypes were grown under irrigation to simulate optimal growing conditions and without supplementary irrigation to implement drought stress. In addition, genotypes were sown at a later time point than usual, leading to higher-than-optimal temperatures throughout the entire crop development period, in order to induce heat stress. This approach allowed for a total of nine different environments in which the different HTP methods could be evaluated. In total, 23 semi-dwarf varieties of durum wheat were included in the trial marketed in Spain over the last 4 decades. While the variation in grain yield between the genotypes was relatively low it gives a good representation of actual variation observed in durum wheat.

In the study, data were obtained by seven different devices, ranging from simple RGB cameras to spectral sensors. To obtain the aerial measurements, an eight-rotor Mikrokopter Oktokopter 6S12 XL (HiSystems GmbH, Moomerland, Germany) was used. The UAV platform was used to obtain RGB, multispectral and thermal images of the trials flying at a height of 50 m (Fig. 6). In post-processing, the multiple images taken during the flight were combined to a single orthomosaic using Agisoft PhotoScan Professional software (Agisoft LLC, St. Petersburg, Russia, www.agisoft.com).

For ground measurements, one image per plot was taken with an RGB camera, with the camera placed 80 cm above the plant canopy in a zenithal plane (Fig. 7).

In addition, the NDVI value of each plot was obtained using the GreenSeeker hand-held crop sensor (Trimble, Sunnyvale, CA, USA), by passing the sensor

Figure 6 From left to right: RGB, thermal and spectral images obtained from a UAV.

Figure 7 (a) Ground-based measurement using the RGB camera. (b) Image obtained using the ground-based method.

over the middle of each plot at a constant height of 0.5 m above the canopy. Canopy temperature data were obtained using a PhotoTempTM MXSTM TD infrared thermometer (Raytek, Santa Cruz, USA), and air temperature across plots was obtained using a thermos-hygrometer (Testo 177-H1 Logger, Lenzkirch, Germany). Furthermore, leaf pigment content was measured using a leaf-clip portable sensor Dualex Force-A (Dualex, Orsay, France).

To extract the single plots from the aerial images and to estimate the various VI, we used a tool developed within our group that performs these steps in a rapid and user-friendly way. The program, called Mosaic tool (Shawn C. Kefauver, https://integrativecropecophysiology.com/software-development/mosaictool/, University of Barcelona, Barcelona, Spain), is a plugin for the open-source image analysis platform FIJI (Fiji is Just ImageJ; http://fiji.sc/Fiji) and is freely available.

In order to consider the influence of the growing stage, measurements were obtained at different time points over the growing season, ranging from anthesis to senescence - phenological stages considered critical for yield formation. The performance of the remote sensing indexes varied considerably across the growing conditions and at different growth stages.

The identified correlations between the remote sensing data and grain yield were in agreement with those described in earlier publications (Yousfi et al., 2016; Crain et al., 2017). For the irrigated trials, the highest correlation with grain yield for the data was found for the RGB image-based green area (GA) index ($R^2 = 0.684$) and the multispectral-based NDVI index and its reformulations ($R^2 < 0.64$) at grain filling obtained via ground measurements. At grain filling, both ground and aerial RGB indices showed correlations very similar to those obtained using anthesis data. In addition, in the irrigated trials, GA and NDVI obtained via ground measurements at anthesis showed a significantly higher heritability than grain yield. This highlights the potential of the two indices for indirect selection at early growth stages in breeding trials. As described in earlier studies, thermal measurements showed a negative correlation with grain yield. It is noteworthy that the correlations between the measured indices and grain yield were higher under water-limited conditions. In the late planting trial, the indices obtained based on ground measurements significantly outperformed those calculated from the aerial HTPP images. However, overall ground and aerial measurements performed similarly in terms of assessing grain yield, with the exception being canopy temperature. As such, the main benefit of aerial platforms is that they cover larger experimental areas in a shorter period of time, thus reducing errors caused by fluctuating environmental factors during measurement. This became most evident in the case of the thermal data, where UAV-derived data significantly outperformed ground-based data as ground-based measurements are much more time consuming and environmental conditions are not constant. Another benefit of aerial measurements is that obtained data are based on the whole plot, reducing bias caused by the heterogeneity of the canopy, while ground-based images only cover part of the full plot. However, this potential cause of bias regarding ground measurements could be removed in future studies when taking several measurements per plot. Despite these benefits of aerial-derived indices, a major point to consider is the higher cost of these platforms. Therefore, ground-based HTP approaches might be a good alternative, being less costly and easier to use while showing comparable performance. Based on the study, we conclude that focusing on VI derived using a simple RGB camera has significant potential for predicting grain yield because despite being a low-cost approach, the indices performed as well as, and sometimes better than, indices derived from more expensive devices (such as multispectral and thermal indexes).

6 Conclusion and future trends

In order to increase grain yield, phenotyping focuses on assessing correlated traits such as plant height, flowering time, aerial biomass and ear density and physiological traits such as stomatal conductance and functional SG. Conventional phenotyping techniques applied to evaluate these traits are generally time consuming, labor intensive and prone to error. Screening of large breeding populations is not feasible with these conventional methods. The implementation of high-throughput phenotyping techniques with high accuracy and repeatability is required to meet the demands of modern breeding programs aimed at improving grain yield. As showcased, HTP for indirect selection for grain yield has great potential to increase the rate of genetic gain achievable by breeders. This is especially true when combining the HTP-derived data with modern genome-based breeding approaches. Rapid progress in the field of HTP goes hand in hand with the accumulation of a wealth of data. With decreasing costs of HTP platforms, the next challenge will be to extract information from this data. New approaches in the field of data science and greater use of machine learning for future extraction will be of great help to discover as yet unknown predictors of grain yield. While these predictors may be determined empirically, as the direct result of data analysis process, the integration of such information into physiological/agronomical models may further fine-tune the use of these secondary selection traits. The development of easy-to-use analysis software will increase the use of HTP methods in the field of plant breeding. The focus should also be on the development of low-cost approaches to enable the use of HTP techniques in small-scale breeding programs.

7 Where to look for further information

- Group website, University of Barcelona: https://integrativecropecophysiology.com/.
- International plant phenotyping network: https://www.plant-phenotyping.org/.
- Optical synergies for spatiotemporal sensing of scalable ecophysiological traits Action: https://www.senseco.eu/about-senseco/.
- Excellence in breeding platform: https://excellenceinbreeding.org/.
- EMPHASIS working group: https://emphasis.plant phenotyping.eu/.
- Index Database, a database for remote sensing indices: https://www.indexdatabase.de/.

8 References

Abdel-Ghani, A. H., Hu, S., Chen, Y., Brenner, E. A., Kumar, B., Blanco, M. and Lübberstedt, T. (2016). Genetic architecture of plant height in maize phenotype-selected introgression families. *Plant Breeding* 135(4), 429-438. https://doi.org/10.1111/pbr.12387.

Acuña, T. L., Condon, A. G., Rebetzke, G. J. and Richards, R. A. (2002). Field evaluation of early vigour for genetic improvement of grain yield in wheat. *Australian Journal of Agricultural Research* 53(10). https://doi.org/10.1071/AR02007.

Alexandratos, N. and Bruinsma, J. (2012). World Agriculture Towards 2030/2050: The 2012 Revision (2521-1838). ESA Working Papers 12-03, Issue. Available at: http://ageconsearch.umn.edu/record/288998/files/a-ap106e.pdf.

Anderegg, J., Yu, K., Aasen, H., Walter, A., Liebisch, F. and Hund, A. (2020). Spectral vegetation indices to track senescence dynamics in diverse wheat germplasm. *Frontiers in Plant Science* 10, 1749.

Araus, J. L. and Cairns, J. E. (2014). Field high-throughput phenotyping: the new crop breeding frontier. *Trends in Plant Science* 19(1), 52-61. https://doi.org/10.1016/j.tplants.2013.09.008.

Araus, J. L., Kefauver, S. C., Zaman-Allah, M., Olsen, M. S. and Cairns, J. E. (2018a). Phenotyping: new crop breeding frontier. In: Meyers, R. A. (Ed.), *Encyclopedia of Sustainability Science and Technology* (pp. 1-11). Springer, New York. https://doi.org/10.1007/978-1-4939-2493-6_1036-1.

Araus, J. L., Kefauver, S. C., Zaman-Allah, M., Olsen, M. S. and Cairns, J. E. (2018b). Translating high-throughput phenotyping into genetic gain. *Trends in Plant Science* 23(5), 451-466. https://doi.org/10.1016/j.tplants.2018.02.001.

Araus, J. L., Buchaillot, M. L. and Kefauver, S. C. (2021). High throughput field phenotyping. In: Reynolds, B. H. (Ed.), *Wheat Breeding Text Book: Underpinning Food Security through Climate Resilience*. Springer.

Araus, J. L., Slafer, G. A., Reynolds, M. P. and Royo, C. (2002). Plant breeding and drought in C3 cereals: what should we breed for? *Annals of Botany* 89(7), 925-940. https://doi.org/10.1093/aob/mcf049.

Asseng, S., Foster, I. A. N. and Turner, N. C. (2011). The impact of temperature variability on wheat yields. *Global Change Biology* 17(2), 997-1012. https://doi.org/10.1111/j.1365-2486.2010.02262.x.

Baenziger, P. S., Dweikat, I., Gill, K., Eskridge, K., Berke, T., Shah, M., Campbell, B. T., Ali, M. L., Mengistu, N., Mahmood, A., Auvuchanon, A., Yen, Y., Rustgi, S., Moreno-Sevilla, B., Mujeeb-Kazi, A. and Morris, M. R. (2011). Understanding grain yield: it is a journey, not a destination. *Czech Journal of Genetics and Plant Breeding* 47(Special Issue), S77-S84.

Baye, A., Berihun, B., Bantayehu, M. and Derebe, B. (2020). Genotypic and phenotypic correlation and path coefficient analysis for yield and yield-related traits in advanced bread wheat (*Triticum aestivum* L.) lines. *Cogent Food and Agriculture* 6(1), 1752603. https://doi.org/10.1080/23311932.2020.1752603.

Bhargava, A. and Srivastava, S. (2019). Plant breeding. In: Bhargava, A. and Srivastava, S. (Eds), *Participatory Plant Breeding: Concept and Applications* (pp. 29-68). Springer, Singapore. https://doi.org/10.1007/978-981-13-7119-6_2.

Bhat, J. A., Ali, S., Salgotra, R. K., Mir, Z. A., Dutta, S., Jadon, V., Tyagi, A., Mushtaq, M., Jain, N., Singh, P. K., Singh, G. P. and Prabhu, K. V. (2016). Genomic selection in the era of next generation sequencing for complex traits in plant breeding. *Frontiers in Genetics* 7, 221.

Boukerrou, L. and Rasmusson, D. D. (1990). Breeding for high biomass yield in spring barley. *Crop Science* 30(1) , 31-35. https://doi.org/10.2135/cropsci1990.0011183X003000010007x.

Caine, R. S., Yin, X., Sloan, J., Harrison, E. L., Mohammed, U., Fulton, T., Biswal, A. K., Dionora, J., Chater, C. C., Coe, R. A., Bandyopadhyay, A., Murchie, E. H., Swarup, R., Quick, W. P. and Gray, J. E. (2019). Rice with reduced stomatal density conserves water and has improved drought tolerance under future climate conditions. *New Phytologist* 221(1), 371-384. https://doi.org/10.1111/nph.15344.

Casadesús, J. and Villegas, D. (2014). Conventional digital cameras as a tool for assessing leaf area index and biomass for cereal breeding. *Journal of Integrative Plant Biology* 56(1), 7-14. https://doi.org/10.1111/jipb.12117.

Chairi, F., Vergara-Diaz, O., Vatter, T., Aparicio, N., Nieto-Taladriz, M. T., Kefauver, S. C., Bort, J., Serret, M. D. and Araus, J. L. (2018). Post-green revolution genetic advance in durum wheat: the case of Spain. *Field Crops Research* 228, 158–169. https://doi.org/10.1016/j.fcr.2018.09.003.

Chapman, E. A., Orford, S., Lage, J. and Griffiths, S. (2020). Capturing and selecting senescence variation in wheat. *Frontiers in Plant Science* 12, 638738. https://doi.org/10.1101/2020.12.06.413641.

Chen, X., Min, D., Yasir, T. A. and Hu, Y.-G. (2012). Evaluation of 14 morphological, yield-related and physiological traits as indicators of drought tolerance in Chinese winter bread wheat revealed by analysis of the membership function value of drought tolerance (MFVD). *Field Crops Research* 137, 195–201. https://doi.org/10.1016/j.fcr.2012.09.008.

Cobb, J. N., DeClerck, G., Greenberg, A., Clark, R. and McCouch, S. (2013). Next-generation phenotyping: requirements and strategies for enhancing our understanding of genotype-phenotype relationships and its relevance to crop improvement. *TAG. Theoretical and Applied Genetics. Theoretische und Angewandte Genetik* 126(4), 867–887. https://doi.org/10.1007/s00122-013-2066-0.

Cockram, J., Jones, H., Leigh, F. J., O'Sullivan, D., Powell, W., Laurie, D. A. and Greenland, A. J. (2007). Control of flowering time in temperate cereals: genes, domestication, and sustainable productivity. *Journal of Experimental Botany* 58(6), 1231–1244. https://doi.org/10.1093/jxb/erm042.

Cogato, A., Meggio, F., De Antoni Migliorati, M. and Marinello, F. (2019). Extreme weather events in agriculture: a systematic review. *Sustainability* 11(9). https://doi.org/10.3390/su11092547.

Condorelli, G. E., Maccaferri, M., Newcomb, M., Andrade-Sanchez, P., White, J. W., French, A. N., Sciara, G., Ward, R. and Tuberosa, R. (2018). Comparative aerial and ground based high throughput phenotyping for the genetic dissection of NDVI as a proxy for drought adaptive traits in durum wheat. *Frontiers in Plant Science* 9, 893.

Crain, J., Reynolds, M. and Poland, J. (2017). Utilizing high-throughput phenotypic data for improved phenotypic selection of stress-adaptive traits in wheat. *Crop Science* 57(2), 648–659. https://doi.org/10.2135/cropsci2016.02.0135.

Davey, J. W., Hohenlohe, P. A., Etter, P. D., Boone, J. Q., Catchen, J. M. and Blaxter, M. L. (2011). Genome-wide genetic marker discovery and genotyping using next-generation sequencing. *Nature Reviews. Genetics* 12(7), 499–510. https://doi.org/10.1038/nrg3012.

Deery, D., Jimenez-Berni, J., Jones, H., Sirault, X. and Furbank, R. (2014). Proximal remote sensing buggies and potential applications for field-based phenotyping. *Agronomy* 4(3), 349–379. https://doi.org/10.3390/agronomy4030349.

Deery, D. M., Rebetzke, G. J., Jimenez-Berni, J. A., Bovill, W. D., James, R. A., Condon, A. G., Furbank, R. T., Chapman, S. C. and Fischer, R. A. (2019). Evaluation of the phenotypic repeatability of canopy temperature in wheat using continuous-terrestrial and airborne measurements. *Frontiers in Plant Science* 10, 875.

Dunn, J., Hunt, L., Afsharinafar, M., Meselmani, M. A., Mitchell, A., Howells, R., Wallington, E., Fleming, A. J. and Gray, J. E. (2019). Reduced stomatal density in bread wheat leads to increased water-use efficiency. *Journal of Experimental Botany* 70(18), 4737–4748. https://doi.org/10.1093/jxb/erz248.

El Sabagh, A. E., Barutçular, C. and Islam, M. S. (2017). Relationships between stomatal conductance and yield under deficit irrigation in maize (*Zea mays* L.). *Journal of*

Experimental Biology and Agricultural Sciences 5(1), 014-021. https://doi.org/10.18006/2017.5(1).014.021.

Falconer, D. S. and Mackay, T. F. C. (1996). *Introduction to Quantitative Genetics*. Longmans Green. ISBN: 9780582243026.

FAO (2020). Available at: http://www.fao.org/faostat/en/#data/QC.

Faralli, M., Matthews, J. and Lawson, T. (2019). Exploiting natural variation and genetic manipulation of stomatal conductance for crop improvement. *Current Opinion in Plant Biology* 49, 1-7. https://doi.org/10.1016/j.pbi.2019.01.003.

Fernandez-Gallego, J., Buchaillot, M., Aparicio Gutiérrez, N., Nieto-Taladriz, M., Araus, J. and Kefauver, S. (2019). Automatic wheat ear counting using thermal imagery. *Remote Sensing* 11(7), 751. https://doi.org/10.3390/rs11070751.

Fernandez-Gallego, J. A., Kefauver, S. C., Gutiérrez, N. A., Nieto-Taladriz, M. T. and Araus, J. L. (2018). Wheat ear counting in-field conditions: high throughput and low-cost approach using RGB images. *Plant Methods* 14(1), 22. https://doi.org/10.1186/s13007-018-0289-4.

Fernandez-Gallego, J. A., Lootens, P., Borra-Serrano, I., Derycke, V., Haesaert, G., Roldán-Ruiz, I., Araus, J. L. and Kefauver, S. C. (2020). Automatic wheat ear counting using machine learning based on RGB UAV imagery. *The Plant Journal: For Cell and Molecular Biology* 103(4), 1603-1613. https://doi.org/10.1111/tpj.14799.

Ferrante, A., Cartelle, J., Savin, R. and Slafer, G. A. (2017). Yield determination, interplay between major components and yield stability in a traditional and a contemporary wheat across a wide range of environments. *Field Crops Research* 203, 114-127. https://doi.org/10.1016/j.fcr.2016.12.028.

Fischer, R. A., Rees, D., Sayre, K. D., Lu, Z.-M., Condon, A. G. and Saavedra, A. L. (1998). Wheat yield progress associated with higher stomatal conductance and photosynthetic rate, and cooler canopies. *Crop Science* 38(6), 1467-1475 https://doi.org/10.2135/cropsci1998.0011183X003800060011x.

Fu, Y., Yang, G., Wang, J., Song, X. and Feng, H. (2014). Winter wheat biomass estimation based on spectral indices, band depth analysis and partial least squares regression using hyperspectral measurements. *Computers and Electronics in Agriculture* 100, 51-59. https://doi.org/10.1016/j.compag.2013.10.010.

Fu, Y. B. (2015). Understanding crop genetic diversity under modern plant breeding. *TAG. Theoretical and Applied Genetics. Theoretische und Angewandte Genetik* 128(11), 2131-2142. https://doi.org/10.1007/s00122-015-2585-y.

Furbank, R. T. and Tester, M. (2011). Phenomics - technologies to relieve the phenotyping bottleneck. *Trends in Plant Science* 16(12), 635-644. https://doi.org/10.1016/j.tplants.2011.09.005.

Gautam, A., Sai Prasad, S. V., Jajoo, A. and Ambati, D. (2015). Canopy temperature as a selection parameter for grain yield and its components in durum wheat under terminal heat stress in late sown conditions. *Agricultural Research* 4(3), 238-244. https://doi.org/10.1007/s40003-015-0174-6.

Ghassemi-Golezani, K. and Tajbakhsh, Z. (2012). Relationship of plant biomass and grain filling with grain yield of maize cultivars. *International Journal of Agriculture and Crop Sciences*. 4(20), 1536-1539.

Golzarian, M. R., Frick, R. A., Rajendran, K., Berger, B., Roy, S., Tester, M. and Lun, D. S. (2011). Accurate inference of shoot biomass from high-throughput images of cereal plants. *Plant Methods* 7(1), 2. https://doi.org/10.1186/1746-4811-7-2.

Gracia Romero, A., Kefauver, S. C., Fernandez-Gallego, J. A., Vergara Diaz, O., Nieto-Taladriz, M. T. and Araus, J. L. (2019). UAV and ground image-based phenotyping: a proof of concept with durum wheat. *Remote Sensing* 11(10), 1244. https://doi.org/10.3390/rs11101244.

Hiltbrunner, J. and Liedgens, M. (2008). Performance of winter wheat varieties in white clover living mulch. *Biological Agriculture and Horticulture* 26(1), 85-101. https://doi.org/10.1080/01448765.2008.9755071.

Hirabayashi, H., Sasaki, K., Kambe, T., Gannaban, R. B., Miras, M. A., Mendioro, M. S., Simon, E. V., Lumanglas, P. D., Fujita, D., Takemoto-Kuno, Y., Takeuchi, Y., Kaji, R., Kondo, M., Kobayashi, N., Ogawa, T., Ando, I., Jagadish, K. S. and Ishimaru, T. (2015). qEMF3, a novel QTL for the early-morning flowering trait from wild rice, Oryza officinalis, to mitigate heat stress damage at flowering in rice, O. sativa. *Journal of Experimental Botany* 66(5), 1227-1236. https://doi.org/doi:10.1093/jxb/eru474.

Holman, F. H., Riche, A. B., Michalski, A., Castle, M., Wooster, M. J. and Hawkesford, M. J. (2016). High throughput field phenotyping of wheat plant height and growth rate in field plot trials using UAV based remote sensing. *Remote Sensing* 8(12). https://doi.org/10.3390/rs8121031.

Hossain, A., Krupnik, T. J., Timsina, J., Mahboob, M. G., Chaki, A. K., Farooq, M., Bhatt, R., Fahad, S. and Hasanuzzaman, M. (2020). Agricultural land degradation: processes and problems undermining future food security. In: Fahad, S., Hasanuzzaman, M., Alam, M., Ullah, H., Saeed, M., Ali Khan, I. and Adnan, M. (Eds), *Environment, Climate, Plant and Vegetation Growth* (pp. 17-61). Springer International Publishing. https://doi.org/10.1007/978-3-030-49732-3_2.

Houle, D., Govindaraju, D. R. and Omholt, S. (2010). Phenomics: the next challenge. *Nature Reviews. Genetics* 11(12), 855-866. https://doi.org/10.1038/nrg2897.

Hunter, M. C., Smith, R. G., Schipanski, M. E., Atwood, L. W. and Mortensen, D. A. (2017). Agriculture in 2050: recalibrating targets for sustainable intensification. *BioScience* 67(4), 386-391. https://doi.org/10.1093/biosci/bix010.

Jannink, J. L., Lorenz, A. J. and Iwata, H. (2010). Genomic selection in plant breeding: from theory to practice. *Briefings in Functional Genomics* 9(2), 166-177. https://doi.org/10.1093/bfgp/elq001.

Jones, H. G. (2004). Irrigation scheduling: advantages and pitfalls of plant-based methods. *Journal of Experimental Botany* 55(407), 2427-2436. https://doi.org/10.1093/jxb/erh213.

Jones, H. G., Serraj, R., Loveys, B. R., Xiong, L., Wheaton, A. and Price, A. H. (2009). Thermal infrared imaging of crop canopies for the remote diagnosis and quantification of plant responses to water stress in the field. *Functional Plant Biology* 36(11), 978-989.

Jung, C. and Müller, A. E. (2009). Flowering time control and applications in plant breeding. *Trends in Plant Science* 14(10), 563-573. https://doi.org/10.1016/j.tplants.2009.07.005.

Kamal, N. M., Alnor Gorafi, Y. S., Abdelrahman, M., Abdellatef, E. and Tsujimoto, H. (2019). Stay-green trait: a prospective approach for yield potential, and drought and heat stress adaptation in globally important cereals. *International Journal of Molecular Sciences* 20(23), 5837. https://doi.org/10.3390/ijms20235837.

Krause, M. R., Mondal, S., Crossa, J., Singh, R. P., Pinto, F., Haghighattalab, A., Shrestha, S., Rutkoski, J., Gore, M. A., Sorrells, M. E. and Poland, J. (2020). Aerial high-throughput phenotyping enabling indirect selection for grain yield at the early-generation seed-limited stages in breeding programs. *bioRxiv*. https://doi.org/10.1101/2020.04.21.054163.

Laidig, F., Piepho, H. P., Rentel, D., Drobek, T., Meyer, U. and Huesken, A. (2017). Breeding progress, variation, and correlation of grain and quality traits in winter rye hybrid and population varieties and national on-farm progress in Germany over 26 years. *TAG. Theoretical and Applied Genetics. Theoretische und Angewandte Genetik* 130(5), 981–998. https://doi.org/10.1007/s00122-017-2865-9.

Lane, H. M., Murray, S. C., Montesinos-López, O. A., Montesinos-López, A., Crossa, J., Rooney, D. K., Barrero-Farfan, I. D., De La Fuente, G. N. and Morgan, C. L. S. (2020). Phenomic selection and prediction of maize grain yield from near-infrared reflectance spectroscopy of kernels. *The Plant Phenome Journal* 3(1), e20002. https://doi.org/10.1002/ppj2.20002.

Li, Y., Li, H., Li, Y. and Zhang, S. (2017). Improving water-use efficiency by decreasing stomatal conductance and transpiration rate to maintain higher ear photosynthetic rate in drought-resistant wheat. *The Crop Journal* 5(3), 231–239. https://doi.org/10.1016/j.cj.2017.01.001.

Liang, Z., Pandey, P., Stoerger, V., Xu, Y., Qiu, Y., Ge, Y. and Schnable, J. C. (2018). Conventional and hyperspectral time-series imaging of maize lines widely used in field trials. *GigaScience* 7(2), 1–11. https://doi.org/10.1093/gigascience/gix117.

Lopes, M. S. and Reynolds, M. P. (2012). Stay-green in spring wheat can be determined by spectral reflectance measurements (normalized difference vegetation index) independently from phenology. *Journal of Experimental Botany* 63(10), 3789–3798. https://doi.org/10.1093/jxb/ers071.

Loss, S. P. and Siddique, K. H. M. (1994). Morphological and physiological traits associated with wheat yield increases in Mediterranean environments. In: Sparks, D. L. (Ed.), *Advances in Agronomy* (vol. 52). Academic Press. https://doi.org/10.1016/S0065-2113(08)60625-2.

Ludovisi, R., Tauro, F., Salvati, R., Khoury, S., Mugnozza Scarascia, G. and Harfouche, A. (2017). UAV-based thermal imaging for high-throughput field phenotyping of black poplar response to drought. *Frontiers in Plant Science* 8, 1681.

Lush, J. L. (1937). *Animal Breeding Plans*. Iowa State College Press.

Madec, S., Jin, X., Lu, H., De Solan, B., Liu, S., Duyme, F., Heritier, E. and Baret, F. (2019). Ear density estimation from high resolution RGB imagery using deep learning technique. *Agricultural and Forest Meteorology* 264, 225–234. https://doi.org/10.1016/j.agrformet.2018.10.013.

Michel, S., Löschenberger, F., Ametz, C., Pachler, B., Sparry, E. and Bürstmayr, H. (2019). Combining grain yield, protein content and protein quality by multi-trait genomic selection in bread wheat. *TAG. Theoretical and Applied Genetics. Theoretische und Angewandte Genetik* 132(10), 2767–2780. https://doi.org/10.1007/s00122-019-03386-1.

Murray, M., Soh, W. K., Yiotis, C., Spicer, R. A., Lawson, T. and McElwain, J. C. (2019). Consistent relationship between field-measured stomatal conductance and theoretical maximum stomatal conductance in C3 woody angiosperms in four major biomes. *International Journal of Plant Sciences* 181(1), 142–154. https://doi.org/10.1086/706260.

Pandey, B. R., Burton, W. A. and Salisbury, P. A. (2016). Non-destructive measurement of canopy cover is an alternative to biomass sampling at anthesis to predict yield of canola-quality Brassica juncea. *Australian Journal of Crop Science* 10(4), 482-489. https://doi.org/10.21475/ajcs.2016.10.04.p7103x.

Paul, K., Pauk, J., Kondic-Spika, A., Grausgruber, H., Allahverdiyev, T., Sass, L. and Vass, I. (2019). Co-occurrence of mild salinity and drought synergistically enhances biomass and grain retardation in wheat. *Frontiers in Plant Science* 10, 501.

Peiffer, J. A., Romay, M. C., Gore, M. A., Flint-Garcia, S. A., Zhang, Z., Millard, M. J., Gardner, C. A., McMullen, M. D., Holland, J. B., Bradbury, P. J. and Buckler, E. S. (2014). The genetic architecture of maize height. *Genetics* 196(4), 1337-1356. https://doi.org/10.1534/genetics.113.159152.

Peng, J., Richards, D. E., Hartley, N. M., Murphy, G. P., Devos, K. M., Flintham, J. E., Beales, J., Fish, L. J., Worland, A. J., Pelica, F., Sudhakar, D., Christou, P., Snape, J. W., Gale, M. D. and Harberd, N. P. (1999). 'Green revolution' genes encode mutant gibberellin response modulators. *Nature* 400(6741), 256-261. https://doi.org/10.1038/22307.

Pietragalla, J. (2012). *Physiological Breeding II: A Field Guide to Wheat Phenotyping*. CIMMYT, Mexico.

Pingali, P. L. (2012). Green revolution: impacts, limits, and the path ahead. *Proceedings of the National Academy of Sciences of the United States of America* 109(31), 12302-12308. https://doi.org/10.1073/pnas.0912953109.

Poland, J., Endelman, J., Dawson, J., Rutkoski, J., Wu, S., Manes, Y., Dreisigacker, S., Crossa, J., Sánchez-Villeda, H., Sorrells, M. and Jannink, J.-L. (2012). Genomic selection in wheat breeding using genotyping-by-sequencing. *The Plant Genome* 5(3). https://doi.org/10.3835/plantgenome2012.06.0006.

Powell, J. P. and Reinhard, S. (2016). Measuring the effects of extreme weather events on yields. *Weather and Climate Extremes* 12, 69-79. https://doi.org/10.1016/j.wace.2016.02.003.

Quarrie, S. A., Pekic Quarrie, S., Radosevic, R., Rancic, D., Kaminska, A., Barnes, J. D., Leverington, M., Ceoloni, C. and Dodig, D. (2006). Dissecting a wheat QTL for yield present in a range of environments: from the QTL to candidate genes. *Journal of Experimental Botany* 57(11), 2627-2637. https://doi.org/10.1093/jxb/erl026.

Rahaman, M. M., Ahsan, M. A., Gillani, Z. and Chen, M. (2017). Digital biomass accumulation using high-throughput plant phenotype data analysis. *Journal of Integrative Bioinformatics* 14(3), 28862986.

Rebetzke, G. J., Condon, A. G., Richards, R. A. and Farquhar, G. D. (2003). Gene action for leaf conductance in three wheat crosses. *Australian Journal of Agricultural Research* 54(4), 381-387. https://doi.org/10.1071/AR02151.

Rebetzke, G. J., Condon, A. G., Richards, R. A. and Read, J. J. (2001). Phenotypic variation and sampling for leaf conductance in wheat (*Triticum aestivum* L.) breeding populations. *Euphytica* 121(3), 335-341. https://doi.org/10.1023/A:1012035720423.

Rebetzke, G. J., Jimenez-Berni, J. A., Bovill, W. D., Deery, D. M. and James, R. A. (2016). High-throughput phenotyping technologies allow accurate selection of stay-green. *Journal of Experimental Botany* 67(17), 4919-4924. https://doi.org/10.1093/jxb/erw301.

Rebetzke, G. J., Rattey, A. R., Farquhar, G. D., Richards, R. A. and Condon, A. T. G. (2012). Genomic regions for canopy temperature and their genetic association with stomatal conductance and grain yield in wheat. *Functional Plant Biology* 40(1), 14-33.

Reeves, T. G. and Cassaday, K. (2002). History and past achievements of plant breeding. *Australian Journal of Agricultural Research* 53(8), 851–863. https://doi.org/10.1071/AR02038.

Roche, D. (2015). Stomatal conductance is essential for higher yield potential of C3 crops. *Critical Reviews in Plant Sciences* 34(4), 429–453. https://doi.org/10.1080/07352689.2015.1023677.

Roy, S. C. and Shil, P. (2020). Assessment of genetic heritability in rice breeding lines based on morphological traits and caryopsis ultrastructure. *Scientific Reports* 10(1), 7830. https://doi.org/10.1038/s41598-020-63976-8.

Rummukainen, M. (2012). Changes in climate and weather extremes in the 21st century. *Wiley Interdisciplinary Reviews* 3(2), 115–129. https://doi.org/10.1002/wcc.160.

Sadeghi-Tehran, P., Sabermanesh, K., Virlet, N. and Hawkesford, M. J. (2017). Automated method to determine two critical growth stages of wheat: heading and flowering. *Frontiers in Plant Science* 8, 252–252. https://doi.org/10.3389/fpls.2017.00252.

Schneider, C. A., Rasband, W. S. and Eliceiri, K. W. (2012). NIH Image to ImageJ: 25 years of image analysis. *Nature Methods* 9(7), 671–675. https://doi.org/10.1038/nmeth.2089.

Sharma, R. C. (1993). Selection for biomass yield in wheat. *Euphytica* 70(1–2), 35–42. https://doi.org/10.1007/BF00029638.

Shi, J., Li, R., Qiu, D., Jiang, C., Long, Y., Morgan, C., Bancroft, I., Zhao, J. and Meng, J. (2009). Unraveling the complex trait of crop yield with quantitative trait loci mapping in Brassica napus. *Genetics* 182(3), 851–861. https://doi.org/10.1534/genetics.109.101642.

Sinclair, T. R., Purcell, L. C. and Sneller, C. H. (2004). Crop transformation and the challenge to increase yield potential. *Trends in Plant Science* 9(2), 70–75. https://doi.org/10.1016/j.tplants.2003.12.008.

Slafer, G. A., Savin, R. and Sadras, V. O. (2014). Coarse and fine regulation of wheat yield components in response to genotype and environment. *Field Crops Research* 157, 71–83. https://doi.org/10.1016/j.fcr.2013.12.004.

Spielmeyer, W., Ellis, M. H. and Chandler, P. M. (2002). Semidwarf (<em>sd-1</em>). *Proceedings of the National Academy of Sciences of the United States of America* 99(13), 9043–9048. https://doi.org/10.1073/pnas.132266399.

Tambussi, E. A., Bort, J., Guiamet, J. J., Nogués, S. and Araus, J. L. (2007). The photosynthetic role of ears in C3 cereals: metabolism, water use efficiency and contribution to grain yield. *Critical Reviews in Plant Sciences* 26(1), 1–16. https://doi.org/10.1080/07352680601147901.

Teich, A. H. (1984). Heritability of grain yield, plant height and test weight of a population of winter wheat adapted to Southwestern Ontario. *TAG. Theoretical and Applied Genetics. Theoretische und Angewandte Genetik* 68(1–2), 21–23. https://doi.org/10.1007/BF00252304.

Thomas, H. and Ougham, H. (2014). The stay-green trait. *Journal of Experimental Botany* 65(14), 3889–3900. https://doi.org/10.1093/jxb/eru037.

Tilly, N., Aasen, H. and Bareth, G. (2015). Fusion of plant height and vegetation indices for the estimation of barley biomass. *Remote Sensing* 7(9), 11449–11480. https://doi.org/10.3390/rs70911449.

Tilman, D., Balzer, C., Hill, J. and Befort, B. L. (2011). Global food demand and the sustainable intensification of agriculture. *Proceedings of the National Academy of Sciences of the United States of America* 108(50), 20260–20264. https://doi.org/10.1073/pnas.1116437108.

Trachsel, S., Sun, D., SanVicente, F. M., Zheng, H., Atlin, G. N., Suarez, E. A., Babu, R. and Zhang, X. (2016). Identification of QTL for early vigor and stay-green conferring tolerance to drought in two connected advanced backcross populations in tropical maize (*Zea mays* L.). *PLoS ONE* 11(3), e0149636. https://doi.org/10.1371/journal.pone.0149636.

Usubiaga-Liaño, A., Mace, G. M. and Ekins, P. (2019). Limits to agricultural land for retaining acceptable levels of local biodiversity. *Nature Sustainability* 2(6), 491–498. https://doi.org/10.1038/s41893-019-0300-8.

Velumani, K., Oude Elberink, S., Yang, M. Y. and Baret, F. (2017). Wheat ear detection in plots by segmenting mobile laser scanner data. *ISPRS Annals of the Photogrammetry, Remote Sensing and Spatial Information Sciences* IV-2/W4, 149–156. https://doi.org/10.5194/isprs-annals-IV-2-W4-149-2017.

Vollset, S. E., Goren, E., Yuan, C. W., Cao, J., Smith, A. E., Hsiao, T., Bisignano, C., Azhar, G. S., Castro, E., Chalek, J., Dolgert, A. J., Frank, T., Fukutaki, K., Hay, S. I., Lozano, R., Mokdad, A. H., Nandakumar, V., Pierce, M., Pletcher, M., Robalik, T., Steuben, K. M., Wunrow, H. Y., Zlavog, B. S. and Murray, C. J. L. (2020). Fertility, mortality, migration, and population scenarios for 195 countries and territories from 2017 to 2100: a forecasting analysis for the global burden of disease study. *The Lancet* 396(10258), 1285–1306. https://doi.org/10.1016/S0140-6736(20)30677-2.

Wang, S. G., Jia, S. S., Sun, D. Z., Wang, H. Y., Dong, F. F., Ma, H. X., Jing, R. L. and Ma, G. (2015). Genetic basis of traits related to stomatal conductance in wheat cultivars in response to drought stress. *Photosynthetica* 53(2), 299–305. https://doi.org/10.1007/s11099-015-0114-5.

Wang, X., Singh, D., Marla, S., Morris, G. and Poland, J. (2018a). Field-based high-throughput phenotyping of plant height in sorghum using different sensing technologies. *Plant Methods* 14(1), 53. https://doi.org/10.1186/s13007-018-0324-5.

Wang, X., Xu, Y., Hu, Z. and Xu, C. (2018b). Genomic selection methods for crop improvement: current status and prospects. *The Crop Journal* 6(4), 330–340. https://doi.org/10.1016/j.cj.2018.03.001.

Wang, X., Zhang, R., Song, W., Han, L., Liu, X., Sun, X., Luo, M., Chen, K., Zhang, Y., Yang, H., Yang, G., Zhao, Y. and Zhao, J. (2019). Dynamic plant height QTL revealed in maize through remote sensing phenotyping using a high throughput unmanned aerial vehicle (UAV). *Scientific Reports* 9(1), 3458. https://doi.org/10.1038/s41598-019-39448-z.

Wang, Y. and Li, J. (2008). Molecular basis of plant architecture. *Annual Review of Plant Biology* 59, 253–279. https://doi.org/10.1146/annurev.arplant.59.032607.092902.

White, E. M. and Wilson, F. E. A. (2006). Responses of grain yield, biomass and harvest index and their rates of genetic progress to nitrogen availability in ten winter wheat varieties. *Irish Journal of Agricultural and Food Research* 45(1), 85–101.

Wiegmann, M., Maurer, A., Pham, A., March, T. J., Al-Abdallat, A., Thomas, W. T. B., Bull, H. J., Shahid, M., Eglinton, J., Baum, M., Flavell, A. J., Tester, M. and Pillen, K. (2019). Barley yield formation under abiotic stress depends on the interplay between flowering time genes and environmental cues. *Scientific Reports* 9(1), 6397. https://doi.org/10.1038/s41598-019-42673-1.

Wu, G., Miller, N. D., de Leon, N., Kaeppler, S. M. and Spalding, E. P. (2019). Predicting *Zea mays* flowering time, yield, and kernel dimensions by analyzing aerial images. *Frontiers in Plant Science* 10, 1251.

Xu, Y., Li, P., Zou, C., Lu, Y., Xie, C., Zhang, X., Prasanna, B. M. and Olsen, M. S. (2017). Enhancing genetic gain in the era of molecular breeding. *Journal of Experimental Botany* 68(11), 2641–2666. https://doi.org/10.1093/jxb/erx135.

Yang, W., Feng, H., Zhang, X., Zhang, J., Doonan, J. H., Batchelor, W. D., Xiong, L. and Yan, J. (2020). Crop phenomics and high-throughput phenotyping: past decades, current challenges, and future perspectives. *Molecular Plant* 13(2), 187–214. https://doi.org/10.1016/j.molp.2020.01.008.

Yousfi, S., Kellas, N., Saidi, L., Benlakehal, Z., Chaou, L., Siad, D., Herda, F., Karrou, M., Vergara, O., Gracia, A., Araus, J. L. and Serret, M. D. (2016). Comparative performance of remote sensing methods in assessing wheat performance under Mediterranean conditions. *Agricultural Water Management* 164, 137–147. https://doi.org/10.1016/j.agwat.2015.09.016.

Zadoks, J. C., Chang, T. T. and Konzak, C. F. (1974). A decimal code for the growth stages of cereals. *Weed Research* 14(6), 415–421. https://doi.org/10.1111/j.1365-3180.1974.tb01084.x.

Zhang, L., Niu, Y., Zhang, H., Han, W., Li, G., Tang, J. and Peng, X. (2019). Maize Canopy temperature extracted from UAV thermal and RGB imagery and its application in water stress monitoring. *Frontiers in Plant Science* 10, 1270.

Zhang, Y., Yu, C., Lin, J., Liu, J., Liu, B., Wang, J., Huang, A., Li, H. and Zhao, T. (2017). OsMPH1 regulates plant height and improves grain yield in rice. *PLoS ONE* 12(7), e0180825. https://doi.org/10.1371/journal.pone.0180825.

Zhao, C., Zhang, Y., Du, J., Guo, X., Wen, W., Gu, S., Wang, J. and Fan, J. (2019a). Crop phenomics: current status and perspectives. *Frontiers in Plant Science* 10, 714.

Zhao, Z., Rebetzke, G. J., Zheng, B., Chapman, S. C. and Wang, E. (2019b). Modelling impact of early vigour on wheat yield in dryland regions. *Journal of Experimental Botany* 70(9), 2535–2548. https://doi.org/10.1093/jxb/erz069.

Zhou, C., Liang, D., Yang, X., Xu, B. and Yang, G. (2018). Recognition of wheat spike from field based phenotype platform using multi-sensor fusion and improved maximum entropy segmentation algorithms. *Remote Sensing* 10(2). https://doi.org/10.3390/rs10020246.

Chapter 13

Automated assessment of plant diseases and traits by sensors: how can digital technologies support smart farming and plant breeding?

Anne-Katrin Mahlein, Institute of Sugar Beet Research, Germany; Jan Behmann, Bayer Crop Science, Germany; David Bohnenkamp, BASF Digital Farming GmbH, Germany; René H. J. Heim, UAV Research Centre (URC), Ghent University, Belgium; and Sebastian Streit and Stefan Paulus, Institute of Sugar Beet Research, Germany

1 Introduction

Technologies in agriculture are undergoing constant development, improvement and adaptation (Jung et al., 2021; Walter et al., 2017). With the rise of digital technologies during the last few decades, inventions such as camera-controlled weeding, GPS-based seeding and the use of drones and robots are being implemented in daily farming practices (Maes and Steppe, 2018). Innovative smart farming approaches and the ability to use sensors in the field to characterize the plant status have made it possible to adapt applications in a site-specific manner (McBratney et al., 2005; Koch and Khosla, 2003). Plant diseases reduce the yield and quality of agricultural products such as food, feed and fibre and are hard to control (Savary et al., 2017; Strange and

http://dx.doi.org/10.19103/AS.2022.0102.17

Scott, 2005). The concept of integrated plant protection and pest management includes the application of fungicides and insecticides as ultima ratio, as well as mechanical and chemical weeding, among others. Within this context, the correct diagnosis and quantification of diseases is necessary to avoid potential yield loss for decision-making or fundamental research aspects. Relevant parameters to be assessed are the type of disease (diagnosis), the disease incidence (proportion of plant units with at least one symptom unit), and/or disease severity (quantification of diseased area) (Madden et al., 2007). Disease assessment and diagnosis has traditionally been carried out by humans, but an increasing number of new approaches based on digital technologies have become available (Bock et al., 2020; Nutter et al., 2006). Since Integrated Plant Management (IPM) is a knowledge-based approach, all strategies to improve the knowledge of a crop stand are greatly appreciated (Maslekar et al., 2020; Oerke, 2020) and new technologies can enable precise management and decision-making (Weiss et al., 2020). Modern inventions such as imaging technologies and their interpretation (Fiorani et al., 2012), small-scale environmental data from sensor networks and digital prognosis models (Khattab et al., 2019), data transmission protocols for sensor and actor communication, and data from drone or satellite recordings summarized in maps to predict potential risks are all becoming integrated into agricultural processes (Mahlein et al., 2018). They all aim at either identifying heterogeneities in crop stands, assessing temporal or spatial patterns, mapping plant traits or detecting plant stress. In addition, multiple studies are being undertaken to demonstrate the potential of phenotyping techniques to retrieve new information on host-pathogen interactions in the area of fundamental and mechanistic research (Arens et al., 2016; Fiorani and Schurr, 2013; Kuska et al., 2018; Leucker et al., 2017; Mahlein et al., 2019). Optical sensors can widen our understanding of crop response to biotic and abiotic stresses by providing additional information and new insights.

2 Digital plant disease detection

2.1 Non-invasive technologies

These technologies perform in a non-invasive manner, provide high speed and coverage and can be automated objectively (Bock et al., 2020). As these technologies come from various areas of research such as surveying, image processing, machine learning and automatization, rather than agriculture, there is a reluctance among farmers to implement them (Pierpaoli et al., 2013). Additionally, some critical studies highlight that precision agriculture is not advancing as fast as expected (McBratney et al., 2005). All these technologies must prove and demonstrate their benefits to support the work of farmers and breeders. It is critical to create trust in new technologies. A complex technology can easily enter the market if it fits well into the workflow of a farm or a breeding company. Therefore, several prerequisites need to be fulfilled; the technology must provide precise results, be user-friendly and come with easy-to-use manuals and accessible software interfaces. Smart plant protection and plant disease phenotyping is currently transitioning from research to practical application, and this process must be supported and guided by reliable demonstrators and robust scientific results.

2.2 Motivation

Digital plant disease detection is crucial for agricultural practice (smart farming) and plant phenotyping (Mahlein, 2016). As plant diseases alter plant physiology and development, these changes can be converted into specific signals by optical sensor systems (Mahlein et al., 2018; Oerke, 2020). Intensive research has been conducted into developing innovative optical methods for the assessment of plant disease symptoms. Studies across scales from the lab to the field have been carried out on diverse crops and their respective diseases, and multiple review articles summarize and present examples for plant disease detection (Bock et al., 2020; Mahlein et al., 2018; Oerke, 2020; Sankaran et al., 2010; West et al., 2003). Different types of sensor data, such as thermal sensors (Belin et al., 2013), red/blue/green (RGB) imaging (Barbedo, 2014) and spectral data (Morel et al., 2018) have shown their potential to provide relevant information. In addition, the assessment of tolerance or resistance reactions is also the focus of research, as the breeding of resistant or tolerant varieties is one of the most sustainable tools in crop cultivation and IPM. Disease resistance is characterized as an observed phenotype in a specific environment on which a pathogen (fungi, bacteria or virus) is not able to cause a disease, whereas tolerance is defined as the ability of a plant to compensate for yield loss due to pathogen infection. The cultivation of resistant or tolerant crops can reduce the application of plant protection products and thereby reduce potential negative externalities of agriculture. Since evaluating disease tolerance and resistance in the breeding process is costly and time-consuming, digital plant phenotyping tools are in high demand (Mahlein et al., 2019; Pieruschka and Schurr, 2019; Walter et al., 2015).

2.3 Optical sensors and the combination with data analysis

These approaches towards smart farming and digital plant phenotyping combine classical phytopathology or microbiology with technology-driven methods such as sensors, robotics, machine learning and artificial intelligence (Bock et al., 2020). In contrast to common visual rating, detection and assessment methods, optical sensors in combination with advanced data analysis methods can retrieve pathogen-induced changes in the physiology of susceptible or resistant plants in a non-invasive and objective way. The field of phytopathometry has been the focus of research for half a century (Madden, 1995), starting with the use of simple RGB images and data from remote sensing and later gaining traction with the invention of digital photography (Bock et al., 2020; Nutter at al., 2006). Nowadays, in the field of optical sensor technology, spectral cameras are of great importance as they extend the visible spectrum (380–780 nm) to light ranges that are not visible to the human eye, such as the near-infrared range (780–1000 nm) (Mahlein et al., 2012). A distinction is made between non-imaging spectrometers (1D) and imaging measuring devices (2D and 3D) (Paulus, 2019; Paulus and Mahlein, 2020). Imaging sensors include digital RGB colour cameras, as well as multispectral and hyperspectral cameras. These cameras measure the reflectance on a line on a chip whose pixels are sensitive to different spectral ranges. Depending on the spectral resolution, this corresponds in practice to up to three channels for RGB cameras; several channels, but usually fewer than 20, for multispectral cameras; and up to several hundred channels for hyperspectral cameras. By cumulating the image channels in their spatial resolution, a complete image is created as a spectral cube (see Fig. 1). The light reflected by a single plant or a crop stand is measured, and multiple factors influence the signal. Both natural sunlight and artificial light sources with defined spectral properties can serve as light sources. Systems with artificial light sources are used in the laboratory and greenhouse, as well as in the field.

2.4 Carrier platforms and different scales

The potential applications are manifold and depend on the sensor technology in combination with carrier platforms (e.g. unmanned aerial vehicles and robots) and measurement scales (e.g. spatial, temporal and spectral) (Bohnenkamp et al., 2019). Spectral changes are assessed at the leaf level or at the field scale for the identification of disease infestation (Mishra et al., 2020). Information in the field can be generated at different scale levels. In addition to the single-plant scale level, vehicle-based sensors (on robots, tractors or other carrier platforms) are used to investigate the crop at close range. Drones can be used to view the crop from 10 m to 100 m above the canopy, and aircraft and satellites can be used to view the crop from several hundred meters to many kilometres above the field. The type of sensor and sensor platform play a decisive role in the intended use. There are cameras that must remain motionless above the crop to capture an image (mirror-based cameras and filter-based systems) and those whose image is only acquired by moving continuously line by line over the crop (line scanners). Various large-scale vegetation patterns can be described

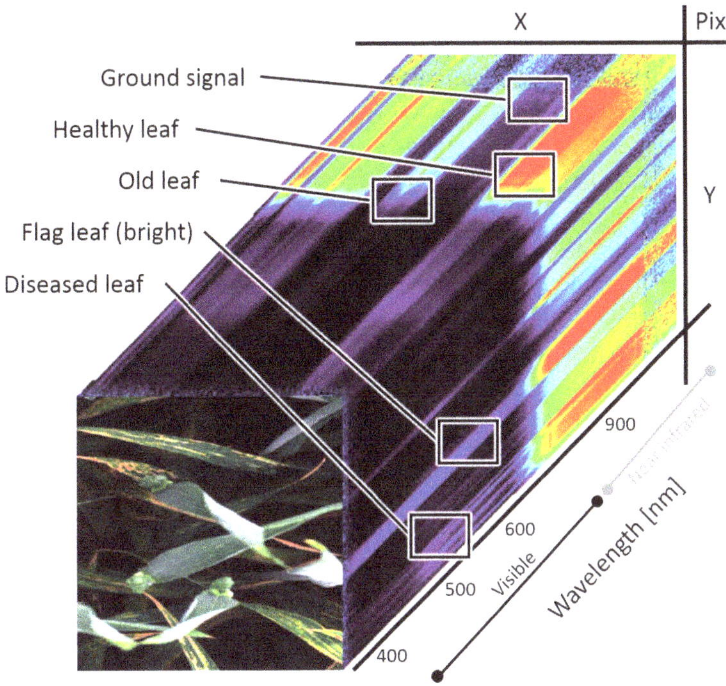

Figure 1 Hyperspectral data cube. Data cube of a hyperspectral image taken in a wheat field with yellow rust infestation. The data cube illustrates the biological heterogeneity that is present in each hyperspectral field image. The spectral information is influenced by ground signals, different leaf altitude signals and disease signals on different leaf altitudes (according to Bohnenkamp, 2019).

with satellite images (e.g. biomass development or crop species). By contrast, a sensor at such a large distance cannot detect a single leaf spot caused by a fungal disease a few millimetres in size. The spatial resolution in combination with the measuring distance becomes the limiting factor for the detection of many features. This is especially relevant for plant disease detection.

3 Complexity of host-pathogen interactions

Plant diseases interact with their host plant in a highly specific way. Multiple factors influence this interaction, such as environmental conditions, crop plant genotype, pathogen genotype, management practices and many more. Plant pathogens influence biochemical processes and the structure of plant leaves following infection. Other characteristics such as the plant architecture or the canopy structure are influenced as well. During host-pathogen interactions, physiological and biochemical changes lead to symptom development in the host tissue. These changes might influence the optical properties in a characteristic way (Mahlein et al., 2010). We know from multiple studies that plant diseases exhibit a highly specific optical response that enables researchers to differentiate healthy from diseased tissue, different types of diseases, and diseases from other types of stresses such as drought or nutrient deficiencies (Delalieux et al., 2009; Gold et al., 2020; Heim et al., 2019b; Mahlein et al., 2012; Wahabzada et al., 2016). The detectability of a disease essentially depends on the stage of pathogenesis, the pathogen's lifestyle and the resulting disease symptoms of the respective disease as shown in Fig. 2 (Bohnenkamp et al., 2021). In plant pathology, pathogens are mainly characterized as pathogens with a necrotrophic lifestyle or pathogens with a biotrophic lifestyle. Necrotrophic pathogens (e.g. *Zymoseptoria tritici* of wheat)

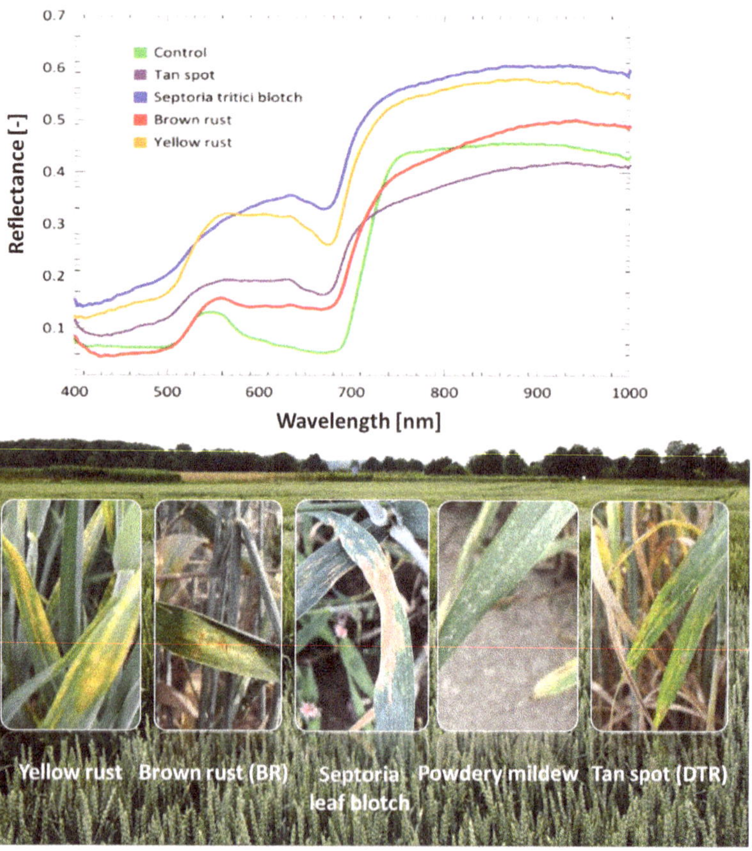

Figure 2 Spectral signatures of wheat diseases. Characteristic spectral signatures of mature symptoms and RGB illustration of foliar wheat diseases. Spectral signatures in the range of 400-1000 nm were assessed with a hyperspectral line scanning camera. Spectral reflectance in the visible range (400-700 nm) and near infrared (700-1000 nm) differs between the individual diseases and compared to the healthy control, respectively.

kill their host cells and feed from the nutrients provided by the dead organic matter to enable fungal proliferation in the host tissue (Steinberg, 2015). Biotrophic fungi, such as mildews and rusts, keep their host's tissue alive as long as possible to derive nutrients from the living photosynthetic active leaf area. Colonization of the host and the formation of haustoria as feeding organs ensure the provision of nutrients (Dracatos et al., 2018). Furthermore, each host-pathogen interaction is highly dynamic in time and space and provides multiple sensor data patterns (Bohnenkamp et al., 2021; Delalieux et al., 2009; Mahlein et al., 2012; Wahabzada et al., 2015). Sensor signals from early-stage symptoms differ from sensor signals from late-stage symptoms. Plant diseases can be detected either indirectly through the detection of structural changes in the host tissue (i.e. loss of water, chlorophyll degradation and cell collapse) or directly through the observation of the formation of fungal structures on or in the host (i.e. pigmented spores or hyphae). Each individual host-parasite interaction has a specific spatial and temporal dynamic, which has an impact on different wavebands of the electromagnetic spectrum during pathogenesis (Wahabzada et al., 2015, 2016). To detect individual symptoms, a camera with high spatial resolution close above the canopy is therefore recommended (Jay et al., 2020; Bohnenkamp et al., 2019). However, large-scale heterogeneities in the plant stand (e.g. nest-wise infestation with nematodes or diseases, soil differences, etc.) can be localized easily by unmanned aerial vehicles and with high-resolution satellite images (Barbedo, 2019). In contrast, a high spectral resolution is often not required and many studies aim to develop plant disease-specific spectral indices for the assessment of plant diseases (Mahlein et al., 2013; Oerke et al., 2016; Zheng et al., 2019). The advantage of using sensors with low spatial resolution and broad spectral bands (i.e. multispectral cameras) is that they are currently commercially available to the public at a relatively low price. Hyperspectral sensors, with high spectral resolution, are still costly and not as straightforward to deploy as snapshot cameras. First hyperspectral snapshot cameras are being developed. Data processing workflows for multispectral imagery are widely available in the form of user-friendly software solutions. Many users prefer their spectral data converted into spectral vegetation indices (SVIs), which are mathematical expressions to extract selected aspects such as chlorophyll or water content by combining the perceived reflectance within the most relevant bands (Thenkabail et al., 2019). A limitation of using SVIs is that the transfer and applicability across environments and data sets is challenging as their output is often generated on empirical data. However, under controlled laboratory conditions and with the appropriate equipment and experimental design, the detection, identification and quantification of plant diseases is possible. Crop- and disease-specific parameters depend on the type of crop and on plant-pathogen interaction. These parameters must be considered for optical measurements in the field (degree of canopy cover, leaf area, etc.). Crop stands of sugar beet are suitable for optical measurements in a different way to wheat. In sugar beet, a large part of the leaf area is at the same height and the illumination of sunlight on the crop is comparatively homogeneous. Wheat forms different leaf layers that must be exposed to varying intensities of light and also vary in their focal distance. In a single image, object sharpness will be different across leaf layers, which will have a direct impact on data measurement and analysis.

4 Complexity in a crop stand

Another source of variability is the complex structure of plants and their different organs which significantly increases imaging complexity. For example, the ear is formed above the canopy. In canopy measurements, the ear can have a negative effect on image quality and image processing, as it increases the complexity of and potentially obscures the plant structures to be classified. Plant diseases also develop on different layers of a crop canopy. The first symptoms of disease often occur on older leaves or leaves closer to the ground. The reasons for this are manifold, but a simple explanation is related to the source of inoculum. Plant diseases, which produce their primary infection from overwintering infectious structures in the soil, tend to produce the first symptoms on lower leaf levels. By contrast, diseases such as powdery mildews or rust diseases, where infectious structures are transported by wind, start with infection on the upper leaves. Ambient conditions and the phenotypic appearance of the crop stand itself often prevent detection under field conditions as shown in Fig. 3 (Bohnenkamp et al., 2019; Jay et al., 2020).

This chapter provides an overview of current developments for automatically detectable plant diseases and associated crop parameters. The goal of using digital technologies is to simplify or replace conventional, manual processes

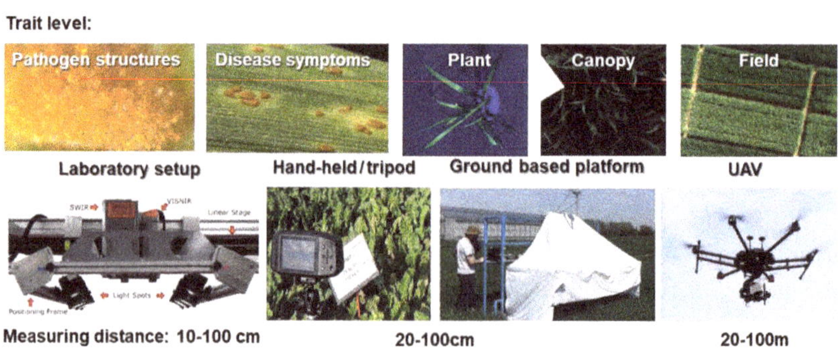

Figure 3 Measurement set-up on different scales. Measurements on different scales demand differing measurement set-ups and sensor platforms. The scale influences the assessable trait level.

of disease monitoring and to provide additional information from sensor data for decision-making in smart disease detection and plant phenotyping. This creates several challenges, showcased in the following case study.

5 Case study: application of deep learning to foliar plant diseases

The application of deep learning algorithms for the detection of foliar plant diseases of wheat on the single plant scale from lab data is presented.

5.1 Description of the leaf diseases

Hyperspectral imaging of individual wheat leaves, inoculated with plant diseases including yellow rust (YR), brown rust (BR) and Septoria tritici blotch (STB), was performed in a laboratory setting with a hyperspectral imaging line scanner (Bohnenkamp et al., 2021). Pathogens causing rust diseases such as YR and BR are biotrophic fungi and colonize the intercellular space and reduce nutrients to build spore clusters with characteristic colour. Necrotrophic diseases such as STB cause chlorotic and necrotic leaf tissue by the release of phytotoxic compounds. All three diseases are highly relevant to wheat production, and early identification could improve the efficiency of plant protection measures.

5.2 Deep learning approach and experimental set-up of the case study

This case study addresses monitoring of the reflectance characteristics of the investigated disease symptoms by optimizing a three-dimensional convolutional neural network (3D-CNN) for evaluation. Hyperspectral images provide significant information on the plant's status and potential disease symptoms, but the interpretation of such data is challenging. The most relevant patterns of plant diseases within the spectral and spatial dimensions are unknown. Deep learning models are known for their adaptation to multidimensional input because the representation of features is optimized during the learning phase. The use of 3D-CNN allows a direct interpretation of hyperspectral data cubes without pre-processing. Details of the selected model architecture and training parameters are explained in the subsequent sections. The model is evaluated on a holdout test data set and its plausibility is verified by class activation maps (CAMs).

In the experimental set-up, inoculation was performed at plant stage BBCH 30 (Meier, 2018). Images were obtained when mature symptoms of each disease were shown, resulting in a different measurement time for each disease. Details of the preparation of the biological material and information on the measurement set-up are provided in Bohnenkamp et al. (2021). The leaves were horizontally fixed in a wireframe to capture hyperspectral images in nadir orientation. The hyperspectral camera ImSpector PFD V10E (Spectral Imaging Ltd., Oulu, Finland) with a spectral range of 400–1000 nm was used in nadir orientation to observe the leaf samples at a distance of 42.5 cm. Spectral raw data were smoothed by employing the Savitzky-Golay (Savitzky and Golay, 1964) filter and noise within the data was reduced by removing the extreme spectral border regions from the data set. Overall, a test set of 500 image samples per disease and 1500 image samples for the healthy control group as well as the holdout test data set were generated (data dimensions 64 px × 64 px × 148 bands). Training and test data were obtained from different leaves to increase the reliability of model evaluation. CNNs using two-dimensional (2D) convolutional filters are well established for image analysis (LeCun et al., 2015; Goodfellow et al., 2016). However, due to the 3D data characteristics of hyperspectral images, 3D-CNNs were applied (see Fig. 4). These use the same methodological frameworks as CNNs but apply 3D filters for convolution. As a result, the spatial-spectral characteristics are not separated but used in a joint fashion as features for the following layers. Tran et al. (2015) introduced 3D-CNNs for action recognition based on video data. Nagasubramanian et al. (2018) used 3D-CNNs for the classification of the disease stem rot and utilized saliency maps for the interpretation and evaluation of the model characteristics. The use of 3D-CNNs is an adaptation of CNNs with 2D convolutional filters. Here, each filter K is a 2D matrix, which is convoluted with the 2D input image 'I' to provide the 2D output image

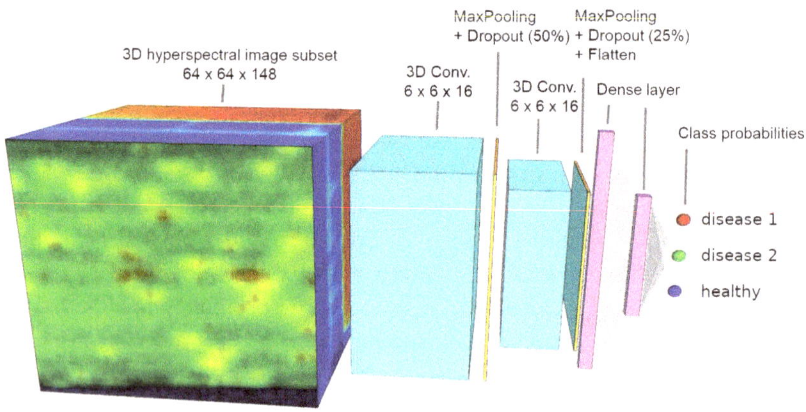

Figure 4 Convolutional neural network (CNN). Structure of a 3D-CNN application on hyperspectral imaging data with two convolutional layers and two fully connected layers with around 500 000 parameters. Three diseases and the healthy control group were used as target classes of the model. The 3D hyperspectral image subset represents a wheat leaf, diseased with brown rust.

'S' for the next layer. The entries of the kernel matrix are optimized during the training phase and shared for the whole image. To deal with RGB images with three channels, the kernels are frequently applied and optimized per channel. For 3D convolutional layers, the kernel is extended to a 3D matrix and the channels or bands are analysed in a joint approach. In image analysis, the combination of convolutional layers with dense layers has been successful (LeCun et al., 2015). Using this approach, the relevant information is compressed in fewer nodes, resulting in more abstract features. The proposed model consists of two 3D convolutional layers (kernel dimensions of 6 × 6 × 16 in both cases) and two fully connected dense layers (50 and 4 neurons). The last layers can be interpreted as non-linear classification. The layers use vector-valued input layers to generate vector-valued output layers. The last dense layer has as many output nodes as classes to predict. Using the activation function softmax, class probabilities are assigned to the output nodes. For the transition of the 3D convolution part of the model to the 1D dense part, a flattening operation is applied that reshapes a 3D input layer to a 1D output layer by simply lining up all entries of the 3D matrix. Because it improves classification accuracy, categorical cross-entropy was used during training by an Adam optimization algorithm (Kingma and Ba, 2014). The training was performed in multiple steps in which the learning rate was adapted from 0.01 to 0.001 to 0.0001. Each of the stages had a duration of 100 epochs consisting of a mini batch size of 64.

Deep learning models are difficult to interpret due to the large number of parameters. A single parameter is generally not related to a specific meaning. Nevertheless, approaches for the visualization of filters and methods of interpretation are the focus of current research (Zhou et al., 2016). CAMs localize the relevance for classification within the data. Averaging the spectral or spatial dimensions across a large data set allows us to extract relevant positions in spectral or spatial domains.

5.3 Results of the optimization phase

The overall training was performed in approximately 23 h of computing time, but training was suspended to adapt the parameters and check the level of convergence. Longer training times, more epochs with a stable learning rate and further training with adjusted learning rate were also tested but did not improve model accuracy. In accordance with Goodfellow et al. (2016), the accuracy of the model during the presented learning structure stagnated after approximately 100 epochs in each phase but improved if the learning rate was changed. Further improvements to the model's accuracy by additional learning rate changes were not successful.

5.4 Improved interpretation by feature relevance

The visualization of relevant regions by CAM within the images allows verification of the plausibility of the model characteristics obtained during the training process (Nagasubramanian et al., 2019). Disease detection models should rely on the specific disease symptoms and not on unrelated leaf tissue or background material (see Fig. 5). For each of the diseases, the specific symptoms were located. For YR the elongated uredinia, for BR the smaller round uredinia, and for STB the locations of the larger areas of necrotic leaf tissue were highlighted. In summary, the trained model had the desired property. Furthermore, the CAM approach may provide support if multiple diseases are present within one image. Using different target classes could allow for localization and differentiation of symptoms of different diseases.

5.5 Classification accuracy and discussion

The application of the optimized model to the holdout test data set reveals a reasonable accuracy of 88.1% (Table 1). Therefore, the accuracy of the classes measured by the F1 score is not homogeneous. The highest F1 is reached by the healthy class with 0.91, followed by BR with 0.89, STB with 0.86 and YR with 0.79. The differences in performance can be explained by the data set composition and symptom characteristics. The healthy class reached the highest accuracy as it is represented by 50% of the training data and is characterized by a low spatial heterogeneity. The BR showed highly characteristic and evenly distributed symptoms on the whole leaf. Image samples showing just a few BR symptoms were very rare. By contrast, YR showed a lower number but larger uredinia of bright yellow colour, similar to chlorotic tissue. Furthermore,

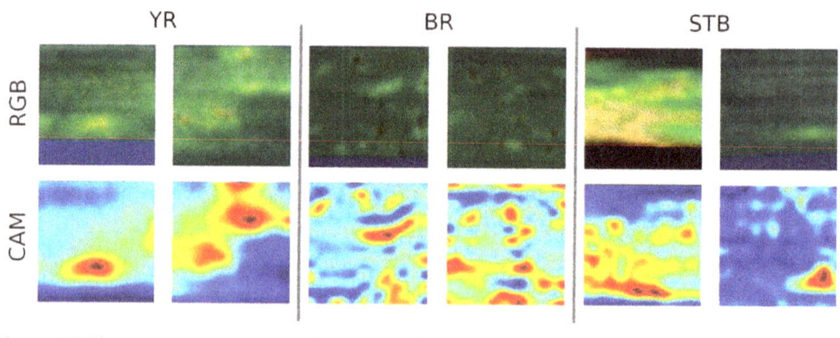

Figure 5 Class activation maps (CAM). Application of 3D-CNN on hyperspectral images of diseased wheat leaves subsamples. The upper line illustrates RGB visualization of hyperspectral data from wheat leaves infested with yellow rust (YR, left), brown rust (BR, centre) and Septoria tritici blotch (STB, right); the lower line shows CAM averaged in the spectral dimension. The colour intensity heat map indicates regions of high relevance for the respective disease with red colour.

Table 1 Confusion matrix of the model performance on a holdout test data set including healthy leaf tissue, yellow rust (YR), brown rust (BR) and Septoria tritici blotch (STB) samples

	Healthy	YR	BR	STB	Precision (%)
Healthy	1392	17	58	33	92.8
YR	87	368	1	44	73.6
BR	50	0	450	0	90.0
STB	15	51	0	434	86.8
Recall (%)	90.1	84.4	88.4	84.9	Accuracy (%) 88.1

YR uredinia were accompanied by larger and more pronounced chlorosis, which is the characteristic symptom of STB. This impedes the separation of YR and STB. The complexity of separating YR and STB in comparison to BR is also reflected in the confusion matrix (Table 1). One aspect that has not been taken into account but may have an impact on the obtained accuracy is optimization of the spatial dimensions of the input samples. Depending on the data set, a specific size is optimal to represent the characteristics of a sample. This number varies between data sets and classes (Chen et al., 2016). Applied to plant diseases, the 64 px × 64 px may represent the BR symptoms much better than the large-scale symptoms of STB and YR. There are two main challenges when training a disease detection model: (i) the heterogeneity of disease severity in test and (ii) training data and the limited amount of training data. Even in highly controlled inoculation experiments, image samples are present with very low or even invisible disease symptoms, which tend to be misclassified to the healthy class. The described label noise has the potential to reduce the overall accuracy significantly (Frénay and Verleysen, 2014). Future studies should homogenize the test data to exclude bias caused by the training data configuration. The ratio of training samples and model complexity is decisive for the optimization process (Foody et al., 1995). Therefore, most of the prediction models in plant contexts are hampered by limited training data because they are based on available models that have been trained on millions of images (Singh et al., 2018; Polder et al., 2019). Training only the last layers and adopting the parameters of the convolutional layer is in many cases sufficient to adapt the model to a new task. In the context of hyperspectral imaging, sophisticated models are not available (Singh et al., 2018), causing the need to learn from scratch with randomized initialization. This may change in future, but for now descriptions of successful configurations should be taken as base. Both challenges can be addressed by the provision of high-quality data sets or alternatively trained models for the interpretation of close-range hyperspectral images.

5.6 Conclusion of the case study

This study presents the application of 3D-CNNs in the novel context of differentiating multiple foliar plant diseases based on close-up hyperspectral images. The model allows hyperspectral images in their raw 3D representation to be classified without neglecting the spatial or spectral context of individual measurements. The straightforward model definition and training process reached an overall accuracy of 88.1%. The plausibility was furthermore verified using CAM analysis. This study provides a working configuration regarding data set characteristics, model architecture, training strategy and evaluation methods. This model is focused on the interpretation of highly controlled close-up images in laboratory set-ups. The applicability or adaptation to an extended application scenario within the greenhouse or even the field must be verified in further studies. However, hyperspectral observations in the field are challenging as illumination is typically less stable and leaf fixation is impossible. Variable leaf angles and observation distances strongly impair the hyperspectral signal (Behmann et al., 2016). Consequently, the sensing approach, as well as training data composition, must reflect this increased natural variability to allow a reliable assessment with significant value for agricultural practice.

6 Summary

From a technical, biological and functional perspective, digital technologies and especially optical sensors combined with artificial intelligence and intuitive robotic platforms are future-oriented technologies for plant disease detection in smart farming and resistance breeding. However, there are still several critical aspects and challenges: (i) The data quality and measurement scale influence the quality of the results significantly. Therefore, tailored measurement settings, depending on the aspect of interest, are required. High precision is often gained by more complex measurement set-ups using an exhaustive approach. Early detection, before disease symptoms are visible under field conditions, is unrealistic and demands further research. (ii) Digital technologies are data intensive - data is often collected at a massive scale, up to several terabytes. Time series measurements are often required, which increases the data sampling intensity. (iii) Manual labelling to create powerful training data is often prevented as subtle changes due to plant disease are hard to locate or their rare occurrence is blurred by natural variability. (iv) To date, no global data models for disease detection are available; algorithms and powerful hardware systems must be tailored to a specific application, as transferring them among and between crop species and pathosystems is not always possible. (v) An interdisciplinary or transdisciplinary approach is needed. Experts with complementary skills and scientific backgrounds such as phytopathology, plant breeding, informatics, electrical engineering, and/or computer vision must combine their knowledge to yield easy to interpret insights from data representations with a biological meaning (Heim et al., 2019a).

We believe there are two approaches to proceed with optical disease detection. In the first approach, all available data will be composed into a large data set. The enormous biological variability will be handled by the huge amount of data provided by multiple research groups. The second approach will rely on a unified sensor set-up that is able to reduce the variability caused by environmental conditions. If such a platform is accepted in the research community, the focus of the obtained data can be broadened, while preserving the transferability of models and datasets. We believe that the second approach is more expedient as the number of research groups working on optical disease detection is still comparatively small and affordable, and user-friendly sensors are not available for imaging data. However, unmanned aerial vehicle (UAV) methods are becoming increasingly available, which will significantly speed up progress.

Furthermore, deep learning is well suited for the interpretation of images and will advance plant disease detection as it has been shown in other fields. As shown in the study by Schramowski and coworkers (2020), a toolbox with many new algorithms and data analysis routines is already available for use by the scientific community. Data mining, machine learning, pattern recognition and computational sustainability already provide solutions for smart farming and plant phenotyping, even in high-throughput set-ups (Kersting et al., 2016; Eberius and Lima-Guerra, 2009; Singh et al., 2016). Recently, fast data-driven approaches for analysing and interpreting sensor data in agriculture and resistance screening have facilitated significant steps to meet the demands of interpretability and scalability (Wahabzada et al., 2016; Coppens et al., 2017). Data-driven approaches search for heterogeneity and patterns within the data space itself. This provides the potential to retrieve relevant information on plant disease occurrence and interaction types while avoiding the risk of losing information and the time-intensive process of data labelling (Wahabzada et al., 2015). In recent years, deep learning models and CNNs have attracted huge attention as they allow valuable information to be extracted without predefined features by learning optimal feature representation during the training process (Bengio et al., 2013). As presented in the case study, deep learning models consist of multiple layers containing data representations with different levels of abstraction. In a classification model, the highest level of abstraction is typically the class probability. As feature representations are optimized during the learning phase, deep learning models can be more flexible regarding feature dimensionality. Traditional approaches represent a sample by a one-dimensional vector for features. In contrast, CNNs make use of 2D input matrices by applying convolutional filters shared between all parts of the input matrix. This reduces the number of model parameters significantly and allows the analysis of images with millions of pixels with regard to the local context. As presented in the case study, the use of 3D-CNNs extends this concept to 3D convolutional filters that allow data cubes to be interpreted directly without neglecting the spatio-spectral context of each single entry.

Future research should integrate sensor data, environmental modelling and genotype-phenotype data for a holistic view on disease development based on digital phenotyping. Results and approaches are of utmost interest; however, so far only a few pilot studies have attempted to include environmental or phenotype to genotype data to explore host-pathogen systems and disease development. In the context of data integration, we expect a significant step forward within the next 5 years.

7 Future trends in research

The applications of sensors and especially hyperspectral cameras shown in the selected proof-of-concept study demonstrate great potential for use in various agronomic issues and for phenotyping plant diseases. However, certain requirements for the measurement system and data processing currently limit the transfer of various techniques from fundamental research to practice. For example, hyperspectral images and sensor data must be corrected for current solar radiation by a reference object in the crop. Leaf geometry also has a major influence on data quality (Behmann et al., 2016). Depending on the leaf and recording angle, the spectrum is distorted. This factor must be considered in the data evaluation and calculated.

Future applications will use spectrally and spatially high-resolution cameras on UAV carrier platforms (Horstrand et al., 2019; Mahlein et al., 2018), which are able to fly over larger fields due to smaller payloads to transfer the potential previously shown at plot level into fields of practical size. Airborne unmanned carrier platforms thus represent the trade-off between slow but very high-resolution measurements on the ground or with moving platforms (West et al., 2003; Mahlein, 2016) and faster but spectrally and spatially low-resolution recordings on flying platforms such as UAVs, aircraft or satellites (Joyce et al., 2021; Mirik et al., 2010).

Manual work will be largely automated in the field in the near future through the use of moving and flying sensor systems. If automated approaches prove to be reliable compared to manual assessments, the temporal frequency of assessments can be increased and measurements can be carried out objectively, independent of human subjectivity in disease rating or the position of the sun. In this way, estimation procedures and extrapolations can be replaced by targeted overflights and, for example, automated counting of plants. Furthermore, the images obtained can be used to extract further crop parameters such as leaf area, abiotic stress, leaf position or possible disease infestation, in addition to the data collected during human assessment. Already-collected data can be analysed in retrospect through the development of new algorithms, thus enabling the observation of the development of new parameters even from past trials.

8 Where to look for further information

8.1 Textbook

- Jaquemoud, S. and Ustin, S. (2019), *Leaf Optical Properties,* 1st ed., Cambridge University Press, Cambridge, UK.

8.2 Review articles

- Mahlein, A.-K., Kuska, M. T., Behmann, J., Polder, G. and Walter A. (2018). 'Hyperspectral sensors and imaging technologies in phytopathology: State of the art', *Annu. Rev. Phytopathol.*, 56, 535-558. doi 10.1146/annurev-phyto-080417-050100.
- Oerke, E. C. (2020). 'Remote Sensing of Diseases', *Annu. Rev. Phytopathol.*, 58, 225-252. doi 10.1146/annurev-phyto-010820-012832.

8.3 Research projects

- Digital trial field 'FarmerSpace': https://www.farmerspace.uni-goettingen.de/, accessed April 30, 2021.
- Cluster of excellence 'Phenorob': http://www.phenorob.de/, accessed April 30, 2021.

9 Acknowledgements

FarmerSpace is supported by funding from the Federal Ministry of Food and Agriculture (BMEL) based on the decision of the Parliament of the Federal Republic of Germany. The Federal Office for Agriculture and Food (BLE) provides coordinating support for digitalization in agriculture as a funding organization, grant number FZK 28DE104A18. The authors are partially funded by the Deutsche Forschungsgemeinschaft (DFG, German Research Foundation) under Germany's Excellence Strategy - EXC 2070 - 390732324. Part of the research was funded by BASF Digital Farming within the project 'Hyperspectral in-field detection'. The authors also thank Sabine von Tiedemann for proofreading and support.

10 References

Arens, N., Backhaus, A., Döll, S., Fischer, S., Seiffert, U. and Mock, H. P. (2016). Non-invasive presymptomatic detection of *Cercospora beticola* infection and identification of early metabolic responses in sugar beet, *Front. Plant Sci.* 7, 1377. doi: 10.3389/fpls.2016.01377.

Barbedo, J. G. A. (2014). An automatic method to detect and measure leaf disease symptoms using digital image processing, *Plant Dis.* 98(12), 1709–1716. doi: 10.1094/PDIS-03-14-0290-RE.

Barbedo, J. G. A. (2019). A review on the use of unmanned aerial vehicles and imaging sensors for monitoring and assessing plant stresses, *Drones* 3(2), 40. doi: 10.3390/drones3020040.

Behmann, J., Acebron, K., Emin, D., Bennertz, S., Matsubara, S., Thomas, S., Bohnenkamp, D., Kuska, M. T., Jussila, J., Salo, H., Mahlein, A. K. and Rascher, U. (2018). Specim IQ: Evaluation of a new, miniaturized handheld hyperspectral camera and its application for plant phenotyping and disease detection, *Sensors (Basel)* 18(2), 441. doi: 10.3390/s18020441.

Behmann, J., Mahlein, A.-K., Paulus, S., Dupuis, J., Kuhlmann, H., Oerke, E.-C. and Plümer, L. (2016). Generation and application of hyperspectral 3D plant models: Methods and challenges, *Mach. Vis. Appl.* 27(5), 611–624. doi: 10.1007/s00138-015-0716-8.

Belin, É., Rousseau, D., Boureau, T. and Caffier, V. (2013). Thermography versus chlorophyll fluorescence imaging for detection and quantification of apple scab, *Comput. Electron. Agr.* 90, 159–163. doi: 10.1016/j.compag.2012.09.014.

Bengio, Y., Courville, A. and Vincent, P. (2013). Representation learning: A review and new perspectives, *IEEE Trans. Pattern Anal. Mach. Intell.* 35(8), 1798–1828. doi: 10.1109/TPAMI.2013.50.

Bock, C. H., Barbedo, J. G. A., Del Ponte, E. M., Bohnenkamp, D. and Mahlein, A.-K. (2020). From visual estimates to fully automated sensor-based measurements of plant disease severity: Status and challenges for improving accuracy, *Phytopathol. Res.* 2(1), 1–36. doi: 10.1186/s42483-020-00049-8.

Bohnenkamp, D., Behmann, J. and Mahlein, A.-K. (2019). In-field detection of yellow rust in wheat on the ground canopy and UAV scale, *Remote Sens.* 11(21), 2495. doi: 10.3390/rs11212495.

Bohnenkamp, D., Behmann, J., Paulus, S., Steiner, U. and Mahlein, A. K. (2021). A hyperspectral library of foliar diseases of wheat, *Phytopathology* 111(9), 1583–1593. doi: 10.1094/PHYTO-09-19-0335-R.

Chen, Y., Jiang, H., Li, C., Jia, X. and Ghamisi, P. (2016). Deep feature extraction and classification of hyperspectral images based on convolutional neural networks, *IEEE Trans. Geosci. Remote Sens.* 54(10), 6232–6251. doi: 10.1109/TGRS.2016.2584107.

Coppens, F., Wuyts, N., Inzé, D. and Dhondt, S. (2017). Unlocking the potential of plant phenotyping data through integration and data-driven approaches, *Curr. Opin. Syst. Biol.* 4, 58–63. doi: 10.1016/j.coisb.2017.07.002.

Delalieux, S., Auwerkerken, A., Verstraeten, W. W., Somers, B., Valcke, R., Lhermitte, S., Keulemans, J. and Coppin, P. (2009). Hyperspectral reflectance and fluorescence imaging to detect scab induced stress in apple leaves, *Remote Sens.* 1(4), 858–874. doi: 10.3390/rs1040858.

Dracatos, P. M., Haghdoust, R., Singh, D. and Park, R. F. (2018). Exploring and exploiting the boundaries of host specificity using cereal rust and mildew models, *New Phytol.* 2, 453–462. doi: 10.1111/nph.15044.

Eberius, M. and Lima-Guerra, J. (2009). High-throughput plant phenotyping - data acquisition transformation, and analysis. In: *Bioinformatics*, Edwards, D., Stajich, J. and Hansen, D. (Eds). Springer, New York, pp. 259–278.

Fiorani, F., Rascher, U., Jahnke, S. and Schurr, U. (2012). Imaging plant dynamics in heterogenic environments, *Curr. Opin. Biotechnol.* 23(2), 227–235. doi: 10.1016/j.copbio.2011.12.010.

Fiorani, F. and Schurr, U. (2013). Future scenarios for plant phenotyping, *Annu. Rev. Plant Biol.* 64, 267–291. doi: 10.1146/annurev-arplant-050312-120137.

Foody, G. M., McCulloch, M. B. and Yates, W. B. (1995). The effect of training set size and composition on artificial neural network classification, *Int. J. Remote Sens.* 16(9), 1707–1723.

Frénay, B. and Verleysen, M. (2014). Classification in the presence of label noise: A survey, *IEEE Trans. Neural Netw. Learn. Syst.* 25(5), 845–869. doi: 10.1109/TNNLS.2013.2292894.

Gold, K. M., Townsend, P. A., Larson, E. R., Herrmann, I. and Gevens, A. J. (2020). Contact reflectance spectroscopy for rapid, accurate, and nondestructive *Phytophthora infestans* clonal lineage discrimination, *Phytopathology* 110(4), 851–862. doi: 10.1094/PHYTO-08-19-0294-R.

Goodfellow, I., Bengio, Y. and Courville, A. (2016). *Deep Learning* (vol. 1), MIT Press, Cambridge, MA. http://www.deeplearningbook.org.

Heim, R. H. J., Carnegie, A. J. and Zarco-Tejada, P. J. (2019a). Breaking down barriers between remote sensing and plant pathology, *Trop. Plant Pathol.* 44(4), 398–400. doi: 10.1007/s40858-019-00300-4.

Heim, R. H. J., Wright, I. J., Allen, A. P., Geedicke, I. and Oldeland, J. (2019b). Developing a spectral disease index for myrtle rust (*Austropuccinia psidii*), *Plant Pathol.* 68(4), 738–745. doi: 10.1111/ppa.12996.

Horstrand, P., Guerra, R., Rodriguez, A., Diaz, M., Lopez, S. and Lopez, J. F. (2019). A UAV platform based on a hyperspectral sensor for image capturing and on-board processing, *IEEE Access* 7, 66919–66938. doi: 10.1109/ACCESS.2019.2913957.

Jay, S., Comar, A., Benicio, R., Beauvois, J., Dutartre, D., Daubige, G., Li, W., Labrosse, J., Thomas, S., Henry, N., Weiss, M. and Baret, F. (2020). Scoring Cercospora leaf spot on sugar beet: Comparison of UGV and UAV phenotyping systems, *Plant Phenom.* 9452123, 1–18. doi: 10.34133/2020/9452123.

Joyce, K. E., Anderson, K. and Bartolo, R. E. (2021). Of course we fly unmanned - we're women!, *Drones* 5(1), 21. doi: 10.3390/drones5010021.

Jung, J., Maeda, M., Chang, A., Bhandari, M., Ashapure, A. and Landivar-Bowles, J. (2021). The potential of remote sensing and artificial intelligence as tools to improve the resilience of agriculture production systems, *Curr. Opin. Biotechnol.* 70, 15–22. doi: 10.1016/j.copbio.2020.09.003.

Kersting, K., Bauckhage, C., Wahabzada, M., Mahlein, A.-K., Steiner, U., Oerke, E.-C., Römer, C. and Plümer, L. (2016). Feeding the world with big data: Uncovering spectral characteristics and dynamics of stressed plants. In: *Computational Sustainability*, Lässig, J., Kersting, K. and Morik, K. (Eds). Springer International Publishing, Cham, Switzerland, pp. 99–120.

Khattab, A. K. F., Habib, S. E. D., Ismail, H., Zayan, S., Fahmy, Y. and Khairy, M. M. (2019). An IoT-based cognitive monitoring system for early plant disease forecast, *Comput. Electron. Agr.* 166. doi: 10.1016/j.compag.2019.105028.

Kingma, D. P. and Ba, J. (2014). Adam: A method for stochastic optimization, arXiv:1412.6980.

Koch, B. and Khosla, R. (2003). The role of precision agriculture in cropping systems, *J. Crop Prod.* 9(1–2), 361–381. doi: 10.1300/J144v09n01_02.

Kuska, M. T., Behmann, J., Großkinsky, D. K., Roitsch, T. and Mahlein, A. K. (2018). Screening of barley resistance against powdery mildew by simultaneous high-throughput enzyme activity signature profiling and multispectral imaging, *Front. Plant Sci.* 9, 1074. doi: 10.3389/fpls.2018.01074.

LeCun, Y., Bengio, Y. and Hinton, G. (2015). Deep learning review, *Nature* 521(7553), 436-444. doi: 10.1038/nature14539.

Leucker, M., Wahabzada, M., Kersting, K., Peter, M., Beyer, W., Steiner, U., Mahlein, A. K. and Oerke, E. C. (2017). Hyperspectral imaging reveals the effect of sugar beet quantitative trait loci on Cercospora leaf spot resistance, *Funct. Plant Biol.* 44(1), 1-9. doi: 10.1071/FP16121.

Madden, L. V. (1995). Epidemiology, crop loss assessment, and phytopathometry: Introduction, *Can. J. Plant Pathol.* 17(2), 95. doi: 10.1080/07060669509500699.

Madden, L. V., Hughes, G. and van den Bosch, F. (2007). *The Study of Plant Disease Spidemics*, APS Press, St. Paul, MN.

Maes, W. H. and Steppe, K. (2018). Perspectives for remote sensing with unmanned aerial vehicles in precision agriculture, *Trends Plant Sci.* 24(2), 152–164. doi: 10.1016/j.tplants.2018.11.007.

Mahlein, A. K. (2016). Plant Disease detection by imaging sensors - parallels and specific demands for precision agriculture and plant phenotyping, *Plant Dis.* 100(2), 241-251. doi: 10.1094/PDIS-03-15-0340-FE.

Mahlein, A. K., Kuska, M. T., Behmann, J., Polder, G. and Walter, A. (2018). Hyperspectral sensors and imaging technologies in phytopathology: State of the art, *Annu. Rev. Phytopathol.* 56(1), 535–558. doi: 10.1146/annurev-phyto-080417-050100.

Mahlein, A. K., Kuska, M. T., Thomas, S., Wahabzada, M., Behmann, J., Rascher, U. and Kersting, K. (2019). Quantitative and qualitative phenotyping of disease resistance of crops by hyperspectral sensors: Seamless interlocking of phytopathology, sensors, and machine learning is needed!, *Curr. Opin. Plant Biol.* 50, 156–162. doi: 10.1016/j.pbi.2019.06.007.

Mahlein, A.-K., Rumpf, T., Welke, P., Dehne, H.-W., Plümer, L., Steiner, U. and Oerke, E.-C. (2013). Development of spectral indices for detecting and identifying plant diseases, *Remote Sens. Environ.* 128, 21–30. doi: 10.1016/j.rse.2012.09.019.

Mahlein, A.-K., Steiner, U., Dehne, H.-W. and Oerke, E.-C. (2010). Spectral signatures of sugar beet leaves for the detection and differentiation of diseases, *Precis. Agric.* 11(4), 413–431. doi: 10.1007/s11119-010-9180-7.

Mahlein, A.-K., Steiner, U., Hillnhütter, C., Dehne, H.-W. and Oerke, E.-C. (2012). Hyperspectral imaging for small-scale analysis of symptoms caused by different sugar beet diseases, *Plant Methods* 8(1), 3. doi: 10.1186/1746-4811-8-3.

Maslekar, N. V., Kulkarni, K. P. and Chakravarthy, A. K. (2020). Application of unmanned aerial vehicles (UAVs) for pest surveillance, monitoring and management. In: Chakravarthy, A. (Ed.) *Innovative Pest Management Approaches for the 21st Century*, Springer, Singapore.

McBratney, A., Whelan, B., Ancev, T. and Bouma, J. (2005). Future directions of precision agriculture, *Precis. Agric.* 6(1), 7–23. doi: 10.1007/s11119-005-0681-8.

Meier, U. (2018). *Growth Stages of Mono- and Dicotyledonous Plants. BBCH Monograph.* Open Agrar Repositorium, Quedlinburg.

Mirik, M., Jones, D. C., Price, J. A., Workneh, F., Ansley, R. J. and Rush, C. M. (2010). Satellite remote sensing of wheat infected by *Wheat streak mosaic virus*, *Plant Dis.* 95(1), 4-12. doi: 10.1094/PDIS-04-10-0256.

Mishra, P., Polder, G. and Vilfan, N. (2020). 'Close range spectral imaging for disease detection in plants using autonomous platforms: A review on recent studies', *Curr.* Robot Rep., *Curr. Robot. Rep.* 1(2), 43–48. doi: 10.1007/s43154-020-00004-7.

Morel, J., Jay, S., Féret, J. B., Bakache, A., Bendoula, R., Carreel, F. and Gorretta, N. (2018). Exploring the potential of PROCOSINE and close-range hyperspectral imaging to study the effects of fungal diseases on leaf physiology, *Sci. Rep. UK* 8(1), 15933. doi: 10.1038/s41598-018-34429-0.

Nagasubramanian, K., Jones, S., Singh, A. K., Sarkar, S., Singh, A. and Ganapathysubramanian, B. (2019). Plant disease identification using explainable 3D deep learning on hyperspectral images, *Plant Methods* 15, 98. doi: 10.1186/s13007-019-0479-8.

Nagasubramanian, K., Jones, S., Singh, A. K., Singh, A., Ganapathysubramanian, B. and Sarkar, S. (2018). Explaining hyperspectral imaging based plant disease identification: 3D CNN and saliency maps, arXiv:1804.08831 [cs.CV].

Nutter, F. W., Esker, P. D. and Netto, R. A. C. (2006). Disease assessment concepts and the advancements made in improving the accuracy and precision of plant disease data, *Eur. J. Plant Pathol.* 115(1), 95–103. doi: 10.1007/s10658-005-1230-z.

Oerke, E. C. (2020). Remote sensing of diseases, *Annu. Rev. Phytopathol.* 58(1), 225–252. doi: 10.1146/annurev-phyto-010820-012832.

Oerke, E. C., Herzog, K. and Toepfer, R. (2016). Hyperspectral phenotyping of the reaction of grapevine genotypes to *Plasmopara viticola*, *J. Exp. Bot.* 67(18), 5529–5543. doi: 10.1093/jxb/erw318.

Paulus, S. (2019). Measuring crops in 3D: Using geometry for plant phenotyping, *Plant Methods* 15, 103. doi: 10.1186/s13007-019-0490-0.

Paulus, S. and Mahlein, A. K. (2020). Technical workflows for hyperspectral plant image assessment and processing on the greenhouse and laboratory scale, *GigaScience* 9(8), giaa090. doi: 10.1093/gigascience/giaa090.

Pierpaoli, E., Carli, G., Pignatti, E. and Canavari, M. (2013). Drivers of precision agriculture technologies adoption: A literature review, *Procedia Technol.* 8, 61–69. doi: 10.1016/j.protcy.2013.11.010.

Pieruschka, R. and Schurr, U. (2019). Plant phenotyping: Past, present, and future, *Plant Phenom.* 7507131, 1–6. doi: 10.34133/2019/7507131.

Polder, G., Blok, P. M., de Villiers, H. A. C., van der Wolf, J. M. and Kamp, J. (2019). *Potato virus Y* detection in seed potatoes using deep learning on hyperspectral images, *Front. Plant Sci.* 10, 209. doi: 10.3389/fpls.2019.00209.

Sankaran, S., Mishra, A., Ehsani, R. and Davis, C. (2010). A review of advanced techniques for detecting plant diseases, *Comput. Electron. Agric.* 72(1), 1–13. doi: 10.1016/j.compag.2010.02.007.

Savary, S., Bregaglio, S., Willocquet, L., Gustafson, D., Mason D'Croz, D., Sparks, A., Castilla, N., Djurle, A., Allinne, C., Sharma, M., Rossi, V., Amorim, L., Bergamin, A., Yuen, J., Esker, P., McRoberts, N., Avelino, J., Duveiller, E., Koo, J. and Garrett, K. (2017). Crop health and its global impacts on the components of food security, *Food Secur.* 9(2), 311–327. doi: 10.1007/s12571-017-0659-1.

Savitzky, A. and Golay, M. J. E. (1964). Smoothing and differentiation of data by simplified least squares procedures, *Anal. Chem.* 36(8), 1627–1639. doi: 10.1021/ac60214a047.

Schramowski, P., Stammer, W., Teso, S., Brugger, A., Herbert, F., Shao, X., Luigs, H.-G., Mahlein, A.-K., and Kersting, K. (2020). Making deep neural networks right for the right scientific reasons by interacting with their explanations, *Nat. Mach. Intell.* 2(8), 476–486. doi: 10.1038/s42256-020-0212-3.

Singh, A., Ganapathysubramanian, B., Singh, A. K. and Sarkar, S. (2016). Machine learning for high-troughput stress phenotyping in plants, *Trends Plant Sci.* 21(2), 110–124. doi: 10.1016/j.tplants.2015.10.015.

Singh, A. K., Ganapathysubramanian, B., Sarkar, S. and Singh, A. (2018). Deep learning for plant stress phenotyping: Trends and future perspectives, *Trends Plant Sci.* 23(10), 883–898. doi: 10.1016/j.tplants.2018.07.004.

Steinberg, G. (2015). Cell biology of *Zymoseptoria tritici*: Pathogen cell organization and wheat infection, *Fungal Genet. Biol.* 79, 17–23. doi: 10.1016/j.fgb.2015.04.002.

Strange, R. N. and Scott, P. R. (2005). Plant disease: A threat to global food Security, *Annu. Rev. Phytopathol.* 43(1), 83–116. doi: 10.1146/annurev.phyto.43.113004.133839.

Thenkabail, P. S., Lyon, J. G. and Huete, A. (2019). *Hyperspectral Remote Sensing of Vegetation* (2nd edn.), Four Volume Set, CRC Press Press, Boca Raton, FL.

Tran, D., Bourdev, L., Fergus, R., Torresani, L. and Paluri, M. (2015). Learning spatiotemporal features with 3D convolutional networks. *Proceedings of the IEEE International Conference on Computer Vision (ICCV)*, December 2015, pp. 4489–4497.

Wahabzada, M., Mahlein, A. K., Bauckhage, C., Steiner, U., Oerke, E. C. and Kersting, K. (2015). Metro maps of plant disease dynamics - Automated mining of differences using hyperspectral images, *PLoS ONE* 10(1), e0116902. doi: 10.1371/journal.pone.0116902.

Wahabzada, M., Mahlein, A. K., Bauckhage, C., Steiner, U., Oerke, E. C. and Kersting, K. (2016). Plant phenotyping using probabilistic topic models: Uncovering the hyperspectral language of plants, *Sci. Rep. UK* 6, 22482. doi: 10.1038/srep22482.

Walter, A., Finger, R., Huber, R. and Buchmann, N. (2017). Opinion: Smart farming is key to developing sustainable agriculture, *Proc. Natl. Acad. Sci. U. S. A.* 114(24), 6148–6150. doi: 10.1073/pnas.1707462114.

Walter, A., Liebisch, F. and Hund, A. (2015). Plant phenotyping: From bean weighing to image analysis, *Plant Methods* 11, 14. doi: 10.1186/s13007-015-0056-8.

Weiss, M., Jacob, F. and Duveiller, G. (2020). Remote sensing for agricultural applications: A meta-review, *Remote Sens. Environ.* 236, 111402, JRC114361. doi: 10.1016/j.rse.2019.111402.

West, J. S., Bravo, C., Oberti, R., Lemaire, D., Moshou, D. and McCartney, H. A. (2003). The potential of optical canopy measurement for targeted control of field crop diseases, *Annu. Rev. Phytopathol.* 41, 593–614. doi: 10.1146/annurev.phyto.41.121702.103726.

Zheng, Q., Huang, W., Cui, X., Dong, Y., Shi, Y., Ma, H. and Liu, L. (2019). Identification of wheat yellow rust using optimal three-band spectral indices in different growth stages, *Sensors (Basel)* 19(1), 35. doi: 10.3390/s19010035.

Zhou, B., Khosla, A., Lapedriza, A., Oliva, A. and Torralba, A. (2016). Learning deep features for discriminative localization. *Proceedings of the IEEE Conference on Computer Vision and Pattern Recognition (CVPR)*, pp. 2921–2929.

Index

www.ingramcontent.com/pod-product-compliance
Lightning Source LLC
Jackson TN
JSHW080849171125
94236JS00015B/15
* 9 7 8 1 7 8 6 7 6 8 5 6 8 *